ESTIMATION AND TRACKING:
PRINCIPLES, TECHNIQUES, AND SOFTWARE

Yaakov Bar-Shalom
Dept. of Electrical and Systems Engineering
University of Connecticut
Storrs, Connecticut

Xiao-Rong Li
Dept. of Electrical Engineering
University of New Orleans
New Orleans, Louisiana

Yaakov Bar-Shalom
Box U-157, Storrs, CT 06269-2157
860-486-4823
ybs@ee.uconn.edu

To the memory of my parents, Moshe and Irit, and to Michael

YBS

To Peizhu

XRL

Lemma.   Make things as simple as posible but not simpler.

A. Einstein

Theorem.   By making things absolutely clear, people will become confused.

A Chinese fortune cookie

Corollary.   We will make things
simple
but not too simple,
clear
but not too clear.

Contents

CONTENTS

PREFACE

FOREWORD

This text — set of lecture notes — presents the material from a second semester graduate level course on Estimation offered in the Department of Electrical and Systems Engineering at the University of Connecticut. This course is a standard requirement in the department's M.S. program in Control and Communication Systems. The prerequisites are a solid knowledge of linear systems and probability theory at the first semester graduate level. These, as well as some additional useful material from Statistics, are summarized for ready reference in Chapter 1.

The main goal of this course is to convey the knowledge necessary for the *evaluation and design of state estimators that operate in a stochastic environment*. The course leads to a major design project that is unique to this text — it combines the basics of standard estimation with the latest adaptive techniques. The emphasis is on mapping the physical quantities of the object of interest and of the sensors — an aircraft in an air traffic control tracking system — into the parameters of the mathematical model, namely, the statistical characterization of the random processes describing the uncertainties of this problem.

The approach is a balanced combination of mathematics — linear systems and probability theory — in order to understand how a state estimator should be designed, with the necessary tools from statistics in order to interpret the results. The use of statistical techniques has been somewhat neglected in the engineering literature pertaining to state estimation but it is necessary for the (nontrivial task of) interpretation of stochastic data and to answer the question whether a design can be accepted as "good."

The material covers the topics usually taught in control-oriented EE/systems and aeronautical engineering programs. The relevance extends to other areas dealing with control in mechanical or chemical engineering. Recently, the state estimation techniques have been gaining wider attention due to their applicability to such new fields as robotics, computer vision for autonomous navigation and image feature extraction with application to medical diagnosis. While the course is mainly directed toward the M.S. students, it is also part of the Ph.D. program at the University of Connecticut, with the intent of providing to the students the knowledge to tackle real-world problems.

The presentation of the material stresses the algorithms, their properties and the understanding of the assumptions behind them. Proofs are given to the extent they are relevant to understanding the results. This will hopefully result in a modest step in bridging the much talked about "gap between theory and practice" — it will illustrate to students the usefulness of state estimation for the real world and provide to engineers and scientists working in industry or laboratories a broader understanding of the algorithms used in practice. It might also avoid the situation summarized by a participant at one of the continuing education courses taught by the first author as follows: "Although I studied Kalman filters when I worked toward my Ph.D. (at one of the major universities), I did not expect that they worked with real data." This happens when, because of the theorems, the students cannot see the forest of applications.

Tuning of a Kalman filter — the choice of its design parameters — is an art. The contribution of this text is to make it less of a black magic technique and more of a systematic approach.

The format of this text is also unique — *textgraph*™ — in that it dares to attempt to accomplish two goals in one format: to serve as a self-contained *concise text*, without excess verbosity, and at the same time, to enable the lecturer to use the pages of this text directly as *viewgraphs for lectures*. Experience over a number of years shows that students with a copy of these notes in front of them can concentrate on the ideas, rather than having to copy a copious amount of equations from the board. Some of the graduate students who took the Estimation course taught using this *textgraph*™ even complained that other graduate courses had not adopted this instruction method.

This format is also a potential way of making lectures more efficient since it makes it possible to convey more material than through the standard lecturing technique where one has to write everything on the board. Last, but not least, the need to spend time on preparing lecture notes, as well as the main hazard of the teaching profession (the inhalation of chalk dust), can be avoided. The ready availability of such class notes also makes it much easier to hold in-house courses in an industrial or government laboratory environment where senior staff — with knowledge of the material but limited time — can serve as lecturers.

It is our hope that this text will contribute to the education of *future practitioners*, as well as be of use to those already involved in the application of state estimation to real-world problems. This textgraph was written with the philosophy that engineers should be able to do more than run-of-the-mill work — they should be prepared to work at the frontiers of knowledge.[1]

---

[1]As the ancient Romans said: "Orchides forum trahite, cordes et mentes veniant."

Many thanks to all the colleagues and friends who helped in putting together the material that eventually became this book: H. Blom, L. Campo, K. C. Chang, E. Daeipour, A. Kumar, C. Jauffret, D. Lerro, P. B. Luh, C. Rago, H. M. Shertukde and C. Yang.

Also, thanks to the sponsors who have supported the research that lead to this book: J. G. Smith and Dr. R. Madan from ONR, Dr. N. Glassman from AFOSR and Dr. R. Baheti from NSF.

A Note on the Printing Style

For the convenience of the reader, all major concepts/terms, when they are introduced or used in a particular application, are in boldface and indexed. Also, all the index terms appear in the text as boldfaced.

Italics are used for emphasized ideas/properties. The main equations of important algorithms are highlighted with boxes.

The Companion Software DynaEst

DynaEst^{TM} is a completely self-explanatory menu-driven interactive design and evaluation tool for Kalman filters and an adaptive interacting multiple model (IMM) estimator. The package provides the following important flexibilities as a design tool:

Model design. The user may specify any linear Gaussian true plant and measurement models, and the plant and measurement models for the filter. The true models and the filter-assumed models can be different. Models are valid for all dimensions. Reduced-order filters may be implemented.

Selection of model parameters. The user may specify any parameter of the true models and/or the filter-assumed models, such as the noise means and covariances.

Statistical testing. Most relevant statistical tests of the Kalman filters may be performed, including the normalized estimation error squared (NEES), the normalized innovation squared (NIS), and the RMS errors.

DynaEst^{TM} is user friendly and has the following features:

- Self-explanatory menu
- Convenient user specification of parameters (on screen or in an online editable file)
- Plotting on screen, to a printer (with high/low resolution option), or to a file
- Printing on screen, to a printer, or to a file
- Automatic (and flexible) installation
- Access to DOS and MATLAB from within *DynaEst*^{TM}
- Easily readable source code transparent to the user (and thus user-modifiable)
- Running on a floppy diskette as well as on a hard disk (so that students can use *DynaEst*^{TM} in a computer lab)
- Knowledge of DOS and MATLAB not required (but helpful, of course).

DynaEst^{TM} is written in MATLAB, which is very powerful, compact, and popular, especially in the science and engineering community. *DynaEst*^{TM} version 1.0 consists of about 100 files, which can be stored in one (1) $5\frac{1}{4}''$ or $3\frac{1}{2}''$ double density diskette. *DynaEst*^{TM} version 2.0 will consist of about 150 files. It will be an interactive software for the design of multiple model adaptive estimators, especially the state-of-the-art Interacting Multiple Model (IMM) estimators, as well as Kalman filters.

DynaEst^{TM} provides valuable hands-on experience on Kalman filters (and other adaptive estimators) for any person who has access to MATLAB, including the student edition. With the help of *DynaEst*^{TM}, the principles/concepts and techniques treated in this book can be better understood and remembered. Some practical problems of estimation and tracking can be handled by *DynaEst*^{TM}. The source code of *DynaEst*^{TM}, which is easy to read, provides a sample implementation of the Kalman filter, as well as the state-of-the-art IMM estimator.

System Requirements

- IBM PC compatible
- At least 320 KB of memory
- DOS 5.0 or higher
- MATLAB (professional, education, or student version)
- A floppy disk drive.

DynaEst 2.11 is availabe from

http://esplab2.ee.uconn.edu/dynaest211.zip

ABBREVIATIONS

cdf	cumulative (probability) distribution function
cpmf	cumulative probability mass function
CLT	central limit theorem
CRLB	Cramer-Rao lower bound
EKF	extended Kalman filter
FIM	Fisher information matrix
GPB	generalized pseudo Bayes
HOT	higher order terms
IE	input estimation
i.i.d.	independent identically distributed
IMM	interacting multiple model
INS	inertial navigation system
KBF	Kalman-Bucy filter (continuous time)
KF	Kalman filter (discrete time)
LF	likelihood function
LG	linear-Gaussian (assumption)
LLN	law of large numbers
LMMSE	linear MMSE
LOS	line of sight
LS	least squares
LSE	least squares estimator (or estimate)
MAP	maximum a posteriori
ML	maximum likelihood
MLE	maximum likelihood estimator (or estimate)
MM	multiple model
MMSE	minimum mean square error
MSE	mean square error
NEES	normalized estimation error squared
NIS	normalized innovation squared
pdf	probability density function
pmf	probability mass function
RMS	root mean square
VSD	variable state dimension

MATHEMATICAL NOTATIONS

arg max	argument that maximizes
arg min	argument that minimizes
χ_n^2	chi-square distribution with n degrees of freedom [Subsection 1.4.17]
$\chi_n^2(Q)$	$100Q$ percentile point of chi-square distribution with n degrees of freedom [Subsection 1.5.4]
col($\cdot$)	column vector [Subsection 1.3.1]
cov	covariance
$\delta(\cdot)$	Dirac (continuous) delta function (impulse function)
δ_{ij}	Kronecker (discrete) delta function
diag($\cdot$)	diagonal or block-diagonal matrix [Subsection 1.3.3, Subsection 3.4.1]
$E[\cdot]$	expectation
$\mathcal{G}(Q)$	$100Q$ percentile point of standard normal (Gaussian) distribution [Subsection 1.5.4]
Λ	likelihood function
$\mu(\cdot)$	pmf
$\mathcal{N}(\bar{x}, \sigma^2)$	normal (Gaussian) distribution with mean $\bar{x}$ and variance σ^2 [Subsection 1.4.13]
$\mathcal{N}(x; \bar{x}, \sigma^2)$	pdf of a normal (Gaussian) random variable x with mean $\bar{x}$ and variance σ^2 [Subsection 1.4.13]
$\mathcal{N}(\bar{x}, P)$	multivariate normal (Gaussian) distribution with mean $\bar{x}$ and covariance P [Subsection 1.4.13]
$\mathcal{N}(x; \bar{x}, P)$	pdf of a normal (Gaussian) random vector x with mean $\bar{x}$ and covariance P [Subsection 1.4.13]
n_x	dimension of x
$O(\cdot)$	order of magnitude of
$p[\cdot]$	pdf [Subsection 1.4.2]
$p(\cdot)$	pdf [Subsection 1.4.2]
$P(k|j)$	conditional covariance matrix of state at time k given observations through time j
P_{xz}	covariance (matrix) of x and z
$P_{x|z}$	conditional covariance (matrix) of x given z
$P[\cdot, \cdot]$	mixed joint probability-pdf
$P\{\cdot\}$	probability of an event
$\tilde{P}$	the spread of the means term [Subsection 1.4.16]
P_D	probability of detection [Subsection 1.5.1]
P_{FA}	false alarm probability [Subsection 1.5.1]
σ_θ	standard deviation of θ
σ_θ^2	variance of θ
tr($\cdot$)	trace (of a matrix) [Subsection 1.3.2]

$\mathcal{U}(a, b)$	uniform distribution over the interval $[a, b]$
$\mathcal{U}(x; a, b)$	pdf of random variable x uniformly distributed over the interval $[a, b]$ [Subsection 1.4.5]
var	variance
Z^k	the sequence $z(j)$, $j = 1, \ldots, k$ [Subsection 2.2.1]
0	scalar zero, zero (null) vector, zero (null) matrix
$\emptyset$	empty set, impossible event [Subsection 1.4.1]
1	unity, or a (column) vector with all elements unity
$1(\cdot)$	unit step function
$'$	transposition (of a matrix or a vector) [Subsection 1.3.1]
$(\cdot)^*$	complex conjugate and transposed [Subsection 1.4.23]
$\bar{x}$	expected value of x [Subsection 1.4.5],
$\hat{x}$	estimate of x [Subsection 2.2.1],
$\tilde{x}$	error corresponding to the estimate of x [Subsection 2.2.1],
$\bar{A}$	complement of set (or event) A [Subsection 1.4.1]
$\forall$	for all
$(\cdot \mid \cdot)$	conditioning (for probabilities, estimates or covariances)
$\mid \cdot \mid$	determinant (of a matrix)
$\mid \cdot \mid$	magnitude (of a scalar)
$\|\cdot\|$	norm of a vector [Subsection 1.3.4]
$\|\cdot\|$	norm of a matrix [Subsection 1.3.6]
$\|\cdot\|$	norm of a random variable [Subsection 3.3.1]
$\perp$	orthogonal vectors [Subsection 1.3.4]
$\perp$	orthogonal random variables [Subsection 3.3.1]
$f(x)\mid_{x=0}$	function evaluated at $x = 0$
$[a, b]$	closed interval between points a and b
$[a_{ij}]$	a matrix whose component at i-row j-column is a_{ij} [Subsection 1.3.1]
$x_{[t_0, t_1]}$	the function $x(t)$ in the corresponding interval [Subsection 4.2.1]
$\{x(j)\}_{j=j_0}^{j_1}$	the sequence $x(j)$, $j = j_0, \ldots, j_1$ [Subsection 4.3.3]
$\{x(j), j = j_0, \ldots, j_1\}$	the set $\{x(j_0), x(j_0 + 1), \ldots, x(j_1)\}$
$\langle \cdot, \cdot \rangle$	inner product of vectors [Subsection 1.3.2]
$\langle \cdot, \cdot \rangle$	inner product of random variables [Subsection 3.3.1]
$\wedge$	the smallest of two variables [Subsection 10.2.4]
$\vee$	the largest of two variables [Subsection 10.2.4]
$\gg$	much larger than
$\triangleq$	equal by definition
$\equiv$	identically equal
$\approx$	approximately equal

$\sim$	distributed as [Subsection 1.4.13]
$\implies$	implies
$\iff$	if and only if — implies and is implied by
$\exists$	there exists
$\ni$	such that
$\in$	element of
$\cup$	logical "OR" operation (set union) [Subsection 1.4.1]
$\cap$	logical "AND" operation (set intersection) [Subsection 1.4.1]
∇_x	gradient with respect to x

Chapter 1

INTRODUCTION

1.1 BACKGROUND

1.1.1 Estimation and Related Areas

Estimation is the process of inferring the value of a quantity of interest from indirect, inaccurate and uncertain observations.

The purpose of estimation can be the:

- Determination of planet orbit parameters — probably the first estimation problem — studied by Laplace, Legendre, and Gauss
- Statistical inference
- Determination of the position and velocity of an aircraft in an air traffic control system — tracking
- Application of control to a plant in the presence of uncertainty (noise and/or unknown parameters) — parameter identification and stochastic control
- Determination of model parameters for prediction of the state of a physical system or forecasting economic or other variables — system identification
- Determination of the characteristics of a transmitted message from noise corrupted observation of the received signal — communication theory
- Decisions regarding resource allocation or scheduling — operations research.

More rigorously, **estimation** can be viewed as the *process of selection of a point from a continuous space* — the "best estimate."

Decision can be viewed as the *selection of one out of a set of discrete alternatives* — the "best choice" from a discrete space. However, one can talk about estimation in a discrete-valued case with the possibility of not making a choice but obtaining some conditional probabilities of the various alternatives. This information can be used without making "hard decisions."

Estimation and decision can, therefore, be seen to be overlapping and techniques from both areas are used simultaneously in many practical problems.

Tracking is the estimation of the state of a moving object based on *remote measurements*. This is done using one or more sensors at fixed locations or on moving platforms.

At first sight, tracking might seem to be a special case of estimation. However, it is wider in scope: not only does it use all the tools from estimation, but it also requires extensive use of statistical decision theory when some of the practical problems (data association — "which is my measurement?" — see, e.g., [BF88]) are considered.

Filtering is the estimation of the (current) state of a dynamic system — the reason for the use of the word "filter" is that the process for obtaining the "best estimate" from noisy data amounts to "filtering out" the noise. The term filtering is thus used in the sense of eliminating an undesired signal, which in this case is the noise.

In control systems, signals are also filtered to obtain the estimate of the state of the (noisy) dynamic system, needed by the controller. Filtering of signals is very commonly used in signal processing — in the frequency domain as well as in the spatial domain. The latter is done to select signals coming from a certain direction.

Navigation is the estimation of the state of the platform ("own ship") on which the sensor is located.

An **optimal estimator** is a computational algorithm that processes observations (measurements) to yield an estimate of a variable of interest, that optimizes a certain criterion.

In general, one can classify the variable that is to be estimated into the following two categories:

- A parameter — a time-invariant quantity (a scalar, a vector, or a matrix);
- The state of a dynamic system (usually a vector).

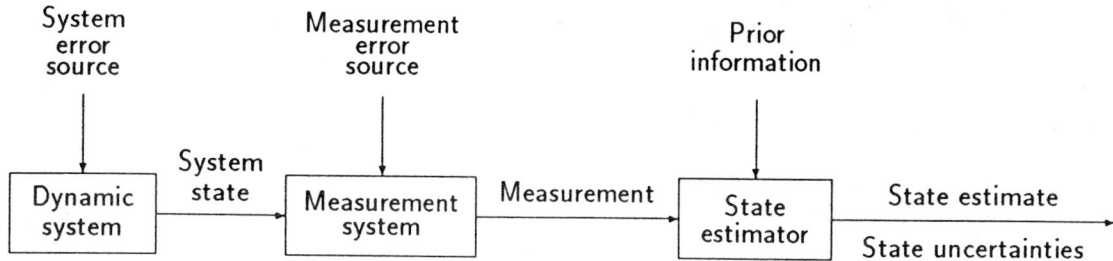

Figure 1.1.1-1: Mathematical view of state estimation.

Figure 1.1.1-1 presents a concise block diagram that illustrates state estimation. In this figure the first two blocks are "black boxes" — there is no access to variables inside them. The only variables to which the estimator has access are the **measurements**, which are affected by the error sources in the form of "noise."

The estimator uses knowledge about

- The evolution of the variable (the system dynamics);
- The probabilistic characterization of the various random factors (disturbances);
- The prior information.

Optimal Estimation — Advantages and Disadvantages

The advantage of an optimal estimator is that it makes the best utilization of the data and the knowledge of the system and the disturbances.

The disadvantages, like for any optimal technique, are that it is possibly sensitive to modeling errors and might be computationally expensive. In view of this, it is very important to have a clear understanding of the assumptions under which an algorithm is optimal and how they relate to the real world.

3

1.1.2 Applications of Estimation

State estimation, in combination with decision theory has a great variety of applications. A partial list of the areas where this has found use is:

- Tracking/surveillance
- Control systems

  - guidance
  - attitude control
  - sensor pointing
  - steel making
  - chemical, nuclear, and industrial processes, etc.

- Power systems
- Failure detection
- Navigation and trajectory determination
- Signal processing
- Image processing
- Communication
- Operations research
- Mapping via remote sensing
- Geophysical problems
- Fluid flow rate measurement
- Econometric systems

  - macroeconomic models
  - microeconomic models

- Demographic systems.

1.2 SCOPE OF THE TEXT

1.2.1 Objectives

The objectives of this text are to present the fundamentals of *state estimation theory* and the tools for the design of state-of-the-art algorithms for *target tracking*.

The text covers the basic concepts and estimation techniques for static and dynamic systems, linear and nonlinear, as well as adaptive estimation. This constitutes a one semester graduate course in estimation theory in an electrical/systems engineering program.

Special emphasis is given to the statistical tools that can be used for the *interpretation of the output from stochastic systems*. These are key tools for the assessment of the performance of state estimation/tracking filters — in simulations as well as in real-time implementation.

The discussion deals mainly with discrete time estimation algorithms, which are natural for digital computer implementation. The basic state estimation algorithm — the Kalman filter — is presented in discrete as well as in continuous time.

The use of the estimation algorithms is illustrated on kinematic motion models because they reveal all the major issues and in particular the subtleties encountered in estimation, and this serves as an introduction to tracking.

Guidelines for *tracking filter design* — selection of the filter design parameters — are given and illustrated in several examples.

Prerequisite Material

The *linear algebra tools* that form the backbone of the state space analysis techniques used in estimation/tracking are presented in Section 1.3 together with some useful results from linear systems. Matrix notation will be used throughout this text — every quantity should be considered a matrix, unless otherwise stated.

An informal review of the *tools from probability theory and random processes* is presented in Section 1.4.

Section 1.5 reviews *statistical hypothesis testing* and presents some useful statistical tables for the chi-square and the normal (Gaussian) distributions. Particular attention is given to the concept of statistical significance, which, while common in the statistics literature, has not received much attention in the engineering literature.

1.2.2 Overview and Chapter Prerequisites

In the following an overview of the text is given with the prerequisite to each chapter.

Brief Review of Background Techniques

- Linear algebra — Section 1.3
- Probability theory and statistics — Sections 1.4, 1.5.

Basic Concepts in Estimation — Chapter 2 (prerequisite: 1)

- Estimation techniques (ML, MAP, LS, MMSE)
- Properties of estimators.

Linear Estimation in Static Systems — Chapter 3 (prerequisite: 2)

- Techniques (LMMSE, recursive LS, prelude to Kalman filtering)
- Application to polynomial fitting.

Linear Dynamic Systems with Random Inputs — Chapter 4 (prerequisite: 1)

- Models — continuous and discrete time state space models
- Response of dynamic systems to random inputs.

State Estimation in Discrete Time Linear Dynamic Systems — Chapter 5 (prerequisites: 3, 4)

- The Kalman filter and its properties
- Consistency of state estimators
- Initialization — in simulations and in practice.

Estimation for Kinematic Models — Chapter 6 (prerequisite: 5)

- Types of kinematic models
- Explicit filters and the target maneuvering index for filter design.

Computational Aspects of Estimation — Chapter 7 (prerequisite: 5)

- The information filter
- Sequential update
- Square-root filtering.

1.2.2 Overview and Chapter Prerequisites

Extensions of Discrete Time Linear Estimation — Chapter 8 (prerequisite: 5)

- Correlated process noise
- Cross-correlated process and measurement noise sequences
- Correlated measurement noise
- Prediction
- Smoothing.

Continuous Time Linear State Estimation — Chapter 9 (prerequisite: 5)

- The Kalman-Bucy filter and its properties
- Prediction
- Duality between estimation and control
- The Wiener-Hopf problem.

State Estimation for Nonlinear Dynamic Systems — Chapter 10 (prerequisite: 5)

- Optimal estimation
- Suboptimal estimator — the extended Kalman filter (EKF)
- Practical issues in implementation
- Trajectory estimation via dynamic programming.

Adaptive Estimation and Maneuvering Targets — Chapter 11 (prerequisite: 5)

- White and colored noise models
- Input estimation
- Variable state dimension filtering
- Comparison of several methods
- The multiple model approach — static and dynamic, the interacting multiple model algorithm (IMM)
- An air traffic control (ATC) tracking algorithm design example
- Use of the EKF for parameter estimation.

At the end of each chapter, a number of problems that enhance the understanding of the theory and the connection of the theoretical material to the real world are given.

7

The sequence of prerequisites is illustrated in Figure 1.2.2-1.

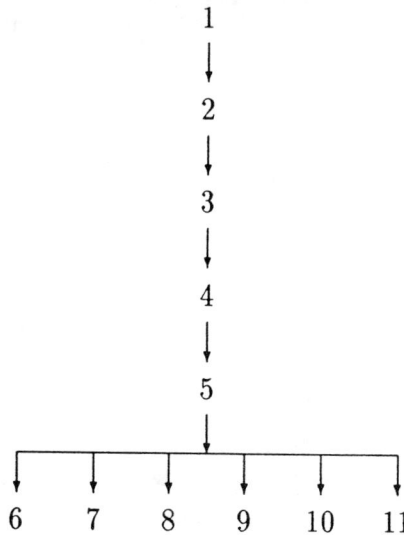

```
                          1
                          ↓
                          2
                          ↓
                          3
                          ↓
                          4
                          ↓
                          5
        ┌─────┬─────┬─────┼─────┬─────┐
        ↓     ↓     ↓     ↓     ↓     ↓
        6     7     8     9    10    11
```

Figure 1.2.2-1: Sequence of chapter prerequisites.

Two major examples of realistic systems are presented:

- A bearings-only target localization from a moving platform based on the Maximum Likelihood technique;
- An ATC tracking problem for maneuvering aircraft using an adaptive estimation technique (the IMM algorithm).

As part of the problems at the end of Chapter 3, a project is given to replicate the first example listed above.

A major term project that involves a similar ATC tracking problem is given at the end of Chapter 11 with two options:

1. With linear measurements;
2. With nonlinear measurements (from a radar — range and azimuth).

The sequences of prerequisites for the term project according to the options are:

1. Chapters 1-6, Section 11.6;
2. Chapters 1-6, Sections 10.3 and 11.6.

1.3 BRIEF REVIEW OF LINEAR ALGEBRA AND LINEAR SYSTEMS

1.3.1 Definitions and Notations

A **matrix** of dimension $n \times m$ is the two-dimensional array

$$A = [a_{ij}] = \begin{bmatrix} a_{11} & \cdots & a_{1m} \\ \vdots & \ddots & \vdots \\ a_{n1} & \cdots & a_{nm} \end{bmatrix} \qquad (1.3.1\text{-}1)$$

The first dimension is the number of rows, the second is the number of columns. The elements of A will be denoted as a_{ij} or as A_{ij}.

An n-dimensional **vector** is the one-dimensional array ($n \times 1$ matrix)

$$a = \begin{bmatrix} a_1 \\ \vdots \\ a_n \end{bmatrix} \triangleq \text{col}(a_i) \qquad (1.3.1\text{-}2)$$

All vectors will be **column vectors**.

Transposition of a matrix or vector will be denoted by an apostrophe. The *transpose* of the matrix (1.3.1-1) is

$$A' = [a_{ji}] = \begin{bmatrix} a_{11} & \cdots & a_{n1} \\ \vdots & \ddots & \vdots \\ a_{1m} & \cdots & a_{nm} \end{bmatrix} \qquad (1.3.1\text{-}3)$$

With this notation one can write the column vector (1.3.1-2) as

$$a = [a_1 \; \cdots \; a_n]' \qquad (1.3.1\text{-}4)$$

The **row vector** obtained from transposing (1.3.1-2) is

$$a' = [a_1 \; \cdots \; a_n] \qquad (1.3.1\text{-}5)$$

A (square) matrix is said to be **symmetric** if

$$A = A' \qquad (1.3.1\text{-}6)$$

or

$$a_{ij} = a_{ji} \qquad \forall i, j \qquad (1.3.1\text{-}7)$$

9

1.3.2 Some Linear Algebra Operations

Addition of matrices and the multiplication of a matrix by a scalar are defined as follows. With α and β scalars, the matrix

$$C = \alpha A + \beta B \qquad (1.3.2\text{-}1)$$

has elements given by

$$c_{ij} = \alpha a_{ij} + \beta b_{ij} \qquad i = 1, \ldots, n; \;\; j = 1, \ldots, m \qquad (1.3.2\text{-}2)$$

where all the matrices have the same dimension $n \times m$.

The product of two matrices

$$C = AB \qquad (1.3.2\text{-}3)$$

has elements

$$c_{ij} = \sum_{k=1}^{m} a_{ik} b_{kj} \qquad i = 1, \ldots, n; \;\; j = 1, \ldots, p \qquad (1.3.2\text{-}4)$$

where A is $n \times m$, B is $m \times p$ and the result C is $n \times p$. The matrix product is, in general, not commutative.

The **transpose of a product** is

$$C' = (AB)' = B'A' \qquad (1.3.2\text{-}5)$$

Thus, if

$$Ab = c \qquad (1.3.2\text{-}6)$$

where A is $n \times m$, b is $m \times 1$ (i.e., an m-vector), c is $n \times 1$ (n-vector), then

$$c' = b'A' \qquad (1.3.2\text{-}7)$$

where c' is $1 \times n$ (n-vector in row form), etc.

The **inner product** of two (real) n-vectors in a Euclidean space is

$$\langle a, b \rangle \;\triangleq\; a'b = \sum_{i=1}^{n} a_i b_i \qquad (1.3.2\text{-}8)$$

The **outer product** of two vectors is the matrix

$$ab' = C = [c_{ij}] \qquad \text{with} \qquad c_{ij} = a_i b_j \qquad (1.3.2\text{-}9)$$

10

The **trace** of an $n \times n$ matrix is the sum of its diagonal elements

$$\text{tr}(A) \triangleq \sum_{i=1}^{n} a_{ii} = \text{tr}(A') \qquad (1.3.2\text{-}10)$$

It can be easily shown[1] that if A is $m \times n$ and B is $n \times m$, then

$$\text{tr}(AB) = \text{tr}(BA) \qquad (1.3.2\text{-}11)$$

For example, if a and b are both n-vectors, then

$$a'b = \text{tr}(a'b) = \text{tr}(ba') \qquad (1.3.2\text{-}12)$$

The first equality above is due to the fact that the trace of a scalar is itself.

It can be shown[2] that for matrices of suitable dimension,

$$\text{tr}(ABC) = \text{tr}(BCA) = \text{tr}(CAB) \qquad (1.3.2\text{-}13)$$

that is, the trace is *invariant under circular permutations* in its argument.

The inner product has a counterpart in probability theory where it is used for random variables, which can be looked upon as vectors in a suitable space. This has applications in linear minimum mean square error estimation.

A (square) matrix is **idempotent** if and only if [3] for all positive integers n

$$A^n = A \qquad (1.3.2\text{-}14)$$

The **square root** of the square matrix A is the (in general, nonunique) matrix $A^{1/2}$, such that

$$A^{1/2} A^{1/2} = A \qquad (1.3.2\text{-}15)$$

---

[1] Also known in mathematical circles as the ICBES argument.

[2] Also known in mathematical circles as the ICBS argument.

[3] The standard shorthand for this in mathematical texts is "**iff**."

1.3.3   Inversion and the Determinant of a Matrix

The **inverse** A^{-1} of the $n \times n$ matrix A is such that

$$A^{-1}A = I \qquad (1.3.3\text{-}1)$$

where

$$I = \begin{bmatrix} 1 & 0 & \cdots & 0 \\ 0 & 1 & \cdots & 0 \\ \vdots & \vdots & \ddots & \vdots \\ 0 & 0 & \cdots & 1 \end{bmatrix}$$
$$\overset{\triangle}{=} \operatorname{diag}(1) \qquad (1.3.3\text{-}2)$$

is the **identity matrix** of the same dimension as A.

The inverse is given by the expression

$$A^{-1} = \frac{1}{|A|}C' \qquad (1.3.3\text{-}3)$$

where C, called the **adjugate** of A, is the matrix of cofactors of A and $|A|$ is the **determinant** of A.

The inverse of a (square) matrix exists if and only if its columns a_i, $i = 1, \ldots, n$ (or its rows) are **linearly independent**, that is,

$$\sum_{i=1}^{n} \alpha_i a_i = 0 \qquad \Longleftrightarrow \qquad \alpha_i = 0 \quad i = 1, \ldots, n \qquad (1.3.3\text{-}4)$$

where 0 denotes the zero vector (of dimension n). This is equivalent to the determinant in (1.3.3-3) being nonzero.

An invertible matrix is also called **nonsingular**.

The determinant of an $n \times n$ matrix multiplied by a scalar is

$$|\alpha A| = \alpha^n |A| \qquad (1.3.3\text{-}5)$$

This is useful in writing the probability density function of Gaussian random vector.

Inversion of a Partitioned Matrix

The inverse of the (nonsingular) $n \times n$ **partitioned matrix**

$$\begin{bmatrix} A & B \\ C & D \end{bmatrix}^{-1} = \begin{bmatrix} E & F \\ G & H \end{bmatrix} \qquad (1.3.3\text{-}6)$$

where A is $n_1 \times n_1$, B is $n_1 \times n_2$, C is $n_2 \times n_1$, D is $n_2 \times n_2$ and $n_1 + n_2 = n$, has the partitions

$$\begin{aligned} E &= A^{-1} + A^{-1}BHCA^{-1} \\ &= (A - BD^{-1}C)^{-1} \end{aligned} \qquad (1.3.3\text{-}7)$$

$$\begin{aligned} F &= -A^{-1}BH \\ &= -EBD^{-1} \end{aligned} \qquad (1.3.3\text{-}8)$$

$$\begin{aligned} G &= -HCA^{-1} \\ &= -D^{-1}CE \end{aligned} \qquad (1.3.3\text{-}9)$$

$$\begin{aligned} H &= (D - CA^{-1}B)^{-1} \\ &= D^{-1} + D^{-1}CEBD^{-1} \end{aligned} \qquad (1.3.3\text{-}10)$$

The Matrix Inversion Lemma

Another useful result is the following identity known as the **matrix inversion lemma**,

$$(P^{-1} + H'R^{-1}H)^{-1} = P - PH'(HPH' + R)^{-1}HP \qquad (1.3.3\text{-}11)$$

where P is $n \times n$, H is $m \times n$, and R is $m \times m$.

An alternative version of the above is

$$(A + BCB')^{-1} = A^{-1} - A^{-1}B(B'A^{-1}B + C^{-1})^{-1}B'A^{-1} \qquad (1.3.3\text{-}12)$$

It can be shown that (1.3.3-6) to (1.3.3-10) and (1.3.3-11) hold by verifying that the corresponding multiplications will yield the identity matrix.

These results have direct application in the derivation of the recursive form of the least squares estimation of parameters as well as in linear estimation for dynamic systems — they yield various forms of the Kalman filter.

1.3.4 Orthogonal Projection of Vectors

The inner product of a vector with itself

$$\langle a, a \rangle \triangleq \|a\|^2 \qquad (1.3.4\text{-}1)$$

is the *squared l_2 norm* of this vector. This applies for the inner product defined in (1.3.2-8) or for any other properly defined inner product.

The **Schwarz inequality** states the following relationship between the magnitude of the inner product of two vectors and their norms

$$|a'b| \leq \|a\| \, \|b\| \qquad (1.3.4\text{-}2)$$

Two vectors are **orthogonal**, which is denoted as

$$a \perp b \qquad (1.3.4\text{-}3)$$

if

$$\langle a, b \rangle = 0 \qquad (1.3.4\text{-}4)$$

The **orthogonal projection** of the vector a on b is

$$\Pi_b(a) = \frac{\langle a, b \rangle}{\|b\|^2} \, b \qquad (1.3.4\text{-}5)$$

It can be easily shown that the difference vector between a and its orthogonal projection on b is orthogonal on b, that is,

$$[a - \Pi_b(a)] \perp b \qquad (1.3.4\text{-}6)$$

The orthogonal projection is used to obtain the linear minimum mean square error (LMMSE) estimates of random variables.

14

1.3.5 The Gradient, Jacobian and Hessian

The **gradient** operator with respect to the n-vector x is defined as

$$\nabla_x = \left[\frac{\partial}{\partial x_1} \cdots \frac{\partial}{\partial x_n} \right]' \tag{1.3.5-1}$$

The gradient of an m-dimensional vector-valued function $f(x)$ is

$$\nabla_x f(x)' = \begin{bmatrix} \dfrac{\partial}{\partial x_1} \\ \vdots \\ \dfrac{\partial}{\partial x_n} \end{bmatrix} [f_1(x) \cdots f_m(x)] = \begin{bmatrix} \dfrac{\partial f_1}{\partial x_1} & \cdots & \dfrac{\partial f_m}{\partial x_1} \\ \vdots & \ddots & \vdots \\ \dfrac{\partial f_1}{\partial x_n} & \cdots & \dfrac{\partial f_m}{\partial x_n} \end{bmatrix} \tag{1.3.5-2}$$

The transpose of the above is the **Jacobian**, an $m \times n$ matrix in this case

$$f_x(x) \triangleq \frac{\partial f}{\partial x} = [\nabla_x f(x)']' = \begin{bmatrix} \dfrac{\partial f_1}{\partial x_1} & \cdots & \dfrac{\partial f_1}{\partial x_n} \\ \vdots & \ddots & \vdots \\ \dfrac{\partial f_m}{\partial x_1} & \cdots & \dfrac{\partial f_m}{\partial x_n} \end{bmatrix} \tag{1.3.5-3}$$

The dimensions m and n are usually (but not necessarily) equal.

It can be easily shown that

$$\nabla_x x' = I \tag{1.3.5-4}$$

and, if A is a symmetric matrix,

$$\nabla_x (x' A x) = 2Ax \tag{1.3.5-5}$$

The **Hessian** of the scalar function $\phi(x)$ with respect to the n-vector x is

$$\phi_{xx}(x) \triangleq \frac{\partial^2 \phi(x)}{\partial x^2} = \nabla_x \nabla_x' \phi(x) = \begin{bmatrix} \dfrac{\partial^2 \phi}{\partial x_1 \partial x_1} & \cdots & \dfrac{\partial^2 \phi}{\partial x_1 \partial x_n} \\ \vdots & \ddots & \vdots \\ \dfrac{\partial^2 \phi}{\partial x_n \partial x_1} & \cdots & \dfrac{\partial^2 \phi}{\partial x_n \partial x_n} \end{bmatrix} \tag{1.3.5-6}$$

which is, obviously, a symmetric $n \times n$ matrix.

These results are used in the linear estimation of the state of nonlinear systems via series expansion.

1.3.6 Eigenvalues, Eigenvectors, and Quadratic Forms

Eigenvalues and Eigenvectors

The **eigenvalues** of an $n \times n$ matrix are the scalars λ_i such that

$$Au_i = \lambda_i u_i \qquad i = 1, \ldots, n \tag{1.3.6-1}$$

where the vectors u_i are the corresponding **eigenvectors**. If the eigenvalues are distinct, then there are n eigenvectors; otherwise the number of the eigenvectors might be less than n.

A matrix is **nonsingular** if and only if all its eigenvalues are nonzero. This follows from the fact that its determinant is the product of its eigenvalues

$$|A| = \prod_{i=1}^{n} \lambda_i \tag{1.3.6-2}$$

The **rank** of a (not necessarily square) matrix is equal to the number of its linear independent rows or columns. For a square ($n \times n$) matrix, this equals the number of its nonzero eigenvalues. In this case its columns (and rows) span the n-dimensional space.

A nonsingular matrix must have **full rank**.

The eigenvalues of a real matrix can be real or complex, however, a *symmetric* real matrix has only real eigenvalues.

It can be shown that the trace of a matrix equals the sum of its eigenvalues

$$\text{tr}(A) = \sum_{i=1}^{n} \lambda_i \tag{1.3.6-3}$$

An **idempotent** matrix, defined in (1.3.2-14), that is, with the property

$$AA = A \tag{1.3.6-4}$$

has eigenvalues that are *either zero or unity*.

Quadratic Forms

The (scalar) function of the real vector x

$$q = x'Ax \qquad (1.3.6\text{-}5)$$

is called a **quadratic form**. It can be easily shown that, without loss of generality, in a quadratic form the matrix A can be considered symmetric.

A **positive (semi)definite quadratic form** is one which is positive (nonnegative) for all nonzero vectors x.

A **positive (semi)definite matrix** is one for which the quadratic form (1.3.6-5) is positive (semi)definite. This can be summarized as follows:

$$A > 0 \qquad \Longleftrightarrow \qquad x'Ax > 0 \quad \forall x \neq 0 \qquad (1.3.6\text{-}6)$$

$$A \geq 0 \qquad \Longleftrightarrow \qquad x'Ax \geq 0 \quad \forall x \neq 0 \qquad (1.3.6\text{-}7)$$

If the matrix A is positive definite, the expression on the right hand side of (1.3.6-5) can also be called the **squared norm with respect to** A of the vector x and denoted sometimes as $\|x\|_A^2$.

A matrix is positive (semi)definite if and only if all its eigenvalues are positive (nonnegative).

The **inequality of two matrices** is defined as follows: the matrix A is smaller (not larger) than the matrix B if and only if the difference $B - A$ is positive (semi)definite.

The outer product of a vector with itself, aa', called a **dyad**, is a positive semidefinite matrix (it has rank one).

The sum of dyads of n-vectors a_i with positive weights

$$\tilde{A} = \sum_{i=1}^{m} \alpha_i a_i a_i' \qquad \alpha_i > 0 \qquad (1.3.6\text{-}8)$$

is (at least) positive semidefinite. If the vectors $a_i, \ldots, a_m$ span the n-dimensional space, in which case one needs $m \geq n$, then (1.3.6-8) is positive definite.

The **spectral representation** of an $n \times n$ matrix that has n eigenvectors is

$$A = \sum_{i=1}^{n} \lambda_i u_i v_i' \qquad (1.3.6\text{-}9)$$

where λ_i are its eigenvalues, u_i are its eigenvectors, and v_i, $i = 1, \ldots, n$, is the **reciprocal basis**, that is,

$$u_i' v_j = \delta_{ij} = \begin{cases} 1 & \text{if } i = j \\ 0 & \text{if } i \neq j \end{cases} \qquad (1.3.6\text{-}10)$$

The vectors v_i are the columns of the inverse of the matrix consisting of the eigenvectors u_i' as rows.

If the $n \times n$ real matrix A is *symmetric*, it will always have n eigenvectors that are *orthogonal*, even if it has multiple eigenvalues (a nonsymmetric matrix with multiple eigenvalues does not necessarily have a full set of eigenvectors). Furthermore, in this case,

$$v_i = u_i \qquad (1.3.6\text{-}11)$$

Thus, the spectral representation of a *symmetric* real $n \times n$ matrix A is

$$A = \sum_{i=1}^{n} \lambda_i u_i u_i' \qquad (1.3.6\text{-}12)$$

The **condition number** of a positive definite symmetric matrix is (usually taken as) the common logarithm of the ratio of its largest to its smallest eigenvalue:

$$\kappa(A) \triangleq \log_{10} \frac{\lambda_{max}}{\lambda_{min}} \qquad (1.3.6\text{-}13)$$

Large condition numbers indicate near singularity (e.g., $\kappa > 6$ for a 32-bit computer indicates an ill-conditioned matrix).

The (induced) **norm of a matrix** is

$$\|A\| \triangleq \max_{x \neq 0} \frac{\|Ax\|}{\|x\|} \qquad (1.3.6\text{-}14)$$

A consequence of the above and the Schwarz inequality is

$$\|Ax\| \leq \|A\| \, \|x\| \qquad (1.3.6\text{-}15)$$

Also

$$\|AB\| \leq \|A\| \, \|B\| \qquad (1.3.6\text{-}16)$$

1.3.7 Continuous Time Linear Dynamic Systems — Controllability and Observability

The State Space Representation

The state space representation of **continuous time linear systems** is

$$\dot{x}(t) = A(t)x(t) + B(t)u(t) \tag{1.3.7-1}$$

where

$x(t)$ is the state vector of dimension n_x,

$u(t)$ is the input (control) vector of dimension n_u,

$A(t)$ is an $n_x \times n_x$ matrix, called the **system matrix**,

$B(t)$ is an $n_x \times n_u$ matrix, called the (continuous time) **input gain**.

Equation (1.3.7-1) is known as the **dynamic equation** or the **plant equation**. The output of the system is, in general, a vector

$$z(t) = C(t)x(t) \tag{1.3.7-2}$$

of dimension n_z and C a known $n_z \times n_x$ matrix, called the **measurement matrix**.

Equation (1.3.7-2) is known as the **output equation** or the **measurement equation**.

Given the initial condition $x(t_0)$ and the input function denoted as

$$u_{[t_0,t]} = \{u(\tau), t_0 \leq \tau \leq t\} \tag{1.3.7-3}$$

one can compute the future output at any time $t > t_0$

$$z(t) = z[x(t_0), u_{[t_0,t]}, t, t_0] \tag{1.3.7-4}$$

The **state** (of a deterministic system) is defined as the smallest vector that *summarizes the past of the system in full.*

A continuous time system is **linear time invariant** if it is described by (1.3.7-1) and (1.3.7-2) with $A(t) = A$, $B(t) = B$, and $C(t) = C$.

19

Controllability

A continuous time (deterministic) system is **completely controllable** if, given an *arbitrary destination point* in the state space, there is an input function that will bring the system from any initial state to this point in a finite time.

For a *linear time-invariant system*, the controllability condition is that the pair $\{A, B\}$ is controllable, that is, the **controllability matrix**

$$\mathcal{Q}_C \triangleq \begin{bmatrix} B & AB & \cdots & A^{n_x-1}B \end{bmatrix} \tag{1.3.7-5}$$

has **full rank**, which in this case is n_x, the lower of its two dimensions (the other is $n_x n_u$).

Observability

A continuous time (deterministic) system is **completely observable** if its initial state can be *fully and uniquely* recovered from its output, observed over a finite time interval, and the knowledge of the input.

Note that since the system is deterministic, knowledge of the initial state is equivalent to knowledge of the state at any time. Thus, using the output (1.3.7-2), which is in general a vector of dimension $n_z < n_x$, and the input, one can then recover the state perfectly.

For a *linear time-invariant system*, the observability condition is that the pair $\{A, C\}$ is observable, that is, the **observability matrix**

$$\mathcal{Q}_O \triangleq \begin{bmatrix} C \\ CA \\ \vdots \\ CA^{n_x-1} \end{bmatrix} \tag{1.3.7-6}$$

has full rank n_x.

1.3.8 Discrete Time Linear Dynamic Systems — Controllability and Observability

The State Space Representation

Deterministic linear dynamic systems can described in discrete time by state equation of the form

$$x(k+1) = F(k)x(k) + G(k)u(k) \qquad (1.3.8\text{-}1)$$

where

$x(k)$ is the state of the system, a vector of dimension n_x,
$F(k)$ is the transition matrix $(n_x \times n_x)$,
$u(k)$ is the input (control), a vector of dimension n_u,
$G(k)$ is the input gain $(n_x \times n_u$ matrix),

all at time k.

The output equation is

$$z(k) = H(k)x(k) \qquad (1.3.8\text{-}2)$$

where

$z(k)$ is the output vector (observation or measurement), of dimension n_z,
$H(k)$ is the measurement matrix $(n_z \times n_x)$.

The **state** of the system is the smallest dimension vector that summarizes the past of the system in full. Then the output at time $j > k$ can be determined fully from $x(k)$ and the intervening inputs

$$z(j) = z[j, x(k), u(k), \dots, u(j-1)] \qquad (1.3.8\text{-}3)$$

A discrete time system is **linear time invariant** if it is described by (1.3.8-1) and (1.3.8-2) with $F(k) = F$, $G(k) = G$, and $H(k) = H$.

Controllability

A discrete time (deterministic) system is **completely controllable** if, given an *arbitrary destination point* in the state space, there is an input sequence that will bring the system from any initial state to this point in a finite number of steps.

For a *linear time-invariant system*, the controllability condition is that the pair $\{F, G\}$ is controllable, that is, the **controllability matrix**

$$\mathcal{Q}_C \triangleq \begin{bmatrix} G & FG & \cdots & F^{n_x-1}G \end{bmatrix} \tag{1.3.8-4}$$

has full rank n_x.

Observability

A (deterministic) system is **completely observable** if its initial state can be *fully and uniquely* recovered from a finite number of observations of its output and the knowledge of its input.

Note that since the system is deterministic, knowledge of the initial state is equivalent to knowledge of the state at any time. Thus, using the input and the output (1.3.8-2), which is in general a vector of dimension $n_z < n_x$, one can *in the absence of noise* recover the state perfectly.

For a *linear time-invariant* system, the observability condition is that the pair $\{F, H\}$ is observable, that is, the **observability matrix**

$$\mathcal{Q}_O \triangleq \begin{bmatrix} H \\ HF \\ \vdots \\ HF^{n_x-1} \end{bmatrix} \tag{1.3.8-5}$$

has full rank n_x.

1.4 BRIEF REVIEW OF PROBABILITY THEORY

1.4.1 Events and the Axioms of Probability

Consider an "experiment", or, in general, a process with random outcomes. An **event** is a collection (set) of such outcomes — it is said to have occurred if the outcome is one of the elements of this set.

Denote by A an event in such an experiment (e.g., "even" or "5" in a die rolling experiment). Let S be the **sure event** in the experiment (e.g., "any number between 1 and 6" in the die rolling). Then the **probability** of an event is a number (**measure**) that satisfies the following three **axioms of probability**:

1. It is nonnegative

$$P\{A\} \geq 0 \qquad \forall A \tag{1.4.1-1}$$

2. It is unity for the sure event

$$P\{S\} = 1 \tag{1.4.1-2}$$

3. It is additive over the union of **mutually exclusive** events, that is, if the events A and B have no common elements (their set intersection is $\emptyset$, the empty set)

$$A \cap B \triangleq \{A \text{ and } B\} \triangleq \{A, B\} = \emptyset \tag{1.4.1-3}$$

then their union (logical "or") has probability

$$P\{A \cup B\} \triangleq P\{A \text{ or } B\} \triangleq P\{A + B\} = P\{A\} + P\{B\} \tag{1.4.1-4}$$

From the above it follows that

$$P\{A\} = 1 - P\{\bar{A}\} \leq 1 \tag{1.4.1-5}$$

where the overbar denotes the **complementary event**.

Since $\emptyset$ is the complement of S, one has

$$P\{\emptyset\} = 0 \tag{1.4.1-6}$$

The event $\emptyset$ is called the **impossible event**.

Extension

The extended version of Axiom 3 such that the probability is additive over the union of an *infinite number of mutually exclusive events* — **countable additivity** — is necessary when an experiment has an infinite number of outcomes. This is a key point in the definition of continuous-valued random variables.

Remarks

Relative-frequency or **measure-of-belief** interpretation of probability are alternative, even though not rigorous, ways of introducing the concept of probability.

In practice one can use the relative frequency or measure of belief to assign, based on intuition (or engineering common sense), probabilities to certain events. This can be of major importance in certain engineering systems.

Note that probability is a scalar quantity that is dimensionless (in a physical sense). This is in contrast to, for instance, a probability density function (pdf) which, while a scalar, has in general a *physical dimension*.

1.4.2 Random Variables and Probability Density Function

A scalar **random variable** is a (real-valued) function that assumes a certain *value* according to the outcome of a *random experiment*. The value taken by a random variable is called its **realization**.

The **probability density function (pdf)** of a scalar continuous-valued random variable x at $x = \xi$ is

$$p_x(\xi) = \lim_{d\xi \to 0} \frac{P\{\xi - d\xi < x \leq \xi\}}{d\xi} \geq 0 \qquad (1.4.2\text{-}1)$$

where $P\{\cdot\}$ is the probability of the event $\{\cdot\}$.

The more common notation

$$p_x(x) = p(x) \qquad (1.4.2\text{-}2)$$

where the argument defines the function, is used instead of (1.4.2-1). Also the term **density** is sometimes used instead of *pdf*.

From (1.4.2-1) and Axiom 3 from Subsection 1.4.1 it follows that

$$P\{\eta < x \leq \xi\} = \int_\eta^\xi p(x)dx \qquad (1.4.2\text{-}3)$$

The function

$$P_x(\xi) = P\{x \leq \xi\} = \int_{-\infty}^\xi p(x)dx \qquad (1.4.2\text{-}4)$$

is called the **cumulative probability distribution function (cdf)** of x at ξ. This is usually referred to as **distribution**.

Since the event $\{x \leq \infty\}$ is the "sure" event, one has

$$P\{x \leq \infty\} = \int_{-\infty}^\infty p(x)dx = 1 \qquad (1.4.2\text{-}5)$$

A pdf has to have the **normalization property** (1.4.2-5) that its total **probability mass** is unity — otherwise it is not a **proper density**.

The relationship between the density and the cumulative distribution is, from (1.4.2-4)

$$p(x) = \frac{d}{d\xi}P_x(\xi)\Big|_{\xi=x} \qquad (1.4.2\text{-}6)$$

if the derivative exists.

An outline of the rigorous way of introducing the pdf is as follows:

1. First define the "basic" events $\{x \leq \xi\}$.
2. Show that from these events one can obtain via (a countable number of) set operations all the events of interest (e.g., $\{a < x \leq \xi\}$, $\{x = a\}$).
3. Define the probabilities of the basic events, from which the probabilities of the other events of interest can be computed using the extended version of the third axiom of probability.
4. Finally, the pdf follows from (1.4.2-6) under suitable conditions of differentiability; note that (1.4.2-6) is equivalent to (1.4.2-1).

Remarks

Note from (1.4.2-1) that with the numerator being dimensionless, the pdf of the random variable x has as its *physical dimension* the inverse of the physical dimension of x. (See also problem 1-5.)

The event $\{x = a\}$ has, for a continuous-valued random variable, probability zero (even though it is *not impossible!*). This follows from (1.4.2-3) when the interval length tends to zero, in which case, as long as the density is finite, the integral is zero. Finite density means "no point masses" — this is discussed in the next section.

Improper densities — which do not integrate to unity (actually, their integral is not defined) — can be used, however, in certain circumstances.

1.4.3 Probability Mass Function

The **probability mass function (pmf)** of a scalar random variable x, which can take values in the set $\{\xi_i, i = 1, \ldots, n\}$ (i.e., it is discrete-valued), is

$$\mu_x(\xi_i) = P\{x = \xi_i\} = \mu_i \qquad i = 1, \ldots, n \qquad (1.4.3\text{-}1)$$

where μ_i are the **point masses**.

Similarly to (1.4.2-5), the requirement for a **proper pmf** is

$$\sum_{i=1}^{n} \mu_i = 1 \qquad (1.4.3\text{-}2)$$

Using the **Dirac (impulse) delta function** defined by

$$\delta(x) = 0 \qquad \forall x \neq 0 \qquad (1.4.3\text{-}3)$$

and

$$\int_{-\infty}^{\infty} \delta(x)dx = 1 \qquad (1.4.3\text{-}4)$$

one can write a pdf corresponding to (1.4.3-1) as

$$p(x) = \sum_{i=1}^{n} \mu_i \delta(x - \xi_i) \qquad (1.4.3\text{-}5)$$

Note that the above satisfies the **normalization property** (1.4.2-5) of a pdf that it integrates to unity.

The distribution corresponding to the above density, called **cumulative probability mass function (cpmf)**, has jumps at ξ_i — it is a staircase function: its derivative is zero everywhere except at the jumps where it is an impulse function.

The expression of the cpmf can be written, in terms of the **unit step function** $1(\cdot)$, as follows

$$P\{x \leq \xi\} = \sum_{i=1}^{n} \mu_i\, 1(\xi - \xi_i) \qquad (1.4.3\text{-}6)$$

1.4.4   Mixed Random Variable and Mixed Probability-PDF

A ***mixed random variable*** or a ***hybrid random variable*** x is one which can take values in a continuous set X as well as over a discrete set of points $\{\xi_i, i = 1, \ldots, n\}$. Such a random variable has a pdf of the form

$$p(x) = p_c(x) + \sum_{i=1}^{n} \mu_i \delta(x - \xi_i) \tag{1.4.4-1}$$

where $p_c(x)$ is the continuous part of the pdf and μ_i are the point masses.

Then

$$\int_{-\infty}^{\infty} p(x)dx = \int_{x \in X} p_c(x)dx + \sum_{i=1}^{n} \mu_i = 1 \tag{1.4.4-2}$$

The ***joint probability-pdf*** *of an event and a random variable* is defined as

$$P_{A,x}[A, \xi] = \lim_{d\xi \to 0} \frac{P\{A, \xi - d\xi < x \le \xi\}}{d\xi} \tag{1.4.4-3}$$

and is denoted without subscripts as $P[A, x]$ where the arguments define the function.

The following notations will be observed [4] in the sequel:

$$P\{\cdot\} = \text{probability of an event} \tag{1.4.4-4}$$

$$p(\cdot) \text{ or } p[\cdot] = \text{pdf} \tag{1.4.4-5}$$

$$\mu(\cdot) = \text{pmf} \tag{1.4.4-6}$$

$$P[\cdot, \cdot] = \text{mixed (joint) probability-pdf} \tag{1.4.4-7}$$

Some of the uncertainties in estimation problems are naturally modeled as mixed random variables.

---

[4]Most of the time.

1.4.5 Expectations and Moments of a Scalar Random Variable

The **expected value** of a scalar random variable, also called its **mean, average**, or **first moment**, is

$$E[x] = \int_{-\infty}^{\infty} xp(x)dx \triangleq \bar{x} \tag{1.4.5-1}$$

The n-th **moment** is

$$E[x^n] = \int_{-\infty}^{\infty} x^n p(x)dx \tag{1.4.5-2}$$

The **second central moment** or **variance** is

$$\begin{aligned}
\text{var}(x) &\triangleq E[(x - \bar{x})^2] = \int_{-\infty}^{\infty} (x - \bar{x})^2 p(x)dx \\
&= E[x^2] - (\bar{x})^2 \triangleq \sigma_x^2
\end{aligned} \tag{1.4.5-3}$$

The square root σ of the variance is called the **standard deviation**.

Note that the second moment, which is the **mean square (MS)** value, is equal to the square of the mean plus the variance

$$E[x^2] = [E(x)]^2 + \text{var}(x) = \bar{x}^2 + \sigma_x^2 \tag{1.4.5-4}$$

For a zero-mean random variable, the standard deviation is the RMS value.[5]

The expected value of a function $g(x)$ of the random variable x is

$$E[g(x)] = \int_{-\infty}^{\infty} g(x)p(x)dx \tag{1.4.5-5}$$

that is, the (Lebesgue) integral with respect to the measure $p(x)dx = dP(x)$.

For example, if x is uniformly distributed in the interval $[a, b]$, that is,

$$p(x) = \mathcal{U}(x; a, b) \triangleq \begin{cases} \dfrac{1}{b - a} & x \in [a, b] \\\\ 0 & \text{elsewhere} \end{cases} \tag{1.4.5-6}$$

then one has

$$E[x] = \frac{b + a}{2} \qquad \text{var}(x) = \frac{(b - a)^2}{12} \tag{1.4.5-7}$$

(See also problem 1-6.)

---

[5]Not to be confused with the RMS value of a waveform.

1.4.6 Joint PDF of Two Random Variables

The **joint pdf** of two random variables x and y is defined in terms of the probability of the following joint event (denoted with the set intersection symbol):

$$p_{x,y}(\xi,\eta) = \lim_{d\xi \to 0, d\eta \to 0} \frac{P\{\{\xi - d\xi < x \le \xi\} \cap \{\eta - d\eta < y \le \eta\}\}}{d\xi d\eta} \qquad (1.4.6\text{-}1)$$

Integrating the joint pdf of two random variables over one of the variables yields the pdf of the other random variable

$$\int_{-\infty}^{\infty} p_{x,y}(\xi,\eta)d\eta = p_x(\xi) \qquad (1.4.6\text{-}2)$$

or, using the simpler notation (1.4.2-2),

$$\int_{-\infty}^{\infty} p(x,y)dy = p(x) \qquad (1.4.6\text{-}3)$$

The resulting pdf, which pertains to a single random variable (introduced in Subsection 1.4.2), is also called **marginal pdf** or **marginal density**.

Similarly to (1.4.2-4), the **joint cdf** is

$$\begin{aligned} P_{x,y}(\xi,\eta) &= P\{x \le \xi, y \le \eta\} \\ &= \int_{x=-\infty}^{\xi} \int_{y=-\infty}^{\eta} p_{x,y}(x,y)dxdy \end{aligned} \qquad (1.4.6\text{-}4)$$

Covariance and Correlation Coefficient

The **covariance of two scalar random variables** x_1 and x_2 with means $\bar{x}_1$ and $\bar{x}_2$, respectively, is

$$
\begin{aligned}
\operatorname{cov}(x_1, x_2) \;&\triangleq\; E[(x_1 - \bar{x}_1)(x_2 - \bar{x}_2)] \\
&=\; \int_{-\infty}^{\infty} \int_{-\infty}^{\infty} (x_1 - \bar{x}_1)(x_2 - \bar{x}_2)p(x_1, x_2)dx_1 dx_2 \\
&\triangleq\; \sigma_{x_1 x_2}^2
\end{aligned}
\tag{1.4.6-5}
$$

The **correlation coefficient** of these two random variables is the normalized quantity

$$
\rho_{12} \triangleq \frac{\sigma_{x_1 x_2}^2}{\sigma_{x_1}\sigma_{x_2}}
\tag{1.4.6-6}
$$

where σ_{x_i} is the standard deviation of x_i.

Due to the normalization, the *magnitude of the correlation coefficient* of any two random variables obeys the following inequality

$$
|\rho_{12}| \leq 1
\tag{1.4.6-7}
$$

It can be shown that this is a consequence of the Schwarz inequality (1.3.4-2) for random variables (where the inner product is the the covariance and the norm squared is the variance — this is discussed in Section 3.3).

Two random variables whose correlation coefficient is zero are said to be **uncorrelated**. The random variables x_1 and x_2 are uncorrelated if and only if

$$
E[x_1 x_2] = E[x_1]E[x_2]
\tag{1.4.6-8}
$$

At the other extreme, if the correlation coefficient of two random variables has magnitude unity, then it can be shown that they are linearly dependent (i.e., one is a linear function of the other).

Two random variables x_1 and x_2 are said to be **linearly dependent** if

$$
a_1 x_1 + a_2 x_2 = 0
\tag{1.4.6-9}
$$

for some $a = [a_1 \quad a_2]' \neq 0$.

1.4.7 Independent Events and Independent Random Variables

Two events are **independent** if the probability of their joint event equals the product of their marginal probabilities

$$P\{A \cap B\} = P\{A, B\} = P\{A\}P\{B\} \tag{1.4.7-1}$$

A set of n events A_i, $i = 1, \ldots, n$ are independent if the (joint) probability of their intersection is equal to the product of the corresponding (marginal) event probabilities

$$P\left\{\bigcap_{i=1}^{n} A_i\right\} = \prod_{i=1}^{n} P\{A_i\} \tag{1.4.7-2}$$

and the same property holds also for any subset of these events.

Similarly, n random variables are **independent** if their joint pdf equals the product of their corresponding marginal densities

$$p(x_1, \ldots, x_n) = \prod_{i=1}^{n} p(x_i) \tag{1.4.7-3}$$

A set of random variables is called **independent, identically distributed (i.i.d.)** if (1.4.7-3) holds and their marginal distributions (or densities) are identical.

The pdf of the sum of two independent random variables

$$y = x_1 + x_2 \tag{1.4.7-4}$$

is the **convolution** of their marginal densities

$$p(y) = \int p_{x_1}(y - x)p_{x_2}(x)dx \tag{1.4.7-5}$$

(See also problem 1-1.)

1.4.8 Vector-Valued Random Variables and Their Moments

The pdf of the **vector-valued random variable**

$$x = [x_1 \ \cdots \ x_n]' \qquad (1.4.8\text{-}1)$$

at

$$\xi = [\xi_1 \ \cdots \ \xi_n]' \qquad (1.4.8\text{-}2)$$

is defined as the **joint density** of its components

$$p_{x_1,\ldots,x_n}(\xi_1,\ldots,\xi_n) \triangleq p_x(\xi) \triangleq \frac{P\left\{\bigcap_{i=1}^n \{\xi_i - d\xi_i < x_i \le \xi_i\}\right\}}{d\xi_1 \cdots d\xi_n} \qquad (1.4.8\text{-}3)$$

where the set intersection symbol $\cap$ is used to denote a joint event

$$A \cap B \triangleq \{A \text{ and } B\} \triangleq \{A, B\} \qquad (1.4.8\text{-}4)$$

The shorter notation $p(x)$ will be used if it does not lead to ambiguities.

The mean of the vector x is the result of the n-fold integration

$$E[x] = \int_{-\infty}^{\infty} \cdots \int_{-\infty}^{\infty} \xi \, p_x(\xi) d\xi_1 \cdots d\xi_n = \int_{-\infty}^{\infty} \cdots \int_{-\infty}^{\infty} x p(x) dx_1 \cdots dx_n \triangleq \bar{x} \qquad (1.4.8\text{-}5)$$

The **covariance matrix** of the n-vector x is obtained from the n-fold integration (written with the short notation and with the limits omitted)

$$\text{cov}(x) \triangleq E[(x - \bar{x})(x - \bar{x})'] = \int (x - \bar{x})(x - \bar{x})' p(x) dx \triangleq P_{xx} \qquad (1.4.8\text{-}6)$$

and is a *symmetric $n \times n$* matrix.

The diagonal elements of the *covariance matrix* are the *variances* of the components of x, while the off-diagonal elements are the *(scalar) covariances* between its components, as in (1.4.6-4).

The *covariance matrix is positive definite* (and thus nonsingular) unless there is a linear dependence among the components of x. If there is such a dependence then the covariance matrix is positive semidefinite.

Note

Covariance matrices will be denoted by the symbol P, with a subscript or argument (in parantheses); the designation of a probability will be $P\{\cdot\}$ where braces are used for the event it pertains to, as in (1.4.4-4).

Characteristic Function and Moments

The **characteristic function** of a vector random variable is defined as the n-fold integral (with the limits, which are as in (1.4.8-5), omitted)

$$
\begin{aligned}
M_x(s) &= E[e^{s'x}] \\
&= \int e^{s'x} p(x)\,dx
\end{aligned}
\tag{1.4.8-7}
$$

which is the (n-dimensional) Fourier transform of the pdf with argument the (purely imaginary) vector s of the same dimension n as that of x.

The first moment of x can be obtained from the characteristic function as

$$
E[x] = \nabla_s M_x(s)|_{s=0}
\tag{1.4.8-8}
$$

that is, its gradient is evaluated at $s = 0$, where ∇_s is the (column) gradient operator

$$
\nabla_s = \left[\frac{\partial}{\partial s_1} \cdots \frac{\partial}{\partial s_n} \right]'
\tag{1.4.8-9}
$$

Similarly,

$$
E[xx'] = \nabla_s \nabla'_s M_x(s)|_{s=0}
\tag{1.4.8-10}
$$

and so forth.

Due to this property, the characteristic function is also called the **moment generating function**.

1.4.9   Conditional Probability and PDF

The **conditional probability** of an event A given B is defined as

$$P\{A|B\} = \frac{P\{A, B\}}{P\{B\}} \tag{1.4.9-1}$$

For independent events, the above becomes the unconditional probability.

Similarly, the **conditional pdf** of one random variable given another random variable is, using the simpler notation as in (1.4.2-2), given by

$$p(x|y) = \frac{p(x, y)}{p(y)} \tag{1.4.9-2}$$

For an event conditioned on a random variable, one has

$$P\{A|x\} = \frac{P[A, x]}{p(x)} \tag{1.4.9-3}$$

The conditional pdf of a random variable x, given an event A, is

$$p(x|A) = \frac{P[A, x]}{P\{A\}} \tag{1.4.9-4}$$

where the numerator above is a mixed probability-pdf, as defined in (1.4.4-3).

If the conditioning event is

$$A = \{x \leq a\} \tag{1.4.9-5}$$

then one gets the **truncated pdf**

$$p(x|x \leq a) = \frac{P[x, x \leq a]}{P\{x \leq a\}} = \begin{cases} \dfrac{p(x)}{P\{x \leq a\}} & \text{if } x \leq a \\ \\ 0 & \text{elsewhere} \end{cases} \tag{1.4.9-6}$$

which is the original one restricted to $x \leq a$ and suitably *renormalized*.

If $z = x + y$ and the conditional pdf of y given x is $p_{y|x}(\cdot)$, then

$$p(z|x) = p_{y|x}(z - x) \tag{1.4.9-7}$$

If, in addition, x and y are independent, then, with $p_y(\cdot)$ denoting the pdf of y, one has

$$p(z|x) = p_y(z - x) \tag{1.4.9-8}$$

(See also problem 1-1.)

1.4.10   The Total Probability Theorem

Let the events B_i, $i = 1, \ldots, n$, be **mutually exclusive**, that is,

$$P\{B_i, B_j\} = 0 \qquad \forall i \neq j \tag{1.4.10-1}$$

and **exhaustive**, that is,

$$\sum_{i=1}^{n} P\{B_i\} = 1 \tag{1.4.10-2}$$

Such a set of events is a *partition of the space*. Then, the **total probability theorem** states that for any event A, its probability can be decomposed in terms of conditional probabilities as follows:

$$
\begin{aligned}
P\{A\} &= \sum_{i=1}^{n} P\{A, B_i\} \\
&= \sum_{i=1}^{n} P\{A|B_i\} P\{B_i\}
\end{aligned}
\tag{1.4.10-3}
$$

The key to the above is the mutual exclusiveness and exhaustiveness of the set of events B_i, $i = 1, \ldots, n$.

Figure 1.4.10-1 illustrates this with the set theory counterpart, namely,

$$A = \bigcup_{i=1}^{n} (A \cap B_i) \tag{1.4.10-4}$$

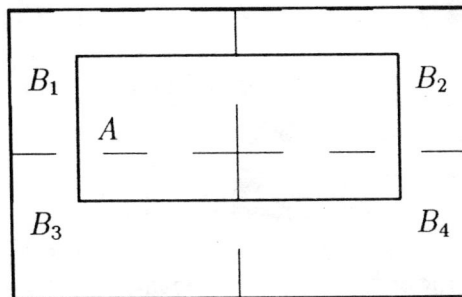

Figure 1.4.10-1: The total probability theorem (set theory counterpart).

1.4.10 The Total Probability Theorem

The version of the total probability theorem for random variables is

$$
\begin{aligned}
p(x) &= \int_{-\infty}^{\infty} p(x,y)dy \\
&= \int_{-\infty}^{\infty} p(x|y)p(y)dy
\end{aligned}
\tag{1.4.10-5}
$$

Note that this is a combination of the definition of the conditional pdf (1.4.9-2) and property (1.4.6-3) that the joint pdf integrated out with respect to one random variable yields the marginal pdf of the other variable.

For mixed event/random variable situations, one has

$$
p(x) = \sum_{i=1}^{n} p(x|B_i)P\{B_i\}
\tag{1.4.10-6}
$$

and

$$
\begin{aligned}
P\{A\} &= \int_{-\infty}^{\infty} P\{A|x\}p(x)dx \\
&= E[P\{A|x\}]
\end{aligned}
\tag{1.4.10-7}
$$

where the expectation operator averages over x.

An additional *common* conditioning in *all* the probabilities is permissible, for instance,

$$
\begin{aligned}
P\{A|C\} &= \sum_{i=1}^{n} P\{A, B_i|C\} \\
&= \sum_{i=1}^{n} P\{A|B_i, C\}P\{B_i|C\}
\end{aligned}
\tag{1.4.10-8}
$$

or

$$
P\{A|y\} = \int_{-\infty}^{\infty} P\{A|x, y\}p(x|y)dx
\tag{1.4.10-9}
$$

Similarly,

$$
\begin{aligned}
p(x|z) &= \int_{-\infty}^{\infty} p(x,y|z)dy \\
&= \int_{-\infty}^{\infty} p(x|y, z)p(y|z)dy
\end{aligned}
\tag{1.4.10-10}
$$

The total probability theorem is the primary tool in obtaining the state estimate in the presence of extraneous uncertainties (e.g., model uncertainties or measurements of uncertain origin).

1.4.11   Bayes' Formula

The probability of an event B_i conditioned on event A can be expressed in terms of the reverse conditioning as follows:

$$P\{B_i|A\} = \frac{P\{A|B_i\}P\{B_i\}}{P\{A\}} \qquad (1.4.11\text{-}1)$$

This is known as **Bayes' formula** or **Bayes' theorem** and is also referred to [6] as **Bayes' rule**.

The conditional of B_i is sometimes referred to as **posterior probability**, while the unconditional one is referred to as **prior probability**.

Using the total probability theorem (1.4.10-3) for the denominator in the above, one also has (if B_j, $j = 1, \ldots, n$ are mutually exclusive and exhaustive)

$$P\{B_i|A\} = \frac{P\{A|B_i\}P\{B_i\}}{\sum_{j=1}^{n} P\{A|B_j\}P\{B_j\}} = \frac{1}{c}P\{A|B_i\}P\{B_i\} \qquad (1.4.11\text{-}2)$$

where now the denominator c appears clearly as the **normalizing constant**, which guarantees that

$$\sum_{i=1}^{n} P\{B_i|A\} = 1 \qquad (1.4.11\text{-}3)$$

Bayes' formula for random variables is written as

$$p(x|y) = \frac{p(y|x)p(x)}{p(y)} = \frac{p(y|x)p(x)}{\int p(y|x)p(x)dx} \qquad (1.4.11\text{-}4)$$

In this case the unconditional pdf $p(x)$ is also called the **prior pdf** and the conditional pdf $p(x|y)$ is also called the **posterior pdf**.

For a mixed case

$$P\{B_i|x\} = \frac{p(x|B_i)P\{B_i\}}{p(x)} = \frac{p(x|B_i)P\{B_i\}}{\sum_{j=1}^{n} p(x|B_j)P\{B_j\}} \qquad (1.4.11\text{-}5)$$

where again the denominator is the normalizing factor.

When there are several conditioning random variables or events, Bayes' formula can be used to "switch" only some of them. For example,

$$P\{B_i|x,y\} = \frac{p(x|B_i,y)P\{B_i|y\}}{p(x|y)} = \frac{p(x|B_i,y)P\{B_i|y\}}{\sum_{j=1}^{n} p(x|B_j,y)P\{B_j|y\}} \qquad (1.4.11\text{-}6)$$

---

[6]**Bayes' decision rule** (or **Bayes' criterion** or **Bayes' principle**) in a decision problem is to minimize the expected value of a cost function, called the **Bayes' risk**.

These equations are the key tools in state estimation for **hybrid systems** — systems with continuous as well as discrete uncertainties.

Remarks

The prior reflects the (possibly subjective) initial **degree of belief**, which, when combined with the **evidence from the data**, quantified by $p(y|x)$, yields the posterior.

Bayes' postulate states that in the absence of prior knowledge, a **uniform prior pdf** should be chosen. While for discrete valued random variables this is reasonable, for continuous valued random variables this would imply a nonuniform pdf on any nonlinear function of this random variable and this has generated a lot of controversy among statisticians.

The use of a uniform prior pdf *over an infinite interval*, called **diffuse** or **noninformative** is discussed in more detail in Subsection 2.3.4.

1.4.12   Conditional Expectations and Their Smoothing Property

The **conditional expectation** is defined similarly to the unconditional expectation (1.4.5-1) but *with respect to a conditional pdf*, that is,

$$E[x|y] = \int_{-\infty}^{\infty} x p(x|y) dx \qquad (1.4.12\text{-}1)$$

Similarly, for a function of the random variable x (and possibly y), one has

$$E[g(x,y)|y] = \int_{-\infty}^{\infty} g(x,y) p(x|y) dx \qquad (1.4.12\text{-}2)$$

Note that (1.4.12-1) and (1.4.12-2) are functions of the conditioning argument y.

The **smoothing property of the expectations** states that the expected value of a conditional expected value is the (unconditional) expected value:

$$
\begin{aligned}
E\{E[x|y]\} &= \int_{y=-\infty}^{\infty} \left[\int_{x=-\infty}^{\infty} x p(x|y) dx \right] p(y) dy \\
&= \int_{x=-\infty}^{\infty} x \left[\int_{y=-\infty}^{\infty} p(x,y) dy \right] dx \\
&= \int_{x=-\infty}^{\infty} x p(x) dx = E[x] \qquad (1.4.12\text{-}3)
\end{aligned}
$$

In the first line of the above, the inside expectation is a function of y, which is "averaged out" (integrated over) by the outside expectation.

Equation (1.4.12-3) is also called the **law of iterated expectations**, summarized as

$$\boxed{E\big[E[x|y]\big] = E[x]} \qquad (1.4.12\text{-}4)$$

The same property holds when the conditioning is on events or mixed random variables.

This is used in the evaluation of the performance of certain estimation algorithms operating in the presence of several types of uncertainties — it allows us to handle different types of uncertainties sequentially rather than simultaneously.

1.4.13 Gaussian Random Variables

The pdf of a (scalar) **Gaussian** or **normal random variable** is

$$
\begin{aligned}
p(x) \;&=\; \mathcal{N}(x; \bar{x}, \sigma^2) \\
&\triangleq\; \frac{1}{\sqrt{2\pi}\,\sigma}\, e^{-\frac{(x-\bar{x})^2}{2\sigma^2}}
\end{aligned}
\tag{1.4.13-1}
$$

where $\mathcal{N}(\cdot)$ denotes the **normal pdf (density)** with argument x, mean $\bar{x}$, and variance σ^2. The first two moments, which fully characterize a Gaussian random variable, are referred to as its **statistics**.

Another notation equivalent to the above is

$$
\boxed{x \sim \mathcal{N}(\bar{x}, \sigma^2)}
\tag{1.4.13-2}
$$

which states that x is *normally distributed* with the corresponding mean and variance. Note the different meanings of $\mathcal{N}$ in (1.4.13-1) and (1.4.13-2) — they are, however, specified by their arguments: three in (1.4.13-1) vs. two in (1.4.13-2).

A **vector-valued Gaussian** random variable has the density

$$
\boxed{\mathcal{N}(x; \bar{x}, P) \triangleq |2\pi P|^{-1/2}\, e^{-\frac{1}{2}(x-\bar{x})'P^{-1}(x-\bar{x})}}
\tag{1.4.13-3}
$$

where

$$
\bar{x} = E[x]
\tag{1.4.13-4}
$$

$$
P = E[(x-\bar{x})(x-\bar{x})']
\tag{1.4.13-5}
$$

are, respectively, the mean and covariance matrix of the vector x.

The determinant in (1.4.13-3) has been written with the factor 2π inside it by making use of (1.3.3-5). This avoids the need to indicate the dimension of the vector x in (1.4.13-3).

The components of a Gaussian distributed vector are said to be **jointly Gaussian**.

If the covariance matrix P is diagonal, that is, the components of the *Gaussian* random vector x are *uncorrelated*, then they are also *independent* because their joint pdf equals the product of the marginals.

An important property of Gaussian random variables is that they stay Gaussian under linear transformations. This latter property is one of the reasons that the Gaussian model is very commonly used in estimation.

If x_1 and x_2 are random vectors of the same dimension and jointly Gaussian distributed, that is,

$$p\left(\begin{bmatrix} x_1 \\ x_2 \end{bmatrix}\right) = \mathcal{N}\left\{\begin{bmatrix} x_1 \\ x_2 \end{bmatrix} ; \begin{bmatrix} \bar{x}_1 \\ \bar{x}_2 \end{bmatrix} , \begin{bmatrix} P_{11} & P_{12} \\ P_{21} & P_{22} \end{bmatrix}\right\} \qquad (1.4.13\text{-}6)$$

then

$$x = x_1 + x_2 \qquad (1.4.13\text{-}7)$$

is Gaussian

$$p(x) = \mathcal{N}(x; \bar{x}, P) \qquad (1.4.13\text{-}8)$$

where

$$\bar{x} = \bar{x}_1 + \bar{x}_2 \qquad (1.4.13\text{-}9)$$

$$P = P_{11} + P_{12} + P_{21} + P_{22} \qquad (1.4.13\text{-}10)$$

Symbolically, one can write, with notation (1.4.13-2),

$$\mathcal{N}(\bar{x}_1, P_{11}) + \mathcal{N}(\bar{x}_2, P_{22}) = \mathcal{N}(\bar{x}_1 + \bar{x}_2, P_{11} + P_{12} + P_{21} + P_{22}) \qquad (1.4.13\text{-}11)$$

Another result of interest is, for x_1 and x_2 independent, the pdf of $x = x_1 + x_2$ conditioned on x_1 is

$$\begin{aligned} p(x|x_1) &= p_{x_2}(x - x_1) \\ &= \mathcal{N}(x - x_1; \bar{x}_2, P_{22}) \\ &= \mathcal{N}(x; x_1 + \bar{x}_2, P_{22}) \end{aligned} \qquad (1.4.13\text{-}12)$$

Note that in (1.4.13-12) x_1 acts as a constant — it shifts the mean — since it is given in the conditioning.

1.4.14 Joint and Conditional Gaussian Random Variables

Two random vectors x and z are **jointly Gaussian** if the **stacked vector**

$$y = \begin{bmatrix} x \\ z \end{bmatrix} \qquad (1.4.14\text{-}1)$$

is Gaussian, that is,

$$p(x,z) = p(y) = \mathcal{N}(y; \bar{y}, P_{yy}) \qquad (1.4.14\text{-}2)$$

The mean and covariance matrix of y in terms of those of x and z are

$$\bar{y} = \begin{bmatrix} \bar{x} \\ \bar{z} \end{bmatrix} \qquad (1.4.14\text{-}3)$$

$$P_{yy} = \begin{bmatrix} P_{xx} & P_{xz} \\ P_{zx} & P_{zz} \end{bmatrix} \qquad (1.4.14\text{-}4)$$

where

$$P_{xx} = \text{cov}(x) = E[(x - \bar{x})(x - \bar{x})'] \qquad (1.4.14\text{-}5)$$

$$P_{zz} = \text{cov}(z) = E[(z - \bar{z})(z - \bar{z})'] \qquad (1.4.14\text{-}6)$$

$$P_{xz} = \text{cov}(x,z) = E[(x - \bar{x})(z - \bar{z})'] = P_{zx}' \qquad (1.4.14\text{-}7)$$

are the blocks of the **partitioned covariance matrix** (1.4.14-4).

The conditional pdf of x given z when they are jointly Gaussian is

$$p(x|z) = \frac{p(x,z)}{p(z)} = \frac{|2\pi P_{yy}|^{-1/2} \, e^{-\frac{1}{2}(y - \bar{y})' P_{yy}^{-1}(y - \bar{y})}}{|2\pi P_{zz}|^{-1/2} \, e^{-\frac{1}{2}(z - \bar{z})' P_{zz}^{-1}(z - \bar{z})}} \qquad (1.4.14\text{-}8)$$

Advantage will be taken of the fact that the above is an exponential whose exponent is the difference of the exponents of the numerator and denominator.

The following notations will be used

$$\xi \overset{\Delta}{=} x - \bar{x} \qquad (1.4.14\text{-}9)$$

$$\zeta \overset{\Delta}{=} z - \bar{z} \qquad (1.4.14\text{-}10)$$

Using the new variables ξ and ζ, the problem for the *nonzero mean random variables* x and y is reduced to that of *zero-mean variables*.

43

The exponent on the right hand side of the conditional density (1.4.14-8) is (after multiplication by -2) the quadratic form

$$
\begin{aligned}
q &= \begin{bmatrix} \xi \\ \zeta \end{bmatrix}' \begin{bmatrix} P_{xx} & P_{xz} \\ P_{zx} & P_{zz} \end{bmatrix}^{-1} \begin{bmatrix} \xi \\ \zeta \end{bmatrix} - \zeta' P_{zz}^{-1} \zeta \\
&= \begin{bmatrix} \xi \\ \zeta \end{bmatrix}' \begin{bmatrix} T_{xx} & T_{xz} \\ T_{zx} & T_{zz} \end{bmatrix} \begin{bmatrix} \xi \\ \zeta \end{bmatrix} - \zeta' P_{zz}^{-1} \zeta
\end{aligned}
\qquad (1.4.14\text{-}11)
$$

The relationships between the partitions of the inverse of the covariance matrix and the partitions of the original matrix are, using (1.3.3-7) to (1.3.3-10), given by

$$
T_{xx}^{-1} = P_{xx} - P_{xz} P_{zz}^{-1} P_{zx} \qquad (1.4.14\text{-}12)
$$

$$
P_{zz}^{-1} = T_{zz} - T_{zx} T_{xx}^{-1} T_{xz} \qquad (1.4.14\text{-}13)
$$

$$
T_{xx}^{-1} T_{xz} = -P_{xz} P_{zz}^{-1} \qquad (1.4.14\text{-}14)
$$

The exponent (1.4.14-11) can be rewritten as

$$
\begin{aligned}
q &= \xi' T_{xx} \xi + \xi' T_{xz} \zeta + \zeta' T_{zx} \xi + \zeta' T_{zz} \zeta - \zeta' P_{zz}^{-1} \zeta \\
&= (\xi + T_{xx}^{-1} T_{xz} \zeta)' T_{xx} (\xi + T_{xx}^{-1} T_{xz} \zeta) + \zeta'(T_{zz} - T_{zx} T_{xx}^{-1} T_{xz}) \zeta - \zeta' P_{zz}^{-1} \zeta \\
&= (\xi + T_{xx}^{-1} T_{xz} \zeta)' T_{xx} (\xi + T_{xx}^{-1} T_{xz} \zeta)
\end{aligned}
\qquad (1.4.14\text{-}15)
$$

where use has been made of (1.4.14-13). The above procedure is called **completion of the squares** (actually, of the quadratic forms).

The result is a quadratic form in x, meaning that the conditional pdf of x given z is also Gaussian. This can be seen as follows: in view of (1.4.14-9), (1.4.14-10), and (1.4.14-14), the expression on the right hand side of (1.4.14-15) is a quadratic form in

$$
\xi + T_{xx}^{-1} T_{xz} \zeta = x - \bar{x} - P_{xz} P_{zz}^{-1} (z - \bar{z}) \qquad (1.4.14\text{-}16)
$$

From this one can recognize the conditional mean of x given z as

$$
\boxed{E(x|z) \stackrel{\Delta}{=} \hat{x} = \bar{x} + P_{xz} P_{zz}^{-1} (z - \bar{z})} \qquad (1.4.14\text{-}17)
$$

The corresponding conditional covariance is

$$
\boxed{\operatorname{cov}(x|z) \stackrel{\Delta}{=} P_{xx|z} = T_{xx}^{-1} = P_{xx} - P_{xz} P_{zz}^{-1} P_{zx}} \qquad (1.4.14\text{-}18)
$$

Note that the conditional mean (1.4.14-17) is linear in the observation z and that the covariance (1.4.14-18) is independent of the observation.

The above are the **fundamental equations of linear estimation**.

1.4.15 Expected Value of Quadratic and Quartic Forms

Consider a vector-valued random variable x with mean $\bar{x} = 0$ and covariance matrix P. Then, the **expected value of a quadratic form** with this random vector can be written as

$$
\begin{aligned}
E[x'Ax] &= E[\mathrm{tr}(x'Ax)] \\
&= E[\mathrm{tr}(Axx')] \\
&= \mathrm{tr}[AE(xx')] \\
&= \mathrm{tr}(AP)
\end{aligned}
\tag{1.4.15-1}
$$

The same result can be obtained for a Gaussian random vector using its characteristic function as follows.

If

$$
x \sim \mathcal{N}(\bar{x}, P)
\tag{1.4.15-2}
$$

then its characteristic (or moment generating) function is

$$
\begin{aligned}
M_x(s) &= E[e^{s'x}] \\
&= e^{\frac{1}{2}s'Ps + s'\bar{x}}
\end{aligned}
\tag{1.4.15-3}
$$

Using the properties of the gradient, one can write

$$
\begin{aligned}
E[x'Ax] &= E\big[(\nabla_s\, e^{s'x})'Ax\big]\Big|_{s=0} \\
&= E\big[\nabla'_s Ax\, e^{s'x}\big]\Big|_{s=0} \\
&= \nabla'_s A\, E\big[x\, e^{s'x}\big]\Big|_{s=0} \\
&= \nabla'_s A\, E\big[\nabla_s\, e^{s'x}\big]\Big|_{s=0} \\
&= \nabla'_s A\, \nabla_s E\big[e^{s'x}\big]\Big|_{s=0} \\
&= \nabla'_s A\, \nabla_s M_x(s)\Big|_{s=0}
\end{aligned}
\tag{1.4.15-4}
$$

Using the characteristic function (1.4.15-3) with $\bar{x} = 0$ in the above yields (1.4.15-1) after the evaluation of the gradient. This technique, even though not the simplest for a quadratic form, can be used conveniently to evaluate the expected value of a quartic form, which is needed for the covariance of quadratic forms.

Note that the expected value of a quadratic form (1.4.15-1) holds regardless of the distribution of x.

45

For the **quartic form** to be considered next, the result will rely on the fact that x is Gaussian.

Analogously to (1.4.15-4), one has the **expected value of a quartic form** written as follows:

$$E[x'Axx'Bx] = \nabla_s' A \, \nabla_s \nabla_s' B \, \nabla_s M_x(s)\big|_{s=0} \qquad (1.4.15\text{-}5)$$

Using (1.4.15-3), it can be shown that after some computations the final result is

$$E[x'Axx'Bx] = \text{tr}(AP)\text{tr}(BP) + 2\text{tr}(APBP) \qquad (1.4.15\text{-}6)$$

The scalar version of the above for $A = B = 1$ and $P = \sigma^2$ is the well-known expression

$$E[x^4] = 3\sigma^4 \qquad (1.4.15\text{-}7)$$

The **covariance of two quadratic forms** is, using (1.4.15-6),

$$
\begin{aligned}
E\big[[x'Ax - E(x'Ax)][x'Bx - E(x'Bx)]\big] &= E[x'Axx'Bx] - E[x'Ax]E[x'Bx] \\
&= 2\text{tr}(APBP) \qquad (1.4.15\text{-}8)
\end{aligned}
$$

These results are used in the state estimation for nonlinear systems where nonlinear functions (system dynamics or measurement equation) are approximated by a series expansion of up to second order.

46

1.4.16 Gaussian Mixture Equations

A **mixture** is a pdf given by a **weighted sum of pdfs** with the weights summing up to unity.

A **Gaussian mixture** is a pdf consisting of a weighted sum of Gaussian densities

$$p(x) = \sum_{j=1}^{n} p_j \mathcal{N}(x; \bar{x}_j, P_j) \tag{1.4.16-1}$$

where, obviously,

$$\sum_{j=1}^{n} p_j = 1 \tag{1.4.16-2}$$

Denote by A_j the event that x is Gaussian distributed with mean $\bar{x}_j$ and covariance P_j, that is,

$$A_j \triangleq \{x \sim \mathcal{N}(\bar{x}_j, P_j)\} \tag{1.4.16-3}$$

With A_j, $j = 1, \ldots, n$, mutually exclusive and exhaustive, and

$$P\{A_j\} = p_j \tag{1.4.16-4}$$

one can rewrite (1.4.16-1) as

$$p(x) = \sum_{j=1}^{n} p(x|A_j) P\{A_j\} \tag{1.4.16-5}$$

The mean of such a mixture is easily seen to be

$$\boxed{\bar{x} = \sum_{j=1}^{n} p_j \bar{x}_j} \tag{1.4.16-6}$$

that is, the weighted sum of the means of the component densities.

The covariance of this mixture is

$$
\begin{aligned}
E[(x - \bar{x})(x - \bar{x})'] &= \sum_{j=1}^{n} E[(x - \bar{x})(x - \bar{x})'|A_j] p_j \\
&= \sum E[(x - \bar{x}_j + \bar{x}_j - \bar{x})(x - \bar{x}_j + \bar{x}_j - \bar{x})'|A_j] p_j \\
&= \sum p_j E[(x - \bar{x}_j)(x - \bar{x}_j)'|A_j] + \sum (\bar{x}_j - \bar{x})(\bar{x}_j - \bar{x})' p_j
\end{aligned}
\tag{1.4.16-7}
$$

which can be written as

$$\boxed{E[(x - \bar{x})(x - \bar{x})'] = \sum_{j=1}^{n} p_j P_j + \tilde{P}} \tag{1.4.16-8}$$

In the above

$$\tilde{P} \triangleq \sum (\bar{x}_j - \bar{x})(\bar{x}_j - \bar{x})' p_j \qquad (1.4.16\text{-}9)$$

is the **spread of the means term**.

An alternative expression for the above is

$$
\begin{aligned}
\tilde{P} &= \sum \bar{x}_j \bar{x}_j' p_j - \bar{x} \sum \bar{x}_j' p_j - \sum \bar{x}_j p_j \bar{x}' + \bar{x}\bar{x}' \sum p_j \\
&= \sum p_j \bar{x}_j \bar{x}_j' - \bar{x}\bar{x}'
\end{aligned}
\qquad (1.4.16\text{-}10)
$$

which yields another form for (1.4.16-8)

$$E[(x - \bar{x})(x - \bar{x})'] = \sum_{j=1}^{n} p_j P_j + \sum_{j=1}^{n} p_j \bar{x}_j \bar{x}_j' - \bar{x}\bar{x}' \qquad (1.4.16\text{-}11)$$

The spread of the means term $\tilde{P}$, defined in (1.4.16-8) is a sum of dyads with positive weightings, and thus, it follows from Subsection 1.3.6 that $\tilde{P}$ is *positive semidefinite*.

Note that (1.4.16-6) and (1.4.16-8), which will be referred to as the **mixture equations**, hold even if the densities in the mixture are *not Gaussian*.

Approximation of a Mixture

A mixture pdf can be approximated by a single Gaussian pdf with moments equal to those of the mixture, given by (1.4.16-6) and (1.4.16-8) — this is called **moment matching**.

Figures 1.4.16-1 and 1.4.16-2 present comparisons between the exact pdf of Gaussian mixtures and the corresponding *moment-matched Gaussian* for several values of the parameters entering into the mixture.

In Figure 1.4.16-1 the difference of the means is 5 while in Figure 1.4.16-2 it is 10; in both cases the (common) standard deviation of the mixture components is $\sqrt{10}$.

The resulting "umbrella" Gaussian appears to be close to the exact pdf of the mixture as long as its components are not too far apart; namely, the distance between the means of the componenets is up to about *two standard deviations*.

This approximate condition is met for the cases depicted in Figure 1.4.16-1, where the match appears good, and it is not met for the cases in Figure 1.4.16-2, where the match does not appear good.

These results are used in obtaining recursive filtering algorithms for systems with hybrid uncertainties — **hybrid systems**. Systems falling into this category are those described by multiple models and/or with measurements of uncertain origin. (See also problem 1-4.)

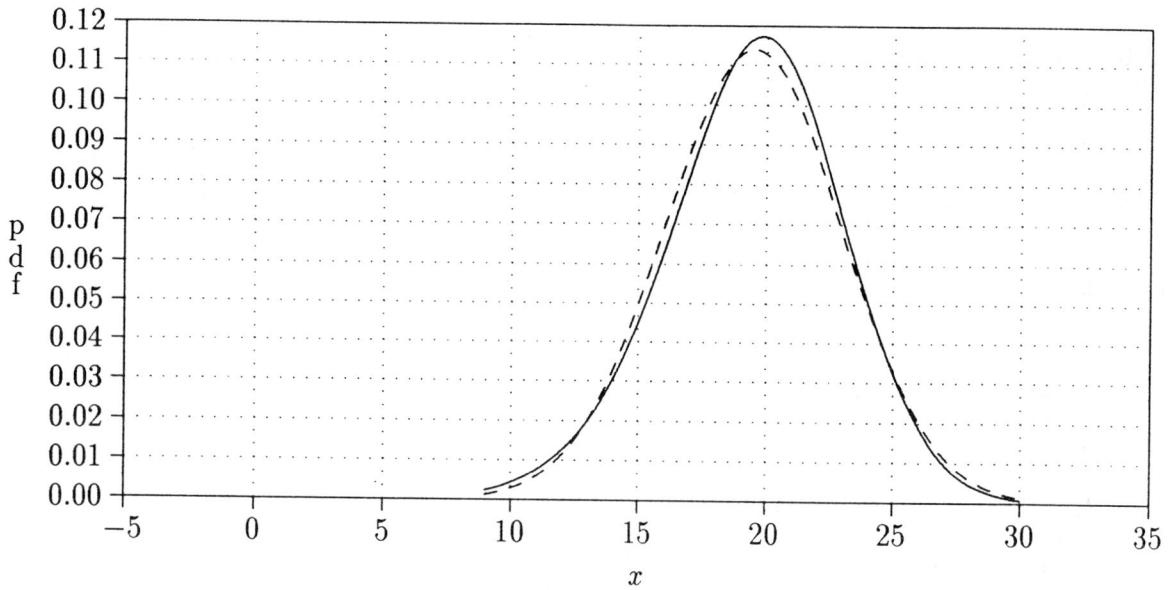

(a) $p_1 = 0.1$, $p_2 = 0.9$

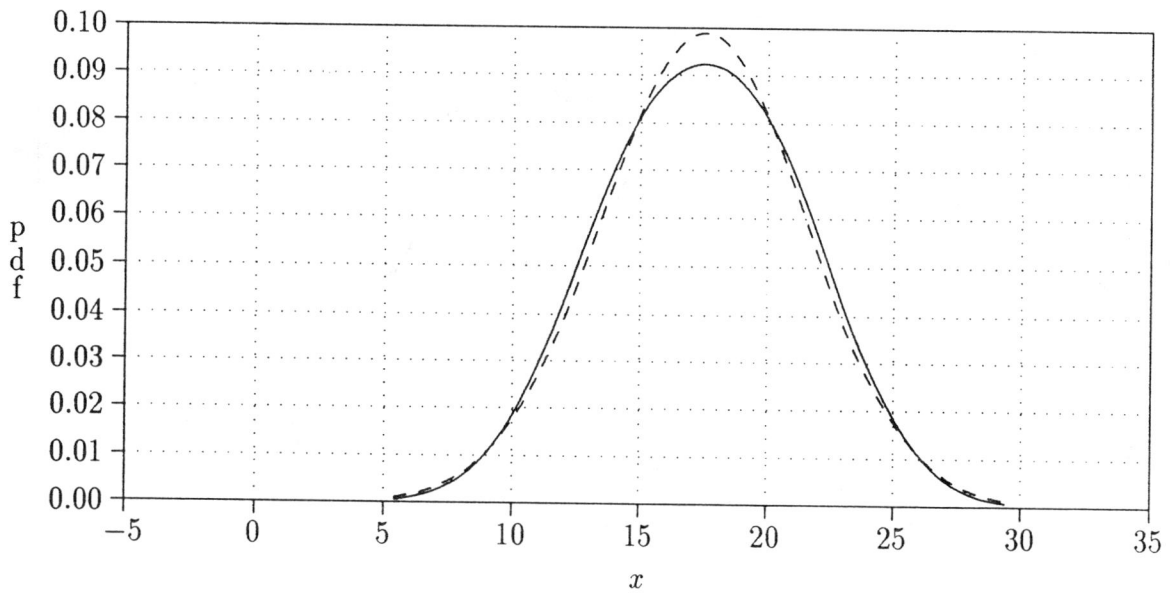

(b) $p_1 = 0.5$, $p_2 = 0.5$

Figure 1.4.16-1: The exact Gaussian mixture pdf (solid line) and the corresponding moment-matched Gaussian (dashed line) for $\bar{x}_1 = 15, \bar{x}_2 = 20, \sigma_1^2 = \sigma_2^2 = 10$.

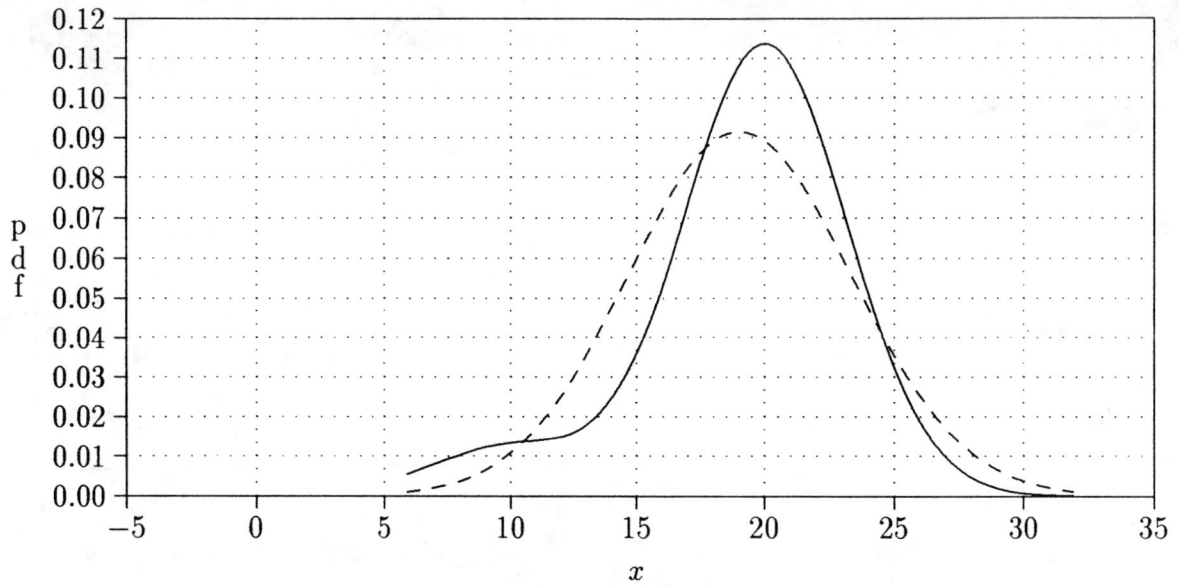

(a) $p_1 = 0.1$, $p_2 = 0.9$

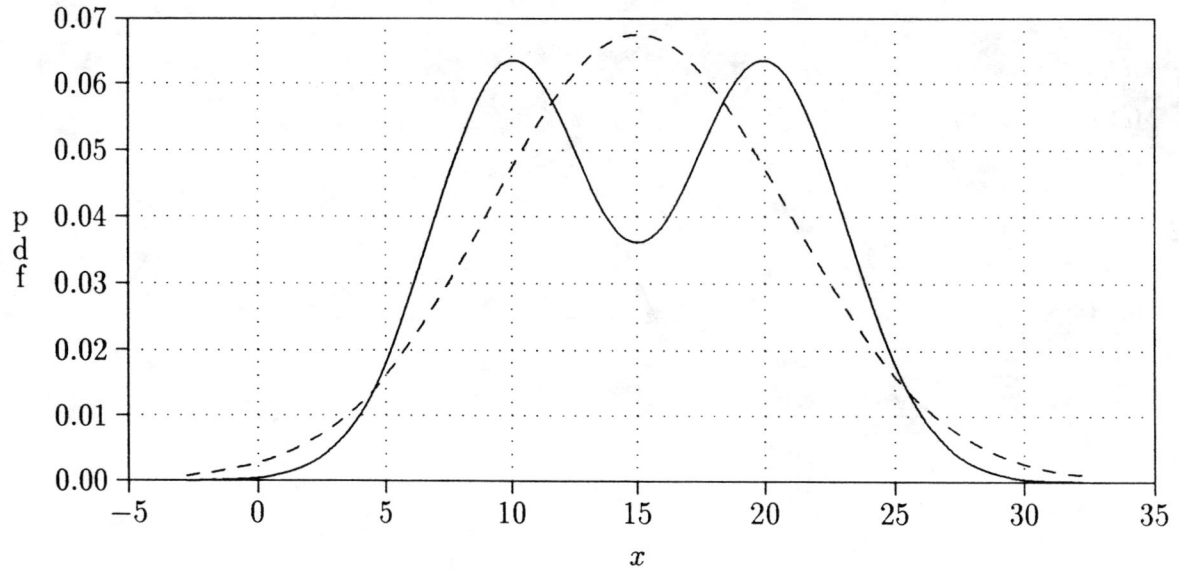

(b) $p_1 = 0.5$, $p_2 = 0.5$

Figure 1.4.16-2: The exact Gaussian mixture pdf (solid line) and the corresponding moment-matched Gaussian (dashed line) for $\bar{x}_1 = 10, \bar{x}_2 = 20, \sigma_1^2 = \sigma_2^2 = 10$.

1.4.17 Chi-Square Distributed Random Variables

If the n-dimensional random vector x is Gaussian, with mean $\bar{x}$ and covariance P, then the (scalar) random variable defined by the quadratic form

$$q = (x - \bar{x})' P^{-1} (x - \bar{x}) \tag{1.4.17-1}$$

can be shown to be the sum of the squares of n independent zero-mean, unity-variance Gaussian random variables. Such a random variable is said to have a **chi-square distribution with n degrees of freedom** (the meaning of degrees of freedom is discussed in Subsection 3.6.2).

This can be seen as follows. Let

$$u \triangleq P^{-1/2} (x - \bar{x}) \tag{1.4.17-2}$$

Then u is Gaussian with

$$E[u] = 0 \tag{1.4.17-3}$$

$$
\begin{aligned}
E[uu'] &= P^{-1/2} E[x - \bar{x}](x - \bar{x})'] P^{-1/2} \\
&= P^{-1/2} P P^{-1/2} \\
&= I
\end{aligned}
\tag{1.4.17-4}
$$

where I is the identity matrix. Therefore, since the covariance matrix of u is diagonal, its components are uncorrelated and, in view of the fact that they are jointly Gaussian, they are also independent.

Thus

$$
\begin{aligned}
q &= u'u \\
&= \sum_{i=1}^{n} u_i^2
\end{aligned}
\tag{1.4.17-5}
$$

where, from (1.4.17-3) and (1.4.17-4) u_i are now **standard Gaussian random variables** (zero mean and unity variance)

$$u_i \sim \mathcal{N}(0, 1) \tag{1.4.17-6}$$

Therefore q is chi-square distributed with n degrees of freedom, which can be written as

$$q \sim \chi_n^2 \tag{1.4.17-7}$$

The mean and variance of the χ_n^2 random variable q are

$$E[q] = E\left[\sum_{i=1}^{n} u_i^2\right] = n \qquad (1.4.17\text{-}8)$$

$$\begin{aligned}
\text{var}(q) &= E\left[\sum_{i=1}^{n}(u_i^2 - 1)\right]^2 = \sum_{i=1}^{n} E[(u_i^2 - 1)^2] \\
&= \sum_{i=1}^{n}\left(E[u_i^4] - 2E[u_i^2] + 1\right) = \sum_{i=1}^{n}(3 - 2 + 1) = 2n \qquad (1.4.17\text{-}9)
\end{aligned}$$

where the fact that the cross-terms are zero mean has been used together with the expression of the fourth moment of a Gaussian variable, as given in (1.4.15-7).

The pdf of q — the chi-square density with n degrees of freedom — is

$$p(q) = \frac{1}{2^{\frac{n}{2}}\Gamma\left(\frac{n}{2}\right)} q^{\frac{n-2}{2}} e^{-\frac{q}{2}} \qquad q \geq 0 \qquad (1.4.17\text{-}10)$$

where Γ is the gamma function, with the following useful properties:

$$\Gamma\left(\frac{1}{2}\right) = \sqrt{\pi} \qquad \Gamma(1) = 1 \qquad \Gamma(m + 1) = m\Gamma(m) \qquad (1.4.17\text{-}11)$$

The square root of q is "chi-distributed." The **chi distributions** for $n = 2$ and 3 are the **Rayleigh distribution** and **Maxwell distribution**, respectively.

Given the *independent* random variables

$$q_1 \sim \chi_{n_1}^2 \qquad\qquad q_2 \sim \chi_{n_2}^2 \qquad (1.4.17\text{-}12)$$

then it can be easily shown that their sum

$$q_3 = q_1 + q_2 \qquad (1.4.17\text{-}13)$$

is chi-square distributed with

$$n_3 = n_1 + n_2 \qquad (1.4.17\text{-}14)$$

degrees of freedom. Symbolically, one can write

$$\chi_{n_1}^2 + \chi_{n_2}^2 = \chi_{n_1+n_2}^2 \qquad (1.4.17\text{-}15)$$

The chi-square distribution is often used to check state estimators for "consistency," that is, whether their actual errors are consistent with the variances calculated by the estimator.

1.4.18 Weighted Sum of Chi-Square Random Variables

Consider the following independent and identically distributed random variables

$$x_i \sim \chi_m^2 \qquad i = 1, \ldots, n \qquad (1.4.18\text{-}1)$$

with m degrees of freedom chi-square distribution.

Then the **weighted sum of chi-square variables** with weights a_i,

$$y_n \triangleq \sum_{i=1}^{n} a_i x_i \qquad (1.4.18\text{-}2)$$

has a distribution which is not chi-square and is very complicated. Its mean is

$$E[y_n] = m \sum_{i=1}^{n} a_i \qquad (1.4.18\text{-}3)$$

and, from (1.4.17-9), its variance is

$$\mathrm{var}(y_n) = 2m \sum_{i=1}^{n} a_i^2 \qquad (1.4.18\text{-}4)$$

The pdf of (1.4.18-2) can be approximated by **moment matching** — equating its first two moments to those of another random variable. In this case, the latter is chosen as a "scaled" chi-square with a number of degrees of freedom n', to be determined together with the scaling factor.

To find a random variable whose moments are matched to those of (1.4.18-2), let

$$v \sim \chi_{n'}^2 \qquad (1.4.18\text{-}5)$$

Then the mean and variance of a "scaled" version of v,

$$w \triangleq cv \qquad (1.4.18\text{-}6)$$

are cn' and $2c^2n'$, respectively. Equating these two moments to (1.4.18-3) and (1.4.18-4) yields the following equations for c and n':

$$m \sum_{i=1}^{n} a_i = cn' \qquad (1.4.18\text{-}7)$$

$$2m \sum_{i=1}^{n} a_i^2 = 2c^2 n' \qquad (1.4.18\text{-}8)$$

The solution of these equations yields the scaling factor as

$$c = \frac{\sum_{i=1}^{n} a_i^2}{\sum_{i=1}^{n} a_i} \qquad (1.4.18\text{-}9)$$

and the number of degrees of freedom of v as

$$n' = \frac{m\left(\sum_{i=1}^{n} a_i\right)^2}{\sum_{i=1}^{n} a_i^2} \qquad (1.4.18\text{-}10)$$

Therefore, the distribution of y_n, defined in (1.4.18-2) is approximately

$$y_n \sim \frac{\sum_{i=1}^{n} a_i^2}{\sum_{i=1}^{n} a_i} \chi_{n'}^2 \qquad (1.4.18\text{-}11)$$

Note that the χ_n^2 density was defined only for integer n; a noninteger n makes it into a case of the gamma density. In practice, one can interpolate from the chi-square tables.

Consider the case where the weights in (1.4.18-2) are exponential, that is,

$$a_i = \alpha^{n-i} \qquad (1.4.18\text{-}12)$$

with $0 < \alpha < 1$. Then the weighted sum

$$z_n \overset{\Delta}{=} \sum_{i=1}^{n} \alpha^{n-i} x_i \qquad (1.4.18\text{-}13)$$

is called the **fading memory sum** of the variables x_i.

Then, for $n \gg 1$ one obtains

$$z_n \sim \frac{1}{1+\alpha} \chi_{n'}^2 \qquad (1.4.18\text{-}14)$$

and

$$n' = m\frac{1+\alpha}{1-\alpha} \qquad (1.4.18\text{-}15)$$

When (1.4.18-13) is normalized by dividing it with the sum of the coefficients, it becomes the **fading memory average**.

These results are used in real-time monitoring of state estimator performance.

1.4.19 Random Processes

A scalar random variable is a (real) number x determined by the outcome ω of a random experiment

$$x = x(\omega) \qquad (1.4.19\text{-}1)$$

A (scalar) **random process** or a **stochastic process** is a *function of time* determined by the outcome of a random experiment

$$x(t) = x(t, \omega) \qquad (1.4.19\text{-}2)$$

This is a family or an **ensemble** of functions of time, in general different for each outcome ω.

The **mean** or **ensemble average** of the random process is

$$
\begin{aligned}
\bar{x}(t) &= E[x(t)] \\
&= \int_{-\infty}^{\infty} \xi \, p_{x(t)}(\xi) d\xi
\end{aligned}
\qquad (1.4.19\text{-}3)
$$

while its **autocorrelation** is defined for a real-valued (scalar) process as

$$
\begin{aligned}
R(t_1, t_2) &\triangleq E[x(t_1)x(t_2)] \\
&= \int_{-\infty}^{\infty} \int_{-\infty}^{\infty} \xi\eta \, p_{x(t_1),x(t_2)}(\xi, \eta) d\xi d\eta
\end{aligned}
\qquad (1.4.19\text{-}4)
$$

The **autocovariance** of this random process is

$$
\begin{aligned}
V(t_1, t_2) &\triangleq E\Big[[x(t_1) - \bar{x}(t_1)][x(t_2) - \bar{x}(t_2)]\Big] \\
&= R(t_1, t_2) - \bar{x}(t_1)\bar{x}(t_2)
\end{aligned}
\qquad (1.4.19\text{-}5)
$$

Note that the autocorrelation of a random process is an (unnormalized) *noncentral moment* — the mean is not subtracted — while the correlation of two random variables is the normalized joint *central moment* (1.4.6-6). If the process is zero mean, the distinction between central and noncentral moments disappears.

For a **vector-valued random process**, (1.4.19-4) and (1.4.19-5) contain the corresponding outer products.

Stationarity

A random process whose mean is time invariant and whose autocorrelation is of the form

$$
\begin{aligned}
R(t_1, t_2) &= R(t_1 - t_2) \\
&= R(\tau)
\end{aligned}
\tag{1.4.19-6}
$$

where

$$
\tau = t_1 - t_2
\tag{1.4.19-7}
$$

that is, its first two moments are invariant with respect to a shift of the time axis, is called **wide sense stationary** or, less rigorously, **stationary**.

As can be seen from (1.4.19-4), for a real-valued process the autocorrelation is symmetric in its two time arguments

$$
R(t_1, t_2) = R(t_2, t_1)
\tag{1.4.19-8}
$$

Thus

$$
R(\tau) = R(-\tau)
\tag{1.4.19-9}
$$

The **power spectrum** or **power spectral density** of a *stationary* random process is the Fourier transform of its autocorrelation

$$
S(\omega) = \int_{-\infty}^{\infty} e^{-j\omega\tau} R(\tau) d\tau
\tag{1.4.19-10}
$$

where ω denotes the (angular) frequency.

Strict stationarity requires that all pdfs (rather than only moments up to second order) be invariant with respect to time shift. In practice wide sense stationarity is about all one can assume (and hope for).

Ergodicity

If a stationary random process is **ergodic**, then **time averages** (of some functions of the process) are equal to the corresponding expected values (*ensemble averages*), given by (1.4.19-3), that is,

$$
\lim_{T \to \infty} \frac{1}{2T} \int_{-T}^{T} x(t) dt = \bar{x}
\tag{1.4.19-11}
$$

Time averages are used in real-time performance monitoring of state estimation filters.

White Noise

A (not necessarily stationary) random process whose autocovariance is zero for any two different times is called **white noise**. In this case

$$
\begin{aligned}
V(t_1, t_2) &= \sigma^2(t_1)\delta(t_1 - t_2) \\
&= S_0(t_1)\delta(t_1 - t_2)
\end{aligned}
\tag{1.4.19-12}
$$

where $\sigma^2(t_1)$ is its "instantaneous variance" and $\delta(\cdot)$ is the **Dirac (impulse) delta function**. The above property is the **wide sense whiteness** as opposed to the **strict sense whiteness**, which is defined by *independence* rather than *uncorrelatedness* as in (1.4.19-12).

The impulse function autocorrelation (1.4.19-12) leads to a power spectral density $S_0(t)$ constant across the frequency spectrum (possibly time-varying); the term **instantaneous variance** is not rigorous in this case since the variance is actually *infinite*.

A **stationary zero-mean white** process has the autocorrelation

$$
\begin{aligned}
R(\tau) &= E[x(t + \tau)x(t)] \\
&= S_0\delta(\tau)
\end{aligned}
\tag{1.4.19-13}
$$

It is possible (and convenient in some applications) to consider a **nonstationary zero-mean white** process $x(t)$. The autocorrelation of such a process is

$$
E[x(t)x(t')] = S_0(t)\delta(t - t')
\tag{1.4.19-14}
$$

where $S_0(t)$ is, with some abuse of language, the "time-varying spectral density" or the "instantaneous variance."

Note on Whiteness

In the sequel, when a result requires whiteness, it usually requires strict sense whiteness. In practice, however, one has only moments (typically up to second order) and thus only wide sense whiteness can be assumed. Then the result is of practical value only when the strict sense whiteness is replaced by wide sense whiteness, although in this case the result is only approximately true in theory.

1.4.20   Random Walk and the Wiener Process

The **Wiener random process** (or Wiener-Levy or Brownian motion) is a limiting form of the **random walk**: the sum of independent steps of size $s \to 0$, equiprobable in each direction, taken at intervals $\Delta \to 0$ such that

$$\frac{s}{\sqrt{\Delta}} \to \sqrt{\alpha} \qquad (1.4.20\text{-}1)$$

where α is a constant. This yields a stochastic process $\mathbf{w}(t)$ with the following pdf [assuming $\mathbf{w}(0) = 0$],

$$p[\mathbf{w}(t)] = \mathcal{N}[\mathbf{w}(t); 0, \alpha t] \qquad (1.4.20\text{-}2)$$

that is, normal, zero mean, and with variance αt.

Note that the Wiener process is *nonstationary*. It relates to the zero-mean white noise, denoted here as $n(t)$, as follows

$$\mathbf{w}(t) = \int_0^t n(\tau)d\tau \qquad (1.4.20\text{-}3)$$

where

$$E[n(t_1)n(t_2)] = \alpha\delta(t_1 - t_2) \qquad (1.4.20\text{-}4)$$

Another way of writing (1.4.20-3) is

$$d\mathbf{w}(t) = n(t)dt \qquad (1.4.20\text{-}5)$$

which shows that the Wiener process is an **independent increment process**. Formally, the white noise is the derivative of the Wiener process; however, this is not rigorous, since the Wiener process is **nowhere differentiable** — its derivative has infinite variance.

The autocorrelation of the Wiener process is

$$E[\mathbf{w}(t_1)\mathbf{w}(t_2)] = \alpha \; \min(t_1, t_2) \qquad (1.4.20\text{-}6)$$

The white noise and the Wiener process are used to model unknown inputs (maneuvers) in state estimation/tracking.

1.4.21 Markov Processes

Markov processes are defined by the following **Markov property**

$$p[x(t)|x(\tau), \tau \leq t_1] = p[x(t)|x(t_1)] \qquad \forall t > t_1 \qquad (1.4.21\text{-}1)$$

that is, the past up to any t_1 is *fully characterized* by the value of the process at t_1.

An equivalent statement to the above is:

"The future is independent of the past if the present is known."

The Wiener process is Markov. This follows from the fact that it is the integral of white noise

$$\mathbf{w}(t) = \mathbf{w}(t_1) + \int_{t_1}^{t} n(\tau)d\tau \qquad (1.4.21\text{-}2)$$

and $n(\tau)$, $\tau \in [t_1, t]$ is independent of $\mathbf{w}(t_1)$.

Furthermore, the state $x(t)$ of a (possibly time-varying) dynamic system driven by white noise $n(t)$,

$$\dot{x}(t) = f[t, x(t), n(t)] \qquad (1.4.21\text{-}3)$$

is a Markov process. In general, both $x(t)$ and $n(t)$ are vector-valued random processes.

Markov Processes with Rational Spectra

Given a *linear time-invariant dynamic system excited by a stationary white noise* $n(t)$ with mean zero

$$\dot{x}(t) = Ax(t) + Bn(t) \qquad (1.4.21\text{-}4)$$

its state is (in steady state — if the system is stable) a **stationary Markov process** with spectrum (power spectral density)

$$S(\omega) = H(j\omega)QH(j\omega)^* \qquad (1.4.21\text{-}5)$$

In the above the asterisk denotes complex conjugate transpose and

$$H(j\omega) = (j\omega I - A)^{-1}B \qquad (1.4.21\text{-}6)$$

is the **transfer function matrix** (from the input to the state) of the system (1.4.21-4) and the autocorrelation of the input (in general, a matrix) is

$$\begin{aligned} R_n(t_1, t_2) &= E[n(t_1)n(t_2)'] \\ &= Q\delta(t_1 - t_2) \end{aligned} \qquad (1.4.21\text{-}7)$$

The matrix Q is sometimes called (nonrigorously) the covariance of $n(t)$; actually it is its *power spectral density*.

Note that (1.4.21-5) is a **rational spectrum** — a ratio of polynomials.

Conversely, one has the following result:

> *Every Markov process with a rational spectrum can be represented as a linear time-invariant system excited by white noise.*

The models used in state estimation are Markov processes — linear or nonlinear systems driven by white noise. If the system is driven by noise that is not white, then it has to undergo **prewhitening**: it is the output of a subsystem driven by white noise.

Prewhitening

The prewhitening is illustrated in Figures 1.4.21-1 and 1.4.21-2. In Figure 1.4.21-1 the system $\mathcal{S}$ with state $x(t)$ is driven by the **autocorrelated noise** (also called **colored noise**) $n(t)$. Since $n(t)$ is not white, $x(t)$ is not a Markov process.

Assume that $n(t)$ can be represented as the output of a system $\mathcal{S}_0$ with white noise input $v(t)$. With the state of the **prewhitening system** or **shaping filter** $\mathcal{S}_0$ denoted as $x_0(t)$, the augmented state of the composite system shown in Figure 1.4.21-2,

$$y(t) \triangleq \left[\begin{array}{c} x(t) \\ x_0(t) \end{array} \right] \tag{1.4.21-8}$$

is then a Markov process.

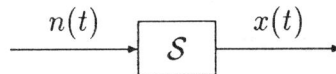

Figure 1.4.21-1: A system driven by autocorrelated noise.

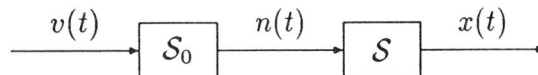

Figure 1.4.21-2: The same system augmented to be driven by white noise.

1.4.22   Random Sequences, Markov Sequences and Markov Chains

A **random sequence**, or a **discrete time stochastic process**, is a time-indexed sequence of random variables

$$X^k = \{x(j)\}_{j=1}^k \qquad k = 1, 2, \ldots \tag{1.4.22-1}$$

Similarly to the continuous-time definition of the Markov property, a random sequence is Markov if

$$p[x(k)|X^j] = p[x(k)|x(j)] \qquad \forall k > j \tag{1.4.22-2}$$

The (real-valued) *zero-mean sequence* $v(j)$, $j = 1, \ldots$, is a **discrete time white noise** (a **white sequence**) if

$$E[v(k)v(j)] = q(k)\delta_{kj} \tag{1.4.22-3}$$

where the **Kronecker delta function**

$$\delta_{kj} = \begin{cases} 1 & \text{if } k = j \\ 0 & \text{if } k \neq j \end{cases} \tag{1.4.22-4}$$

is used and $q(k)$ denotes its variance. If $q(k) = q$, that is, the variance is time invariant, then this is a **stationary white sequence**.

The sequence with property (1.4.22-3) is actually only uncorrelated, or weakly independent — strong independence requires (1.4.7-3) to be satisfied. However, in practice one does not have pdfs but only moments up to second order and, thus, the usual assumption is the **weak independence** (**uncorrelatedness**) indicated in (1.4.22-3).

The state $x(k)$ of a dynamic system excited by white noise $v(k)$

$$x(k+1) = f[k, x(k), v(k)] \tag{1.4.22-5}$$

is a **discrete time Markov process** or **Markov sequence**. In general, both $x(k)$ and $v(k)$ are vector valued.

The state of a linear dynamic system excited by white Gaussian noise

$$x(k+1) = Fx(k) + v(k) \tag{1.4.22-6}$$

is a **Gauss-Markov process**.

The reasons for this are: because of the linearity, $x(k)$ is Gaussian (assuming the initial condition Gaussian); and because of the whiteness of the driving ("process") noise it is Markov.

A special case of (1.4.22-6), for a scalar x, is

$$x(k+1) = x(k) + v(k) \tag{1.4.22-7}$$

in which case x becomes the *integral (sum)* of the white noise sequence terms, and is called a **discrete time Wiener process**.

Markov Chains

A **Markov chain** is a special case of a Markov sequence, in which the state space is finite:

$$x(k) \in \{x_i,\ i = 1, \ldots, n\} \tag{1.4.22-8}$$

Its characterization is given in full by the **transition probabilities**

$$P\{x(k) = x_j \mid x(k-1) = x_i\} \triangleq \pi_{ij} \qquad i, j = 1, \ldots, n \tag{1.4.22-9}$$

and the initial probabilities.

Define the vector

$$\mu(k) \triangleq [\mu_1(k), \ldots, \mu_n(k)]' \tag{1.4.22-10}$$

where

$$\mu_i(k) \triangleq P\{x(k) = x_i\} \tag{1.4.22-11}$$

which describes the pmf of the state of the chain. The evolution in time of (1.4.22-11) is then given by

$$\mu_i(k+1) = \sum_{j=1}^{n} \pi_{ji} \mu_j \qquad i = 1, \ldots, n \tag{1.4.22-12}$$

It can be easily shown that the above can be written in vector form with notation (1.4.22-10) as

$$\mu(k+1) = \Pi' \mu(k) \tag{1.4.22-13}$$

where

$$\Pi = [\pi_{ij}] \tag{1.4.22-14}$$

is the **transition matrix of the Markov chain**.

1.4.23 The Law of Large Numbers and the Central Limit Theorem

The Law of Large Numbers

The **law of large numbers (LLN)** states loosely that the sum of a large number of random variables tends, under some fairly nonrestrictive conditions, to its expected value. One of the versions of the LLN is the following.

Given a stationary sequence of random variables x_i, $i = 1, \ldots$, that are *asymptotically uncorrelated*, that is,

$$E[x_i] = \bar{x} \qquad (1.4.23\text{-}1)$$

$$E[(x_i - \bar{x}_i)(x_j - \bar{x}_j)] = \sigma^2 \rho(i - j) \qquad (1.4.23\text{-}2)$$

where the correlation coefficients are such that

$$\lim_{|i-j| \to \infty} \rho(i - j) = 0 \qquad (1.4.23\text{-}3)$$

that is, the correlations tend to zero. If the correlations in (1.4.23-3) tend to zero "sufficiently fast" (e.g., exponential decay), then the **sample average**

$$y_n \triangleq \frac{1}{n} \sum_{i=1}^{n} x_i \qquad (1.4.23\text{-}4)$$

converges, as $n \to \infty$, in the mean square sense to its expected value

$$
\begin{aligned}
\bar{y}_n &= \frac{1}{n} \sum_{i=1}^{n} \bar{x} \\
&= \bar{x} \qquad (1.4.23\text{-}5)
\end{aligned}
$$

In other words, the variance of y_n in (1.4.23-4) tends to zero as n increases.

If the stronger condition of uncorrelatedness holds, then the convergence of y_n given by (1.4.23-4) to $\bar{y}_n$ given by (1.4.23-5) is quite obvious — it can be easily shown that in this case the mean square value of their difference is σ^2/n.

The Central Limit Theorem

The **central limit theorem (CLT)** states that if the sequence x_i, $i = 1, \ldots$, consists of *independent* random variables, then under some reasonably mild conditions the pdf of the sum

$$z_n = \frac{1}{\sqrt{n}} \sum_{i=1}^{n} x_i \qquad (1.4.23\text{-}6)$$

will tend to a Gaussian pdf as $n \to \infty$.

Since y_n defined in (1.4.23-4) is a scaled version of z_n, the central limit theorem holds for y_n as well. Thus, if the random variables x_i are **independent and identically distributed (i.i.d.)**, then for large n,

$$y_n \sim \mathcal{N}(\bar{y}_n, \sigma_{y_n}^2) \qquad (1.4.23\text{-}7)$$

where

$$\bar{y}_n = \bar{x} \qquad (1.4.23\text{-}8)$$

$$\begin{aligned} \sigma_{y_n}^2 &= \frac{1}{n^2} \sum_{i=1}^{n} \sigma^2 \\ &= \frac{\sigma^2}{n} \end{aligned} \qquad (1.4.23\text{-}9)$$

are the mean and variance of y_n.

The independence requirement can be weakened — there are versions of the central limit theorem that allow dependence, as long as the random variables become "asymptotically independent," in a manner somewhat similar to the asymptotic uncorrelatedness property (1.4.23-3).

Note on the CLT

The CLT has a very important role in characterizing many real-world sources of uncertainty: it is used as the justification/excuse to make the omnipresent Gaussian assumption. For example, the thermal noise in electronic devices, as the sum of many "small contributions," is indeed close to Gaussian.

1.5 BRIEF REVIEW OF STATISTICS

1.5.1 Hypothesis Testing

Consider the **hypothesis testing** problem between the following "simple" hypotheses — each is defined by a point (a certain value of a parameter).

The **null hypothesis** is

$$H_0: \quad \theta = \theta_0 \tag{1.5.1-1}$$

while the **alternate hypothesis** is

$$H_1: \quad \theta = \theta_1 \tag{1.5.1-2}$$

where θ is a certain parameter whose value is equal to a certain value $\theta_0(\theta_1)$ under $H_0(H_1)$.

The **type I error** probability is defined as

$$
\begin{aligned}
P_{e_I} &\triangleq P\{``H_1"|H_0\} \\
&\triangleq P\{\text{accept } H_1|H_0 \text{ true}\}
\end{aligned}
\tag{1.5.1-3}
$$

while the **type II error** probability is

$$
\begin{aligned}
P_{e_{II}} &\triangleq P\{``H_0"|H_1\} \\
&\triangleq P\{\text{accept } H_0|H_1 \text{ true}\}
\end{aligned}
\tag{1.5.1-4}
$$

In signal detection, if H_0 stands for "signal equal to zero" (i.e., absent) and H_1 stands for "signal present," then the type I error is a **false alarm** while the type II error is a **miss**.

The **power of the test** is

$$\pi \triangleq P\{``H_1"|H_1\} = 1 - P_{e_{II}} \tag{1.5.1-5}$$

and it measures the test's capability of discerning H_1 when it is true — the **detection probability**.

The decision as to which hypothesis to accept is made based on a set of observations, Z, whose pdfs conditioned on H_0 and H_1 are known.

According to the **Neyman-Pearson Lemma**, the optimal decision, in the sense of *minimizing the probability of type II error* (or maximizing the power of the test), *subject to a given (maximum) probability of type I error* is as follows.

66

The test, based on the **likelihood ratio**, is

$$\Lambda(H_1, H_0) = \frac{p(Z|H_1)}{p(Z|H_0)} \underset{\text{``}H_0\text{''}}{\overset{\text{``}H_1\text{''}}{\gtrless}} \Lambda_0 \tag{1.5.1-6}$$

that is, "accept H_1" if Λ exceeds the threshold Λ_0 and "accept H_0" if Λ is below this threshold. The threshold Λ_0 is such that

$$P\{\Lambda(H_1, H_0) > \Lambda_0 | H_0\} = P_{e_I} \tag{1.5.1-7}$$

The pdf of the observations, $p(Z|H_j)$, is called the **likelihood function** of H_j.

Example

Consider the test between the hypotheses (1.5.1-1) and (1.5.1-2) with $\theta_0 = 0$ and $\theta_1 > 0$, based on the single observation

$$z \sim \mathcal{N}(\theta, \sigma^2) \tag{1.5.1-8}$$

that is, we are testing whether the mean of z is zero or positive.

The likelihood ratio is

$$\begin{aligned} \Lambda(H_1, H_0) &= \frac{p(z|H_1)}{p(z|H_0)} = e^{-\frac{(z-\theta_1)^2 - z^2}{2\sigma^2}} \\ &= e^{\frac{2z\theta_1 - \theta_1^2}{2\sigma^2}} \end{aligned} \tag{1.5.1-9}$$

Instead of comparing (1.5.1-9) to a threshold, it is convenient to take its logarithm, incorporate the various constants into the threshold, and then determine it.

Since $\theta_1 > 0$, comparing (1.5.1-9) to a threshold is equivalent to

$$z \underset{\text{``}H_0\text{''}}{\overset{\text{``}H_1\text{''}}{\gtrless}} \lambda_0 \tag{1.5.1-10}$$

where λ_0 is the new threshold to be determined.

In view of (1.5.1-8), the pdf of the observation under H_0 is

$$p(z|H_0) = \mathcal{N}(z; 0, \sigma^2) = \frac{1}{\sqrt{2\pi}\sigma} e^{-\frac{z^2}{2\sigma^2}} \tag{1.5.1-11}$$

The threshold λ_0 for **false alarm probability** $P_{FA} = \alpha$ follows from

$$P_{FA} \overset{\Delta}{=} P_{e_I} = P\{z > \lambda_0 | H_0\} = \int_{\lambda_0}^{\infty} \mathcal{N}(z; 0, \sigma^2) dz = \alpha \tag{1.5.1-12}$$

67

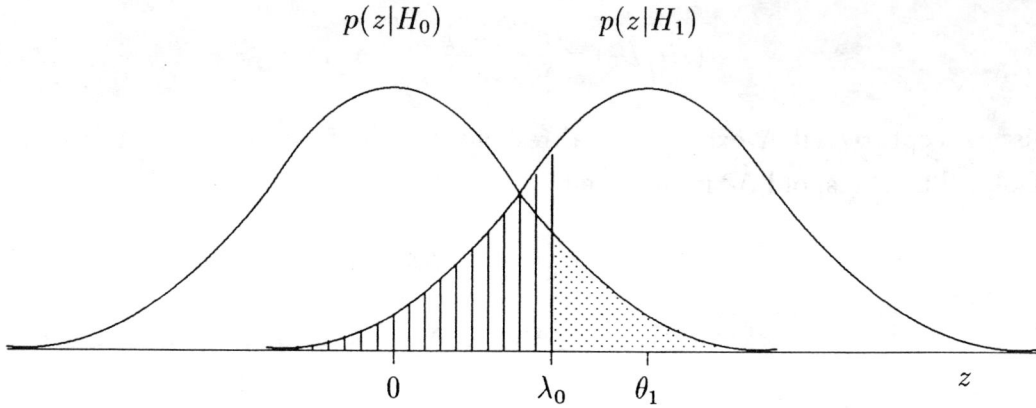

Figure 1.5.1-1: Test between two simple hypotheses.

The **tail mass** or **tail probability** α from (1.5.1-12) is shaded in Figure 1.5.1-1.

Under H_0 there is only a small probability α that z falls in the shaded tail region of the likelihood function of H_0; if it does fall there, then it is deemed unlikely that H_0 is true and is rejected. In this case it is said that the **level of significance** of H_0 is low: α or less. The latter holds if $z = z_1 > \lambda_0$ because such an observation belongs to an even smaller tail mass.

The type II error probability — the **miss probability** — of the test (1.5.1-10) is

$$
\begin{aligned}
P_{e_{II}} &\triangleq P\{z < \lambda_0 | H_1\} \\
&= \int_{-\infty}^{\lambda_0} \mathcal{N}(z; \theta_1, \sigma^2) dz
\end{aligned}
\tag{1.5.1-13}
$$

This corresponds to the striped area of Figure 1.5.1-1.

The power of this test — the **probability of detection** — also denoted as P_D is

$$
\begin{aligned}
P_D &\triangleq \pi \\
&= P\{z > \lambda_0 | H_1\} \\
&= \int_{\lambda_0}^{\infty} \mathcal{N}(z; \theta_1, \sigma^2) dz \\
&= 1 - P_{e_{II}}
\end{aligned}
\tag{1.5.1-14}
$$

and, obviously, it depends on θ_1 — the farther it is from 0, the more powerful the test will be.

Example

For $\alpha = 5\%$, the threshold is obtained (see Subsection 1.5.4) based on the one-sided tail of the Gaussian pdf as $\lambda_0 = 1.64\sigma$. Table 1.5.1-1 shows the power of this test as a function of the separation between the two means (in units of σ, the common standard deviation for both hypotheses).

θ_1/σ	2	3	4	5
Power π	0.641	0.913	0.991	0.9996

Table 1.5.1-1: Power of the test between simple hypotheses as a function of their separation (threshold set for $\alpha = 5\%$).

1.5.2 Confidence Regions and Significance

Assume one desires to test the null hypothesis

$$H_0: \quad \theta = 0 \tag{1.5.2-1}$$

versus the **one-sided alternative hypothesis**

$$H_1': \quad \theta > 0 \tag{1.5.2-2}$$

based upon a set of observations

$$Z = \{z_i, i = 1, \ldots, n\} \tag{1.5.2-3}$$

where

$$z_i = \theta + w_i \tag{1.5.2-4}$$

and the "noises" w_i are independent and identically distributed:

$$w_i \sim \mathcal{N}(0, \sigma^2) \tag{1.5.2-5}$$

Note that H_1' is a **composite hypothesis** — it is defined by more than one point (an interval in this case).

The test is subject to

$$P_{e_I} = \alpha \tag{1.5.2-6}$$

The likelihood ratio for hypothesis H_1' versus H_0 is

$$\begin{aligned}
\Lambda(H_1', H_0) &= \frac{p(Z|\theta)}{p(Z|\theta = 0)} = e^{\frac{1}{2\sigma^2} \sum_{i=1}^{n} [z_i^2 - (z_i - \theta)^2]} \\
&= e^{\frac{1}{2\sigma^2} \sum_{i=1}^{n} (2z_i\theta - \theta^2)}
\end{aligned} \tag{1.5.2-7}$$

The comparison of the above to a threshold is equivalent to comparing the sample mean of the observations to another threshold:

$$\bar{z} \overset{\Delta}{=} \frac{1}{n} \sum_{i=1}^{n} z_i \underset{"H_0"}{\overset{"H_1'"}{\gtrless}} \lambda_1 \tag{1.5.2-8}$$

Equation (1.5.2-8) follows by taking the logarithm of (1.5.2-7) and lumping θ and σ into the (yet undetermined) threshold; the division by n is done for the convenience of normalization — this yields $\bar{z}$ in the same range regardless of the number of samples n.

For this problem $\bar{z}$ is the **test statistic** — the function of the observations used in the test.

The threshold λ_1 is obtained by noting that

$$p(\bar{z}|H_0) = \mathcal{N}(\bar{z}; 0, \frac{\sigma^2}{n}) \tag{1.5.2-9}$$

and putting the condition

$$
\begin{aligned}
P\{``H_1'"|H_0\} &= P\{\bar{z} > \lambda_1|H_0\} \\
&= 1 - \int_{-\infty}^{\lambda_1} \mathcal{N}(\bar{z}; 0, \frac{\sigma^2}{n})d\bar{z} \\
&= \alpha
\end{aligned}
\tag{1.5.2-10}
$$

If the one-sided alternate hypothesis (1.5.2-2) is replaced by the **two-sided alternate hypothesis**, denoted as H_1,

$$H_1: \quad \theta \neq 0 \tag{1.5.2-11}$$

then it can be shown that (1.5.2-8) is to be replaced by

$$|\bar{z}| \underset{``H_0"}{\overset{``H_1"}{\gtrless}} \lambda \tag{1.5.2-12}$$

The threshold λ is obtained by putting the condition

$$
\begin{aligned}
P\{``H_1"|H_0\} &= P\{|\bar{z}| > \lambda|H_0\} \\
&= 1 - \int_{-\lambda}^{\lambda} \mathcal{N}(\bar{z}; 0, \frac{\sigma^2}{n})d\bar{z} \\
&= \alpha
\end{aligned}
\tag{1.5.2-13}
$$

In other words, λ is such that the **acceptance region** for H_0, which is the interval $[-\lambda, \lambda]$, contains $1 - \alpha$ probability mass for the pdf (1.5.2-9).

For $\alpha = 0.05$, one obtains from tables of the normal distribution (see Subsection 1.5.4)

$$\lambda = 1.96\frac{\sigma}{\sqrt{n}} \tag{1.5.2-14}$$

The power of the test depends on the specific value of θ under the alternate hypothesis H_1.

$$p(\bar{z}|H_0)$$

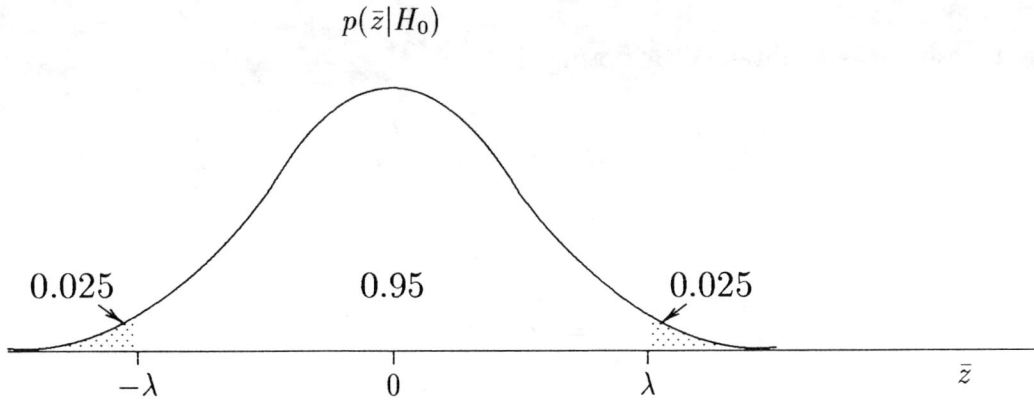

Figure 1.5.2-1: The test between hypotheses H_0 and H_1.

The sample mean $\bar{z}$ is also the **maximum likelihood estimate** (and the **least squares estimate**) of the unknown parameter θ

$$\bar{z} = \arg\max_{\theta} p(Z|\theta) = \arg\min_{\theta} \sum_{i=1}^{n}(z_i - \theta)^2 \qquad (1.5.2\text{-}15)$$

since $p(Z|\theta)$ is the likelihood function of the parameter θ.

Figure 1.5.2-1 illustrates the acceptance region for H_0 — the *two-sided* 95% **probability region** for $\bar{z}$ under hypothesis H_0 — the interval $[-\lambda, \lambda]$. The shaded areas represent the tails on the two sides of $p(\bar{z}|H_0)$, whose total probability mass is $\alpha = 0.05$; this is the region of rejection of H_0 (acceptance of H_1).

Since the difference between the sample mean $\bar{z}$ and the true mean θ (whatever its value) is

$$\bar{z} - \theta \sim \mathcal{N}(0, \frac{\sigma^2}{n}) \qquad (1.5.2\text{-}16)$$

one can say that the true mean lies within the interval $[\bar{z} - \lambda, \bar{z} + \lambda]$ with "confidence" $1 - \alpha$ (since θ is not a random variable, no probabilistic statement about it can be made). In view of this, such an interval is called the **confidence region** for θ.

Another way of interpreting test (1.5.2-12) is the following: if the null hypothesis value $\theta = 0$ falls within the confidence region $[\bar{z} - \lambda, \bar{z} + \lambda]$, then H_0 is accepted. This is equivalent to $\bar{z}$ falling within the acceptance region $[-\lambda, \lambda]$.

If the estimate $\hat{\theta} = \bar{z}$ of θ falls outside the interval $[-\lambda, \lambda]$, then it is said to be a **significant estimate** — hypothesis H_1 is accepted because H_0 is **insignificant**.

Note that, while $\bar{z}$ is the **point estimate** of θ, the confidence region around $\bar{z}$ can be seen as an **interval estimate** of the parameter θ.

Figure 1.5.2-2 illustrates these concepts.

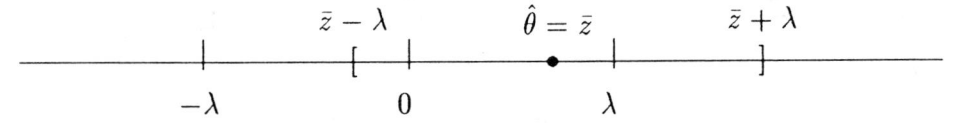

(a) $\left\{ \bar{z} \in [-\lambda, \lambda] \iff 0 \in [\bar{z} - \lambda, \bar{z} + \lambda] \right\} \implies H_1 : \theta = 0$ accepted

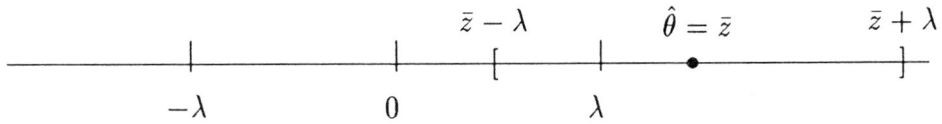

(b) $\left\{ \bar{z} \notin [-\lambda, \lambda] \iff 0 \notin [\bar{z} - \lambda, \bar{z} + \lambda] \right\} \implies H_1 : \theta \neq 0$ accepted

Figure 1.5.2-2: Parameter estimate. (a) $\hat{\theta}$ insignificant; (b) $\hat{\theta}$ significant.

Thus the estimate of a parameter that might be zero is accepted as *significant* (at a "level" of 5% for the null hypothesis) if

$$\frac{|\hat{\theta}|}{\sigma_{\hat{\theta}}} > 1.96 \qquad (1.5.2\text{-}17)$$

where the threshold has been determined under the normal assumption and

$$\sigma_{\hat{\theta}} = \frac{\sigma}{\sqrt{n}} \qquad (1.5.2\text{-}18)$$

is the standard deviation of the estimate, or its **standard error**.

Defining the **estimation error** as

$$\tilde{\theta} \triangleq \theta - \hat{\theta} \qquad (1.5.2\text{-}19)$$

one can see that

$$
\begin{aligned}
p(\hat{\theta}|\theta) &= \mathcal{N}(\hat{\theta}; \theta, \sigma_{\hat{\theta}}^2) = \mathcal{N}(\hat{\theta} - \theta; 0, \sigma_{\hat{\theta}}^2) \\
&= \mathcal{N}(\tilde{\theta}; 0, \sigma_{\hat{\theta}}^2) \triangleq p(\tilde{\theta}) = \mathcal{N}(\tilde{\theta}; 0, \sigma_{\tilde{\theta}}^2)
\end{aligned} \qquad (1.5.2\text{-}20)
$$

Therefore (1.5.2-18), which is the **standard deviation of the estimate** (about the true mean), $\sigma_{\hat{\theta}}$, is the same as $\sigma_{\tilde{\theta}}$, the **standard deviation of the estimation error** (which is zero mean), that is,

$$\sigma_{\hat{\theta}} = \sigma_{\tilde{\theta}} \qquad (1.5.2\text{-}21)$$

The (statistical) significance of parameter estimates is used in choosing the order of models to describe system equations (e.g., for target motion). (See also problem 1-2.)

73

1.5.3 Monte Carlo Runs and Comparison of Algorithms

The performance of an estimation (or control) algorithm is usually the expected value of a "cost" function

$$J = E[C] \tag{1.5.3-1}$$

If C is, for instance, the squared error in the estimation of a certain variable, then J is the corresponding mean square error.

In many situations, the performance of an algorithm of interest cannot be evaluated analytically. [7] In such a case, **Monte Carlo simulations (runs)** are made to obtain an estimate of J from a sample average of *independent* realizations C_i, $i = 1, \ldots, N$, of the cost C. The larger the number of such runs, the smaller is the variability (error) of the resulting estimate. Also, a larger number of runs increases the power of the hypothesis testing used in comparing different algorithms.

The estimate of the performance from N independent runs is the **sample average** (or **sample mean**) of the N realizations of the cost

$$\bar{C} = \frac{1}{N} \sum_{i=1}^{N} C_i \tag{1.5.3-2}$$

with the associated standard deviation — the **standard error**

$$\sigma_{\bar{C}} = \sqrt{\frac{1}{N^2} \sum_{i=1}^{N} (C_i - \bar{C})^2} \tag{1.5.3-3}$$

The above follows from the fact that the variance of the sample mean $\bar{C}$ is the variance of C divided by N; since this variance is not known, it is replaced by the sample variance, which has another N in the denominator. This is the reason for having N^2 in the denominator of (1.5.3-3).

These can be used to obtain a confidence region for the performance assuming that its distribution is, in view of the central limit theorem, approximately normal. Since the pdf of (1.5.3-2) is in general skewed, confidence regions based on the normal assumption might not be accurate unless N is very large — possibly of the order of thousands.

---

[7]Those who can, do. Those who cannot, simulate.

Algorithm Comparison as a Hypothesis Testing Problem

When two algorithms are compared based on simulations, this can be formulated as a hypothesis testing problem as follows.

Assume one has the sample mean of the performance (e.g., MSE, or some other cost to be minimized) for algorithm j from N independent runs

$$\bar{C}^{(j)} = \frac{1}{N} \sum_{i=1}^{N} C_i^{(j)} \qquad j = 1, 2 \qquad (1.5.3\text{-}4)$$

where $C_i^{(j)}$ is the performance of algorithm j in run i and $C_i^{(j)}$ is independent of $C_k^{(j)}$, $\forall k \neq i$. Each pair of runs i uses the *same random variables for the two algorithms* $j = 1, 2$. As shown in problem 1-3, it is *beneficial* to use the same random variables for both algorithms in the same run.

Some Simple Comparison Techniques

Assume that from the simulations one has

$$\bar{C}^{(1)} < \bar{C}^{(2)} \qquad (1.5.3\text{-}5)$$

This does not necessarily imply that algorithm 1 is better than algorithm 2. Any such statement must be qualified by a probability α of error of type I.

Comparison technique #0, based on (1.5.3-5) without any statistical analysis of the individual run outcomes, is clearly *naive*.

Comparison technique #1 — the simplest (but still simplistic) statistical approach — consists of the following:

1. Calculation of confidence regions around the two sample means (based on (1.5.3-2) and (1.5.3-3)).
2. If they do not overlap, to declare that the algorithm with the smaller sample mean is the superior one.

The last step above implicitly assumes the two sample means to be uncorrelated. Note that this is *incorrect*, since each pair of runs uses the same random numbers.

75

The Optimal Comparison Technique

Since $C_i^{(1)}$ is correlated with $C_i^{(2)}$ (because they use the same random variables), the optimal test, which will be based on *independent samples*, is as follows.

The hypothesis testing problem is:

$$H_0 : \ \Delta = J^{(2)} - J^{(1)} \leq 0 \quad \text{(algorithm 1 not better than 2)} \tag{1.5.3-6}$$

versus

$$H_1 : \ \Delta = J^{(2)} - J^{(1)} > 0 \quad \text{(algorithm 1 better than 2)} \tag{1.5.3-7}$$

subject to

$$P\{\text{accept } H_1 | H_0 \text{ true}\} = \alpha \quad \text{(level of significance (of hypothesis } H_0\text{))} \tag{1.5.3-8}$$

where

$$J^{(j)} = E[C^{(j)}] \qquad j = 1, 2 \tag{1.5.3-9}$$

are the *true expected values of the cost functions (true performance)*.

The decision whether H_1 ("1" better than "2") should be accepted is made based upon the **sample performance differences**

$$\boxed{\Delta_i = C_i^{(2)} - C_i^{(1)}} \tag{1.5.3-10}$$

Note that Δ_i is independent of Δ_k, $\forall k \neq i$.

Then, H_1 is accepted if

$$\boxed{\mu \stackrel{\Delta}{=} \frac{\bar{\Delta}}{\sigma_{\bar{\Delta}}} > \mu_0} \tag{1.5.3-11}$$

where

$$\boxed{\bar{\Delta} = \frac{1}{N} \sum_{i=1}^{N} \Delta_i} \tag{1.5.3-12}$$

and

$$\boxed{\sigma_{\bar{\Delta}} = \sqrt{\frac{1}{N^2} \sum_{i=1}^{N} (\Delta_i - \bar{\Delta})^2}} \tag{1.5.3-13}$$

are the *sample mean* of the differences (1.5.3-10) and the *standard error* of this sample mean, respectively.

Assuming the error in $\bar{\Delta}$ to be normal, the threshold μ_0 is based on the *upper tail* of the normal density: $\mu_0 = 1.64$ for $\alpha = 5\%$, $\mu_0 = 2.33$ for $\alpha = 1\%$, and so on — (see Subsection 1.5.4). This follows from the fact that the test is for *positive mean (H_1) versus zero or negative mean* and H_1 is accepted if and only if the sample mean (1.5.3-12) is positive and statistically significant (1.5.3-11); α is the probability of accepting H_1 (positive mean) when the true mean is zero.

This procedure is summarized below:

$C_1^{(1)}$	$C_1^{(2)}$	Δ_1
$\vdots$	$\vdots$	$\vdots$
$C_N^{(1)}$	$C_N^{(2)}$	Δ_N
$\bar{C}^{(1)}$	$\bar{C}^{(2)}$	$\bar{\Delta}, \sigma_{\bar{\Delta}}$

Remarks

The optimal algorithm comparison is based on the significance test of the *sample mean of the differences* (the mean of the terms in the last column above) instead of the naive comparison of the performance estimates from the first two columns (technique #0) or using two confidence regions around these estimates (technique #1).

The applicability of the CLT is much more realistic on the average difference (1.5.3-12) than on the average outcomes (1.5.3-4) since the differences are much less skewed (and independent from run to run).

The fact that the same random numbers are used for the two algorithms leads to a *positive correlation* between the two sample means, and the optimal test presented takes advantage of this. (See also problem 1-3.)

Section 11.5 presents a major example of the application of this optimal comparison technique.

1.5.4 Tables of the Chi-Square and Gaussian Distributions

Table 1.5.4-1 presents the points x on the chi-square distribution for a given **upper tail probability**

$$Q = P\{y > x\} \tag{1.5.4-1}$$

where

$$y \sim \chi_n^2 \tag{1.5.4-2}$$

and n is the number of degrees of freedom. This tabulated function is also known as the **complementary distribution**.

An alternative way of writing (1.5.4-1) is

$$x(1 - Q) \triangleq \chi_n^2(1 - Q) \tag{1.5.4-3}$$

which indicates that to the left of the point x the probability mass is $1 - Q$. This is the $100(1 - Q)$ **percentile point**.

The 95% probability region for a χ_2^2 variable can be taken as the **one-sided probability region** (cutting off the 5% **upper tail**)

$$[0, \chi_2^2(0.95)] = [0, 5.99] \tag{1.5.4-4}$$

or the **two-sided probability region** (cutting off both 2.5% tails)

$$[\chi_2^2(0.025), \chi_2^2(0.975)] = [0.05, 7.38] \tag{1.5.4-5}$$

For a χ_{100}^2 variable, the two-sided 95% probability region is

$$[\chi_{100}^2(0.025), \chi_{100}^2(0.975)] = [74, 130] \tag{1.5.4-6}$$

Note the skewedness of the chi-square distribution: the above two-sided regions are not symmetric about the corresponding means (2 and 100, respectively, in the above two equations).

For degrees of freedom above 100, the following approximation of the points on the chi-square distribution can be used:

$$\chi_n^2(1 - Q) = \frac{1}{2}\left[\mathcal{G}(1 - Q) + \sqrt{2n - 1}\right]^2 \tag{1.5.4-7}$$

For example, $\chi_{400}^2(0.025) = 346$, $\chi_{400}^2(0.975) = 457$.

$n \backslash Q$	0.99	0.975	0.95	0.90	0.75	0.5	0.25	0.10	0.05	0.025	0.01	5E-3	1E-3
1	2E-4	.001	.003	.016	.102	.455	1.32	2.71	3.84	5.02	6.63	7.88	10.8
2	.020	.051	.103	.211	.575	.139	.277	4.61	5.99	7.38	9.21	10.6	13.8
3	.115	.216	.352	.584	1.21	2.37	4.11	6.25	7.81	9.35	11.3	12.8	16.3
4	.297	.484	.711	1.06	1.92	3.36	5.39	7.78	9.49	11.1	13.3	14.9	18.5
5	.554	.831	1.15	1.61	2.67	4.35	6.63	9.24	11.1	12.8	15.1	16.7	20.5
6	.872	1.24	1.64	2.20	3.35	5.35	7.84	10.6	12.6	14.4	16.8	18.5	22.5
7	1.24	1.69	2.17	2.83	4.25	6.35	9.04	12.0	14.1	16.1	18.5	20.3	24.3
8	1.65	2.18	2.73	3.49	5.07	7.34	10.2	13.4	15.5	17.5	20.1	22.0	26.1
9	2.09	2.70	3.33	4.17	5.90	8.34	11.4	14.7	17.0	19.0	21.7	23.6	27.9
10	2.56	3.25	3.94	4.87	6.74	9.34	12.5	16.0	18.3	20.5	23.2	25.2	29.6
11	3.05	3.82	4.57	5.58	7.58	10.3	13.7	17.3	19.7	22.0	24.7	26.8	31.3
12	3.57	4.40	5.23	6.30	8.44	11.3	14.8	18.5	21.0	23.3	26.2	28.3	32.9
13	4.11	5.01	5.90	7.04	9.30	12.3	16.0	19.8	22.4	24.7	27.7	29.8	34.5
14	4.66	5.63	6.57	7.79	10.2	13.3	17.1	21.1	23.7	26.1	29.1	31.3	36.1
15	5.23	6.26	7.26	8.55	11.0	14.3	18.2	22.3	25.0	27.5	30.6	32.8	37.7
16	5.81	6.91	7.96	9.31	11.9	15.3	19.4	23.5	26.3	28.8	32.0	34.3	39.3
17	6.41	7.56	8.67	10.1	12.8	16.3	20.5	24.8	27.6	30.2	33.4	35.7	40.8
18	7.01	8.23	9.40	10.9	13.7	17.3	21.6	26.0	28.9	31.5	34.8	37.2	42.3
19	7.63	8.91	10.1	11.7	14.6	18.3	22.7	27.2	30.1	32.9	36.2	38.6	43.8
20	8.26	9.60	10.9	12.4	15.5	19.3	23.8	28.4	31.4	34.2	37.6	40.0	45.3
25	11.5	13.1	14.6	16.5	19.9	24.3	29.3	34.4	37.7	40.6	44.3	46.9	52.6
30	15.0	16.8	18.5	20.6	24.5	29.3	34.8	40.3	43.8	47.0	50.9	53.7	59.7
40	22.2	24.4	26.5	29.1	33.7	39.3	45.6	51.8	55.8	59.3	63.7	66.8	73.4
50	29.7	32.4	34.8	37.7	43.0	49.3	56.3	63.2	67.5	71.4	76.2	79.5	86.7
60	37.5	40.5	43.2	46.5	52.3	59.3	67.0	74.4	79.1	83.3	88.4	92.0	99.6
70	45.4	48.8	51.7	55.3	61.7	69.3	77.6	85.5	90.5	95.0	100	104	112
80	53.5	57.2	60.4	64.2	71.1	79.3	88.1	96.6	102	107	112	116	125
90	61.8	65.6	69.1	73.3	80.6	89.3	98.6	108	113	118	124	128	137
100	70.1	74.2	77.9	82.4	90.1	99.3	109	118	124	130	136	140	149
$\mathcal{G}()$	-2.33	-1.96	-1.64	-1.28	-.675	0	.675	1.28	1.64	1.96	2.33	2.58	3.09

Table 1.5.4-1: Tail probabilities of the chi-square and normal densities.

The last line of Table 1.5.4-1 shows the points x on the **standard** (zero mean and unity variance) **normal (Gaussian)** distribution for the same tail probabilities. In this case

$$y \sim \mathcal{N}(0,1) \tag{1.5.4-8}$$

and, with

$$Q = P\{y > x\} \tag{1.5.4-9}$$

these points will be denoted as

$$x(1-Q) \triangleq \mathcal{G}(1-Q) \tag{1.5.4-10}$$

Thus, with this notation the 95% **two-sided probability region** for an $\mathcal{N}(0,1)$ random variable is

$$[\mathcal{G}(0.025), \mathcal{G}(0.975)] = [-1.96, 1.96] \tag{1.5.4-11}$$

In terms of the cumulative distribution function (cdf) of a standard Gaussian random variable

$$\begin{aligned} P_{\mathcal{G}}(x) &\triangleq P\{y \le x\} \\ &= \int_{-\infty}^{x} \mathcal{N}(y; 0, 1) dy \\ &= \frac{1}{\sqrt{2\pi}} \int_{-\infty}^{x} e^{-\frac{y^2}{2}} dy \end{aligned} \tag{1.5.4-12}$$

one can write (1.5.4-11) as

$$P_{\mathcal{G}}(1.96) = 0.975 \tag{1.5.4-13}$$

$$P_{\mathcal{G}}(-1.96) = 0.025 \tag{1.5.4-14}$$

Note that $P_{\mathcal{G}}$ defined in (1.5.4-12) is the inverse function of $\mathcal{G}$ defined in (1.5.4-10), that is, $\mathcal{G}$ is the inverse cdf.

The 95% **one-sided probability region** for a standard Gaussian random variable is given by its 95% point:

$$(-\infty, \mathcal{G}(0.95)] = (-\infty, 1.64] \tag{1.5.4-15}$$

1.6 NOTES AND PROBLEMS

1.6.1 Bibliographical Notes

An excellent historical survey of the developments of estimation theory, starting from the least squares technique of Gauss and Legendre, the maximum likelihood technique of R. A. Fisher, the Wiener-Kolmogorov filtering for signals in noise, and the emergence of digital computer oriented recursive algorithms developed by Swerling, Kalman, Bucy, and others, can be found in [Sor85].

For background material, further references on linear algebra as well as linear systems are [ZD63, FH77, Che84]. The existing literature on this topic consists of tens of books and most of them are equally good. A succinct summary of this material can be found in [Gel74]. The probabilistic tools are covered in [Pap84] and succinctly in [Gel74]. The statistical tools (hypothesis testing) can be found, for example, in [MC78, Joh72], among others.

For the more specialized target tracking topics — in particular tracking in with uncertain origin observations and multitarget tracking, the reader is referred to [Bla86, BF88].

The first author's short course notes *Multitarget-Multisensor Tracking: Principles and Techniques* (1993 edition) are the most recent compilation of algorithms in this area, with suitable explanations and design guidelines.

1.6.2 Problems

1-1 **Independence versus conditional independence.** Given x and y independent Gaussian random variables with zero mean and unity variance.

1. For an arbitrary new random variable does the following hold

$$p(x, y|z) = p(x|z)p(y|z) \quad ?$$

2. Let $z = x + y$. Write the explicit expression of the pdf $p(x, y|z)$. Use the Dirac delta function if necessary.

1-2 **Monte Carlo runs for low probability events.** Given an experiment with binary outcome x, assume your colleague did a theoretical calculation that indicates that

$$P\{x = 1\} = \hat{p} = 10^{-4}$$

However, you do not believe this colleague and plan to conduct N Monte-Carlo (i.i.d.) repetitions of the experiment. The outcomes are x_i, $i = 1, \ldots, N$. Estimate, using confidence regions, the number N of experiments you have to carry out to confirm or refute your colleague's theoretical calculation.

1-3 **Monte Carlo runs for algorithm comparison.** Two estimation algorithms are to be compared based on Monte Carlo runs. Assume their average performances (MSE) were obtained as $\bar{C}^{(1)} = 5$ and $\bar{C}^{(2)} = 8$ from $N = 100$ runs.

1. Can one say that algorithm 1 is superior to algorithm 2?

2. Assume the standard deviations associated with these sample means are

$$\sigma_{\bar{C}^{(j)}} = \left[\frac{1}{N^2} \sum_{i=1}^{N} \left[C_i^{(j)} - \bar{C}^{(j)} \right]^2 \right]^{1/2} = 1 \qquad j = 1, 2$$

Using the above information answer question 1.

3. Denote the correlation coefficient between the outcomes of the runs with the two algorithms as ρ, that is,

$$E \left[(C_i^{(1)} - J^{(1)})(C_i^{(2)} - J^{(2)}) \right] = \rho \sigma_{C^{(1)}} \sigma_{C^{(2)}}$$

Show that, if $\rho = 0.5$, the rigorous statistical test will yield a conclusive answer.

1-4 **Covariance of a mixture.** Prove that an equivalent expression of the "spread of the means" term (1.4.16-9) in the covariance of a mixture is

$$\tilde{P} = \sum_i \sum_{j<i} p_i p_j (\bar{x}_i - \bar{x}_j)(\bar{x}_i - \bar{x}_j)'$$

where the summation is over all the pairs of indices without repetition.

1-5 **Dimension of a pdf.**

1. Find the physical dimension of the pdf of the random vector x when x is
   a. The position of a point in an n-dimensional Euclidean space.
   b. The state of a constant acceleration point moving along, say, the x-axis.
2. Find the units of the pdf of a velocity measured in furlongs/fortnight.
3. Can one compare the pdf of a random variable whose dimension is length to the joint pdf of two such random variables?

1-6 **Log-normal random variable.** Given the random variable $y \sim \mathcal{N}(\mu, \sigma)$, then $x \triangleq e^y$ is called a log-normal random variable. Find its mean and variance.

1-7 **Moments of linear transformation of random variables.** Given the random variables x and y of dimensions n_x and n_y, with means $\bar{x}$ and $\bar{y}$, respectively, and with covariances P_{xx}, P_{yy}, P_{xy}:

1. Find the mean and covariance of the n_z-dimensional vector

$$z = Ax + By + c$$

where A, and B are matrices of appropriate dimensions.

2. Indicate the dimensions of A, B, and c.

1-8 Chapman-Kolmogorov equation. Prove that the following equation holds for a (discrete time) Markov process

$$\int p[x(k)|x(k-1)]p[x(k-1)|x(k-2)]dx(k-1) = p[x(k)|x(k-2)]$$

1-9 Hypothesis testing with correlated noise. Given the hypotheses

$$H_0 : \theta = 0 \qquad\qquad H_1 : \theta \neq 0$$

and the observations

$$z_i = \theta + w_i \qquad i = 1, \dots, n$$

with w_i zero-mean jointly Gaussian but not independent. Denoting

$$w = [w_1 \cdots w_n]'$$

one has the covariance matrix (assumed given)

$$E[ww'] = P$$

For the above

1. Specify the optimal hypothesis test for false alarm probability α.

2. Solve explicitly for $n = 2$, $P = \begin{bmatrix} 1 & 0.5 \\ 0.5 & 1 \end{bmatrix}$ and $\alpha = 1\%$.

1-10 Partial derivative with respect to a matrix. The partial derivative of a scalar q with respect to the matrix $A = [a_{ij}]$ is defined as

$$\frac{\partial q}{\partial A} \triangleq \left[\frac{\partial q}{\partial a_{ij}} \right]$$

Prove that

1. For B symmetric, $\frac{\partial}{\partial A}\text{tr}[ABA'] = 2AB$.

2. For B not symmetric, $\frac{\partial}{\partial A}\text{tr}[AB] = B'$.

1-11 Fourth joint moment of Gaussian random variables. Show that, if the zero-mean scalar random variables x_i, $i = 1, \dots, 4$, are jointly Gaussian, then

$$E[x_1 x_2 x_3 x_4] = E[x_1 x_2]E[x_3 x_4] + E[x_1 x_3]E[x_2 x_4] + E[x_1 x_4]E[x_2 x_3]$$

Hint: Use the characteristic function (1.4.15-3) and take a suitable fourth derivative of it.

1-12 Wiener process increments.

1. Using (1.4.20-3) show that

$$E[d\mathbf{w}(t)]^2 = \alpha dt \qquad \text{and} \qquad d\mathbf{w}(t) \sim \mathcal{N}(0, \alpha dt)$$

2. Using the above, show that the limit of $\frac{d\mathbf{w}(t)}{dt}$ tends to ∞. Hint: find the "order of magnitude" of $d\mathbf{w}(t)$ in terms of dt (a zero-mean random variable is of the order of its standard deviation).

83

1-13 **Conditional pdf of the sum of two Gaussian random variables.** Find the pdf (1.4.13-12) in the general case (1.4.13-6).

1-14 **Probability matrix.** Find $\sum_{j=1}^{n} \pi_{ij}$ for (1.4.22-9).

Chapter 2

BASIC CONCEPTS IN ESTIMATION

2.1 INTRODUCTION

2.1.1 Outline

This chapter introduces some of the basic techniques of estimation that provide the foundation for state estimation and tracking.

The problem of parameter estimation is defined in Section 2.2, where the two models for unknown parameters most commonly used (nonrandom and random) are also described. Section 2.3 deals with the maximum likelihood (ML) and maximum a posteriori (MAP) estimates. The least squares (LS) and minimum mean square error (MMSE) estimates are presented in Section 2.4.

The remaining sections deal with various "measures of quality" of estimators. Section 2.5 discusses unbiasedness, Section 2.6 the variances of estimators. The consistency of estimators is discussed in Section 2.7, together with the Cramer-Rao lower bound, the Fisher information, and estimator efficiency.

2.1.2   Basic Concepts – Summary of Objectives

Distinguish between

- Random parameters   *Bayesian approach*
- Nonrandom parameters.

Define the following estimates

- Maximum likelihood
- Maximum a posteriori
- Least squares
- Minimum mean square error.

Present "measures of quality" of estimators

- Unbiasedness
- Variance
- Consistency
- The Cramer-Rao lower bound, Fisher information
- Efficiency.

2.2 THE PROBLEM OF PARAMETER ESTIMATION

2.2.1 Definitions

The term **parameter** is used to designate a quantity (scalar or vector valued) that is assumed to be *time invariant*. If it does change with time it can be designated (with a slight abuse of language) as a "time-varying parameter," but its time variation must be slow compared to the state variables of a system.

The problem of estimating a (time invariant) parameter x is the following. Given the measurements

$$z(j) = h[j, x, w(j)] \qquad j = 1, \ldots, k \qquad (2.2.1\text{-}1)$$

made in the presence of the disturbances (noises) $w(j)$, find a function of the k observations

$$\boxed{\hat{x}(k) \triangleq \hat{x}[k, Z^k]} \qquad (2.2.1\text{-}2)$$

where these observations are denoted compactly as

$$Z^k \triangleq \{z(j)\}_{j=1}^k \qquad (2.2.1\text{-}3)$$

that estimates the value of x in some sense.

The function (2.2.1-2) is called the **estimator**. The value of this function is the **estimate**. These terms, while not the same, will be used (sometimes) interchangeably.

The **estimation error** corresponding to the estimate $\hat{x}$ is

$$\boxed{\tilde{x} \triangleq x - \hat{x}} \qquad (2.2.1\text{-}4)$$

An alternate notation instead of (2.2.1-2) that will be used when k is fixed (and can be omitted) is

$$\hat{x}(Z) \triangleq \hat{x}[k, Z^k] \qquad (2.2.1\text{-}5)$$

where Z is the set of observations.

2.2.2 Models for Estimation of a Parameter

There are two models one can use in the estimation of a (time invariant) parameter:

1. Nonrandom ("unknown constant"): There is an unknown true value x_0. This is also called the **non-Bayesian** or **Fisher approach**.

2. Random: The parameter is a random variable with a *prior* (or *a priori*) pdf $p(x)$ — a *realization* (see Subsection 1.4.2) of x according to $p(x)$ is assumed to have occurred; this value then stays constant during the measurement process. This is also called the **Bayesian approach**.

The Bayesian Approach

In the **Bayesian approach**, one starts with the **prior** pdf of the parameter from which one can obtain its **posterior** pdf (or **a posteriori** pdf) using **Bayes' formula**:

$$\boxed{p(x|Z) = \frac{p(Z|x)p(x)}{p(Z)} = \frac{1}{c}p(Z|x)p(x)}$$

(2.2.2-1)

where c is the **normalization constant** (does not depend on x).

The posterior pdf can be used in several ways to estimate x.

The Non-Bayesian (Likelihood Function) Approach

In contrast to the above, in the **non-Bayesian approach** there is no prior pdf associated with the parameter and thus one cannot define a posterior pdf for it.

In this case one has the pdf of the measurements conditioned on the parameter, called the **likelihood function (LF)** of the parameter

$$\boxed{\Lambda_Z(x) \triangleq p(Z|x)}$$

(2.2.2-2)

or

$$\Lambda_k(x) \triangleq p(Z^k|x)$$

(2.2.2-3)

as a measure of how "likely" a parameter value is for the observations that are made. The likelihood function serves as a measure of the **evidence from the data**.

The use of the LF (2.2.2-2) and a similar usage of the posterior pdf (2.2.2-1) are discussed in the next section.

2.3 MAXIMUM LIKELIHOOD AND MAXIMUM A POSTERIORI ESTIMATORS

2.3.1 Definitions of ML and MAP Estimators

Maximum Likelihood Estimator

A common method of estimation of nonrandom parameters is the **maximum likelihood method** that maximizes the likelihood function (2.2.2-2). This yields the **maximum likelihood estimator (MLE)**

$$\boxed{\hat{x}^{\mathrm{ML}}(Z) = \arg\max_x \Lambda_Z(x) = \arg\max_x p(Z|x)}$$ (2.3.1-1)

Note that, while x is an unknown constant, $\hat{x}^{\mathrm{ML}}(Z)$, being a function of the set of random observations Z, is a random variable.

The MLE is the solution of the **likelihood equation**

$$\frac{d\Lambda_Z(x)}{dx} = \frac{dp(Z|x)}{dx} = 0$$ (2.3.1-2)

Maximum A Posteriori Estimator

The corresponding estimate for a random parameter is the **maximum a posteriori (MAP)** estimator, which follows from the maximization of the posterior pdf (2.2.2-1):

$$\boxed{\hat{x}^{\mathrm{MAP}}(Z) = \arg\max_x p(x|Z) = \arg\max_x [p(Z|x)p(x)]}$$ (2.3.1-3)

The last equality above follows from the fact that, when using Bayes' formula (2.2.2-1), the normalization constant is irrelevant for the maximization.

The MAP estimate, which depends on the observations Z, and through them on the realization of x is, obviously, a random variable.

2.3.2   MLE versus MAP Estimator with Gaussian Prior

Consider the single measurement

$$z = x + w \tag{2.3.2-1}$$

of the unknown parameter x in the presence of the additive measurement noise w, assumed to be a normally (Gaussian) distributed random variable with mean zero and variance σ^2, that is,

$$w \sim \mathcal{N}(0, \sigma^2) \tag{2.3.2-2}$$

First assume that x is an unknown constant (no prior information about it). The likelihood function of x (denoted here without a subscript, for simplicity) is

$$
\begin{aligned}
\Lambda(x) &= p(z|x) \\
&= \mathcal{N}(z; x, \sigma^2) \\
&= \frac{1}{\sqrt{2\pi}\sigma} e^{-\frac{(z-x)^2}{2\sigma^2}}
\end{aligned} \tag{2.3.2-3}
$$

Then

$$\hat{x}^{\mathrm{ML}} = \arg \max_x \Lambda(x) = z \tag{2.3.2-4}$$

since the peak or **mode** of (2.3.2-3) occurs at $x = z$.

Next assume that the prior information about the parameter is that x is Gaussian with mean $\bar{x}$ and variance σ_0^2, that is,

$$p(x) = \mathcal{N}(x; \bar{x}, \sigma_0^2) \tag{2.3.2-5}$$

It is also assumed that x is independent of w.

Then the posterior pdf of x conditioned on the observation z is

$$
\begin{aligned}
p(x|z) &= \frac{p(z|x)p(x)}{p(z)} \\
&= \frac{1}{c} e^{-\frac{(z-x)^2}{2\sigma^2} - \frac{(x-\bar{x})^2}{2\sigma_0^2}}
\end{aligned} \tag{2.3.2-6}
$$

where

$$c = 2\pi\sigma\sigma_0 p(z) \tag{2.3.2-7}$$

is the normalization constant independent of x. This normalization constant, which guarantees that the pdf integrates to unity, is given explicitly next, after rearranging the exponent in (2.3.2-6).

90

After rearranging the exponent in the above by completing the squares in x, it can be easily shown in a manner similar to the one used in Subsection 1.4.14 that the posterior pdf of x is

$$p(x|z) = \mathcal{N}[x; \xi(z), \sigma_1^2] = \frac{1}{\sqrt{2\pi}\sigma_1} e^{-\frac{[x-\xi(z)]^2}{2\sigma_1^2}} \qquad (2.3.2\text{-}8)$$

i.e., Gaussian, where

$$\xi(z) \triangleq \frac{\sigma^2}{\sigma_0^2 + \sigma^2}\bar{x} + \frac{\sigma_0^2}{\sigma_0^2 + \sigma^2}z = \bar{x} + \frac{\sigma_0^2}{\sigma_0^2 + \sigma^2}(z - \bar{x}) \qquad (2.3.2\text{-}9)$$

and (the "parallel resistors formula")

$$\sigma_1^2 \triangleq \frac{\sigma_0^2 \sigma^2}{\sigma_0^2 + \sigma^2} \qquad (2.3.2\text{-}10)$$

The maximization of (2.3.2-8) with respect to x yields immediately

$$\hat{x}^{\text{MAP}} = \xi(z) \qquad (2.3.2\text{-}11)$$

that is, $\xi(z)$ given by (2.3.2-9) is the *maximum a posteriori estimator* for the random parameter x with the prior pdf (2.3.2-5).

Note that the MAP estimator (2.3.2-9) for this (purely Gaussian) problem is a weighted combination of

1. z, the MLE, which is the peak (or **mode**) of the likelihood function;
2. $\bar{x}$, which is the peak of the prior pdf of the parameter to be estimated.

Equation (2.3.2-9) can be rewritten as follows:

$$\begin{aligned}
\hat{x}^{\text{MAP}} &= (\sigma_0^{-2} + \sigma^{-2})^{-1}\sigma_0^{-2}\bar{x} + (\sigma_0^{-2} + \sigma^{-2})^{-1}\sigma^{-2}z \\
&= (\sigma_0^{-2} + \sigma^{-2})^{-1}\left[\frac{\bar{x}}{\sigma_0^2} + \frac{z}{\sigma^2}\right]
\end{aligned} \qquad (2.3.2\text{-}12)$$

which indicates that the weightings of the prior mean and the measurement are *inversely proportional to their variances*.

Similarly, (2.3.2-10) can be rewritten as follows:

$$\sigma_1^{-2} = \sigma_0^{-2} + \sigma^{-2} \qquad (2.3.2\text{-}13)$$

which shows that the *inverse variances* (also called **information** — this will be discussed in more detail in Subsection 3.4.2) are *additive*. This additivity property of information holds in general when the information sources are *independent*. (See also problem 2-8.)

2.3.3 MAP Estimator with One-Sided Exponential Prior

Consider the same problem as before except that the prior pdf of x is a **one-sided exponential pdf**

$$p(x) = ae^{-ax} \qquad x \geq 0 \qquad (2.3.3\text{-}1)$$

This can model, for instance, the arrival time in a stochastic process where the number of arrivals is Poisson distributed.

The ML estimate is the same as before in (2.3.2-4), that is,

$$\hat{x}^{\text{ML}} = z \qquad (2.3.3\text{-}2)$$

The posterior pdf of x is now

$$p(x|z) = c(z)e^{-\frac{(z-x)^2}{2\sigma^2} - ax} \qquad x \geq 0 \qquad (2.3.3\text{-}3)$$

Since the exponent is quadratic in x, the above posterior pdf is Gaussian but truncated due to the fact that x cannot be negative as modeled by the prior given in (2.3.3-1).

The maximizing argument of (2.3.3-3) is, in view of the fact that it cannot be negative, given by

$$\hat{x}^{\text{MAP}} = \max(z - \sigma^2 a, 0) \qquad (2.3.3\text{-}4)$$

Note that the MAP estimate (2.3.3-4) in this case will always be smaller than the MLE (2.3.3-2) as long as the latter is not negative because the prior (2.3.3-1) attaches higher probability to smaller values of x.

2.3.4 MAP Estimator with Diffuse Prior

While $\hat{x}^{\text{ML}}$ is based on a non-Bayesian approach and $\hat{x}^{\text{MAP}}$ is based on the Bayesian approach, the latter will coincide with the former for a certain prior pdf, called a **diffuse pdf**.

This can be seen by rewriting the denominator in Bayes' formula

$$p(x|Z) = \frac{p(Z|x)p(x)}{p(Z)} \tag{2.3.4-1}$$

with the total probability theorem

$$p(Z) = \int_{-\infty}^{\infty} p(Z|x)p(x)dx \tag{2.3.4-2}$$

and assuming a **diffuse uniform prior pdf** for the parameter

$$p(x) = \epsilon \qquad \text{for } |x| < \frac{1}{2\epsilon} \tag{2.3.4-3}$$

over a "sufficiently large" region of length $1/\epsilon$ where $\epsilon > 0$ but small. Using (2.3.4-3) in (2.3.4-2) yields

$$p(Z) = \epsilon \int_{-\infty}^{\infty} p(Z|x)dx = \epsilon g(Z) \tag{2.3.4-4}$$

where g does not depend on x.

Then, inserting (2.3.4-4) into Bayes' formula (2.3.4-1) yields

$$p(x|Z) = \frac{p(Z|x)\epsilon}{\epsilon g(Z)} = \frac{p(Z|x)}{g(Z)} = \frac{1}{c}p(Z|x) \tag{2.3.4-5}$$

since $\epsilon \neq 0$.

This diffuse pdf is also called **improper pdf** because at the limit as $\epsilon \to 0$, it does not integrate to unity as a **proper pdf** does. Another name for it is **noninformative pdf** because it carries no information about the parameter: uniform distribution in an infinite interval at the limit.

A diffuse prior causes the posterior pdf of x to be proportional to its likelihood function and, thus, the MAP estimate to coincide with the MLE.

Bayesian versus Non-Bayesian Philosophies

The non-Bayesian MLE is, in view of the above discussion, nothing but the Bayesian MAP estimate with complete prior ignorance, reflected by the diffuse prior (2.3.4-3). *This provides a philosophically unifying view of the Bayesian and non-Bayesian approaches to estimation.*

Remark

In spite of the fact that the diffuse prior (2.3.4-3) does not integrate to unity and has no moments (i.e., it is *not a proper pdf*), the posterior pdf of x will be, in general, proper.

Example

Consider the problem of Subsection 2.3.2 where the prior (2.3.2-5) is made diffuse by making $\sigma_0 \to \infty$. A Gaussian pdf with very large variance becomes flat and at the limit looks like a uniform pdf over the whole real line.

When $\sigma_0 \to \infty$ it can be seen from (2.3.2-9) that

$$\lim_{\sigma_0 \to \infty} \xi(z) = z \qquad (2.3.4\text{-}6)$$

that is, $\hat{x}^{\text{MAP}}$ coincides with $\hat{x}^{\text{ML}}$. This occurs regardless of the value of $\bar{x}$, which becomes irrelevant when $\sigma_0 \to \infty$ (i.e., this is a *noninformative prior*).

The Philosophical Meaning of the Prior

The prior pdf assumed in a problem is in many cases the subjective assessment of certain phenomena. The uniform prior assumes Nature as "indifferent." In game theory Nature is assumed to be opposed to our interests. While neither of these two extreme points of view are correct, it is useful to keep in mind the well-known concept of *"perversity of inanimate objects."*[1]

---

[1]As mentioned by Richard Bellman [Bel61], this has been established by a number of experiments, The most conclusive of these involved dropping a piece of buttered toast on a rug. In 79.3% of the trials the toast fell buttered side down. For a detailed discussion on priors, see [RS72].

2.3.5 The Sufficient Statistic and the Likelihood Equation

If the likelihood function of a parameter can be decomposed as follows

$$
\begin{aligned}
\Lambda(x) &\triangleq p(Z|x) \\
&= f_1[g(Z), x]f_2(Z)
\end{aligned}
\tag{2.3.5-1}
$$

then it is clear that the maximum likelihood estimate of x depends only on the function $g(Z)$, called the **sufficient statistic**, rather than on the entire data set Z.

The sufficient statistic *summarizes the information contained in the entire data about* x.

Example

Given the scalar measurements

$$
z(j) = x + w(j) \qquad j = 1, \ldots, k
\tag{2.3.5-2}
$$

If the noises $w(j)$, $j = 1$, ..., k, are independent and identically distributed zero-mean Gaussian random variables with variance σ^2, that is,

$$
w(j) \sim \mathcal{N}(0, \sigma^2)
\tag{2.3.5-3}
$$

then

$$
z(j) \sim \mathcal{N}(x, \sigma^2)
\tag{2.3.5-4}
$$

and, conditioned on x, the observations $z(j)$ are mutually independent.

Thus, the **likelihood function** of x in terms of

$$
Z^k \triangleq \{z(j), j = 1, \ldots, k\}
\tag{2.3.5-5}
$$

is then

$$
\begin{aligned}
\Lambda_k(x) &\triangleq p(Z^k|x) \\
&\triangleq p[z(1), \ldots, z(k)|x] \\
&= \prod_{j=1}^{k} \mathcal{N}[z(j); x, \sigma^2] \\
&= ce^{-\frac{1}{2\sigma^2}\sum_{j=1}^{k}[z(j)-x]^2}
\end{aligned}
\tag{2.3.5-6}
$$

The likelihood function (2.3.5-6) can be rewritten into the product of two functions as in (2.3.5-1) as follows:

$$
\begin{aligned}
\Lambda_k(x) &= c\, e^{-\frac{1}{2\sigma^2}\sum_{j=1}^{k} z(j)^2 + \frac{1}{2\sigma^2}2\sum_{j=1}^{k} z(j)x - \frac{1}{2\sigma^2}kx^2} \\
&= c\, e^{-\frac{1}{2\sigma^2}\sum_{j=1}^{k} z(j)^2}\, e^{-\frac{1}{2\sigma^2}kx\left[x-\frac{2}{k}\sum_{j=1}^{k} z(j)\right]} \\
&\triangleq f_2(Z)f_1[g(Z),x]
\end{aligned}
\qquad (2.3.5\text{-}7)
$$

where

$$
f_2(Z) \triangleq ce^{-\frac{1}{2\sigma^2}\sum_{j=1}^{k} z(j)^2} \qquad (2.3.5\text{-}8)
$$

$$
f_1[g(Z),x] \triangleq e^{-\frac{1}{2\sigma^2}kx[x-2\bar{z}]} \qquad (2.3.5\text{-}9)
$$

$$
g(Z) \triangleq \frac{1}{k}\sum_{j=1}^{k} z(j) \triangleq \bar{z} \qquad (2.3.5\text{-}10)
$$

Thus, according to the definition (2.3.5-1), $\bar{z}$ is the *sufficient statistic* for estimating x.

The Likelihood Equation

To maximize the likelihood function (2.3.5-7) one sets its derivative with respect to x to zero. Since f_2 is independent of x, the **likelihood equation** is

$$
\frac{d\Lambda_k(x)}{dx} = 0 \qquad \Longleftrightarrow \qquad \frac{df_1[g(Z),x]}{dx} = 0 \qquad (2.3.5\text{-}11)
$$

Since the logarithm is a monotonic transformation, one can use equivalently the derivative of the **log-likelihood function**

$$
\frac{d\ln\Lambda_k(x)}{dx} = \frac{d\ln f_1[g(Z),x]}{dx} = -\frac{k}{2\sigma^2}2(x-\bar{z}) = 0 \qquad (2.3.5\text{-}12)
$$

which yields

$$
\hat{x}^{\mathrm{ML}} = \bar{z} \qquad (2.3.5\text{-}13)
$$

The concept of sufficient statistic carries over to the MAP procedure in a completely analogous manner.

2.4 LEAST SQUARES AND MINIMUM MEAN SQUARE ERROR ESTIMATION

2.4.1 Definitions of LS and MMSE Estimators

The LS Estimator

Another common estimation procedure for nonrandom parameters is the **least squares (LS) method**. Given the (scalar and nonlinear) measurements

$$z(j) = h(j, x) + w(j) \qquad j = 1, \ldots, k \qquad (2.4.1\text{-}1)$$

the **least squares estimator (LSE)** of x is, with notation (2.2.1-2),

$$\hat{x}^{\text{LS}}(k) = \arg\min_x \left\{ \sum_{j=1}^{k} [z(j) - h(j, x)]^2 \right\} \qquad (2.4.1\text{-}2)$$

This is the **nonlinear LS problem** — if the function h is linear in x, then one has the **linear LS problem**. The linear LS problem is considered in more detail for the vector case in Section 3.4.

The criterion in (2.4.1-2) makes no assumptions about the "measurement errors" or "noises" $w(j)$. If these are independent and identically distributed zero-mean Gaussian random variables, that is,

$$w(j) \sim \mathcal{N}(0, \sigma^2) \qquad (2.4.1\text{-}3)$$

then the LSE (2.4.1-2) coincides with the MLE under these assumptions. In this case,

$$z(j) \sim \mathcal{N}[h(j, x), \sigma^2] \qquad j = 1, \ldots, k \qquad (2.4.1\text{-}4)$$

The likelihood function of x is then

$$\begin{aligned}
\Lambda_k(x) &\triangleq p(Z^k | x) \\
&\triangleq p[z(1), \ldots, z(k) | x] \\
&= \prod_{j=1}^{k} \mathcal{N}[z(j); h(j, x), \sigma^2] \\
&= c e^{-\frac{1}{2\sigma^2} \sum_{j=1}^{k} [z(j) - h(j,x)]^2} \qquad (2.4.1\text{-}5)
\end{aligned}$$

and the minimization (2.4.1-2) is equivalent to the maximization of (2.4.1-5), that is, *the LS method is a "disguised" ML approach.*

The MMSE Estimator

For random parameters, the counterpart of the above is the **minimum mean square error (MMSE) estimator**

$$\hat{x}^{\text{MMSE}}(Z) = \arg \min_{\hat{x}} E[(\hat{x} - x)^2 | Z] \tag{2.4.1-6}$$

The solution to (2.4.1-6) is the **conditional mean** of x

$$\boxed{\hat{x}^{\text{MMSE}}(Z) = E[x|Z] \triangleq \int_{-\infty}^{\infty} x p(x|Z) dx} \tag{2.4.1-7}$$

where the expectation is with respect to the conditional pdf (2.2.2-1).

The above follows by setting to zero the derivative of (2.4.1-6) with respect to $\hat{x}$:

$$\frac{d}{d\hat{x}} E[(\hat{x} - x)^2 | Z] = E[2(\hat{x} - x)|Z] = 2(\hat{x} - E[x|Z]) = 0 \tag{2.4.1-8}$$

For vector random variables, (2.4.1-7) is obtained similarly by setting to zero the gradient of the mean of the squared norm of the error

$$\nabla_{\hat{x}} E[(\hat{x} - x)'(\hat{x} - x)|Z] = 2(\hat{x} - E[x|Z]) = 0 \tag{2.4.1-9}$$

from which (2.4.1-7) follows immediately.

Remarks

1. With x being an unknown constant (nonrandom) and the noises in (2.4.1-1) modeled as random (not necessarily Gaussian), the LSE is a random variable.

2. The MMSE estimate (2.4.1-7) is a random variable that depends on the observations Z and, through them, on (the realization of) x. Also, for a *given* Z, x is a random variable with a conditional pdf (2.2.2-1).

3. The MMSE estimation problem (2.4.1-6) is a particular case of Bayesian estimation where the expected value of a (positive definite) **cost function** $C(\hat{x} - x)$ is to be minimized. The MMSE cost function is a quadratic. The widespread use of the quadratic criterion is due primarily to the (relative) ease of obtaining the solution. (See also problem 2-3.)

2.4.2 Some LS Estimators

LS Estimator from a Single Measurement

For the problem of a single measurement of the unknown parameter x,

$$z = x + w \tag{2.4.2-1}$$

the least squares criterion leads to

$$\hat{x}^{\text{LS}} = \arg\min_x [(z - x)^2] = z \tag{2.4.2-2}$$

which is the same result as $\hat{x}^{\text{ML}}$ *if w* is zero-mean Gaussian. This is due to the fact that maximizing the likelihood function, which is a Gaussian pdf, is equivalent to minimizing the square in its exponent.

LS Estimator from Several Measurements

Assume now that k measurements are made

$$z(j) = x + w(j) \qquad j = 1, \ldots, k \tag{2.4.2-3}$$

where $w(j)$ are idependent, identically distributed, normal, zero mean, and with common variance σ^2.

The likelihood function is, as in (2.4.1-5),

$$\Lambda_k(x) = ce^{-\frac{1}{2\sigma^2}\sum_{j=1}^{k}[z(j)-x]^2} \tag{2.4.2-4}$$

As before, the ML and LS estimates coincide and it can be easily shown that they are given by the following expression:

$$
\begin{aligned}
\hat{x}^{\text{ML}}(k) &= \hat{x}^{\text{LS}}(k) \\
&= \frac{1}{k}\sum_{j=1}^{k} z(j) \\
&= \bar{z} \tag{2.4.2-5}
\end{aligned}
$$

This estimate is known as the **sample mean** or **sample average**, since it estimates the unknown mean x of the k random variables from (2.4.2-3).

2.4.3 MMSE versus MAP Estimator in Gaussian Noise

In the single measurement example with a prior pdf on the parameter to be estimated, discussed in Subsection 2.3.2, the posterior pdf of x was obtained in (2.3.2-8) as

$$p(x|z) = \frac{1}{\sqrt{2\pi}\sigma_1} e^{-\frac{[x-\xi(z)]^2}{2\sigma_1^2}} \tag{2.4.3-1}$$

It is apparent by inspection that the mean of this Gaussian pdf is $\xi(z)$, which is also the **mode** (peak) of this pdf.

Thus

$$\begin{aligned}
\hat{x}^{\text{MMSE}} &= E[x|z] \\
&= \xi(z) \\
&= \hat{x}^{\text{MAP}}
\end{aligned} \tag{2.4.3-2}$$

i.e., the MMSE estimator (the conditional mean) *coincides* with the MAP estimator.

This is due to the fact that the *mean* and the *mode* of a Gaussian pdf coincide.

Note that, in view of (2.4.3-2), equation (2.4.3-1) can be also written as

$$\begin{aligned}
p(x|z) &= \mathcal{N}(x; \hat{x}^{\text{MMSE}}, \sigma_1^2) \\
&= \mathcal{N}(x; \hat{x}^{\text{MAP}}, \sigma_1^2)
\end{aligned} \tag{2.4.3-3}$$

2.5 UNBIASED ESTIMATORS

2.5.1 Definition

Non-Bayesian Case

For a nonrandom parameter, an estimator is said to be **unbiased** if

$$E[\hat{x}(k, Z^k)] = x_0 \tag{2.5.1-1}$$

where x_0 is the true value of the parameter. The expectation in (2.5.1-1) is over the estimate, which is a random variable since it is a function of the measurements (2.2.1-3), and is taken with respect to the conditional pdf $p(Z^k | x = x_0)$.

Bayesian Case

If x is a random variable with a prior pdf $p(x)$, then the unbiasedness property is written as

$$E[\hat{x}(k, Z^k)] = E[x] \tag{2.5.1-2}$$

where the expectation on the left hand side above is with respect to the joint pdf $p(Z^k, x)$ and the one on the right hand side is with respect to $p(x)$.

General Definition

The above unbiasedness requirements can be unified by requiring that the **estimation error**

$$\tilde{x} \triangleq x - \hat{x} \tag{2.5.1-3}$$

be zero mean, that is,

$$\boxed{E[\tilde{x}] = 0} \tag{2.5.1-4}$$

Equation (2.5.1-4) covers both cases, with the expectation being taken over Z^k in the first case and over Z^k *and* x in the second case.

An estimator is unbiased if (2.5.1-4) holds for all k and is **asymptotically unbiased** if it holds in the limit as $k \to \infty$.

2.5.2   Unbiasedness of an ML and a MAP Estimator

Consider the ML estimator (2.3.2-4) of the parameter x with true value x_0

$$\hat{x}^{\mathrm{ML}} = z \qquad (2.5.2\text{-}1)$$

from the single measurement

$$z = x + w \qquad (2.5.2\text{-}2)$$

Its mean is

$$
\begin{aligned}
E[\hat{x}^{\mathrm{ML}}] &= E[z] \\
&= E[x_0 + w] \\
&= x_0 + E[w] \\
&= x_0 \qquad (2.5.2\text{-}3)
\end{aligned}
$$

since the mean of the Gaussian random variable w is zero.

For the MAP estimate (2.3.2-11) of x modeled as a Gaussian random variable with prior mean $\bar{x}$ and prior variance σ^2,

$$
\begin{aligned}
\hat{x}^{\mathrm{MAP}} &= \xi(z) \\
&\overset{\Delta}{=} \frac{\sigma^2}{\sigma_0^2 + \sigma^2}\bar{x} + \frac{\sigma_0^2}{\sigma_0^2 + \sigma^2}z \qquad (2.5.2\text{-}4)
\end{aligned}
$$

one has

$$
\begin{aligned}
E[\hat{x}^{\mathrm{MAP}}] &= E[\xi(z)] \\
&= \frac{\sigma^2}{\sigma_0^2 + \sigma^2}\bar{x} + \frac{\sigma_0^2}{\sigma_0^2 + \sigma^2}E[z] \\
&= \frac{\sigma^2}{\sigma_0^2 + \sigma^2}\bar{x} + \frac{\sigma_0^2}{\sigma_0^2 + \sigma^2}[\bar{x} + E(w)] \\
&= \bar{x} = E[x] \qquad (2.5.2\text{-}5)
\end{aligned}
$$

Thus, both of these estimates are seen to be unbiased.

2.5.3 Bias in the ML Estimation of Two Parameters

Consider the problem of estimating the unknown mean x of a set of k measurements as in (2.4.2-2), with the additional parameter to be estimated being the variance σ^2, now also assumed to be unknown.

The likelihood function for the unknown parameters x and σ is

$$
\begin{aligned}
\Lambda_k(x, \sigma) &= p[z(1), \ldots, z(k)|x, \sigma] \\
&= \frac{1}{(2\pi)^{k/2}\sigma^k} e^{-\frac{1}{2\sigma^2}\sum_{j=1}^{k}[z(j)-x]^2}
\end{aligned}
\tag{2.5.3-1}
$$

To maximize the above, one writes the **likelihood equation** by setting to zero the derivatives of Λ or, more conveniently, of $\ln \Lambda$, with respect to x and σ

$$
\frac{\partial \ln \Lambda_k}{\partial x} = \frac{1}{\sigma^2} \sum_{j=1}^{k} [z(j) - x] = 0
\tag{2.5.3-2}
$$

$$
\frac{\partial \ln \Lambda_k}{\partial \sigma} = -\frac{k}{\sigma} + \frac{1}{\sigma^3} \sum_{j=1}^{k} [z(j) - x]^2 = 0
\tag{2.5.3-3}
$$

The Solution of the Likelihood Equation

The first equation yields $\hat{x}^{\mathrm{ML}}$ as before in (2.4.2-5), that is, the sample mean — in this problem the estimate of x is not affected at all by the fact that σ is also unknown. Substituting this into the second equation yields

$$
\frac{\partial \ln \Lambda_k}{\partial \sigma} = -\frac{k}{\sigma} + \frac{1}{\sigma^3} \sum_{j=1}^{k} [z(j) - \hat{x}^{\mathrm{ML}}]^2 = 0
\tag{2.5.3-4}
$$

The resulting estimate, known as the **sample variance** based on k observations, is

$$
\begin{aligned}
[\hat{\sigma}^{\mathrm{ML}}(k)]^2 &= \frac{1}{k} \sum_{j=1}^{k} [z(j) - \hat{x}^{\mathrm{ML}}]^2 \\
&= \frac{1}{k} \sum_{i=1}^{k} \left[z(j) - \frac{1}{k} \sum_{i=1}^{k} z(i) \right]^2
\end{aligned}
\tag{2.5.3-5}
$$

The Means of the Sample Mean and Sample Variance

Denote by x_0 and σ_0 the true values of the parameters. The expected value of the sample mean is

$$
\begin{aligned}
E[\hat{x}^{\mathrm{ML}}(k)] &= E\left[\frac{1}{k}\sum_{j=1}^{k} z(j)\right] \\
&= x_0
\end{aligned}
\tag{2.5.3-6}
$$

that is, the sample mean estimator (2.4.2-5) is unbiased.

The expected value of the sample variance (2.5.3-5) is

$$
\begin{aligned}
E\left\{[\hat{\sigma}^{\mathrm{ML}}(k)]^2\right\} &= E\left\{\frac{1}{k}\sum_{i=1}^{k}\left[z(j) - \frac{1}{k}\sum_{i=1}^{k} z(i)\right]^2\right\} \\
&= \frac{1}{k}\sum_{j=1}^{k} E\left\{\left[w(j) - \frac{1}{k}\sum_{i=1}^{k} w(i)\right]^2\right\} \\
&= \frac{1}{k^3}\sum_{j=1}^{k} E\left\{\left[(k-1)w(j) - \sum_{\substack{i=1 \\ i\neq j}}^{k} w(i)\right]^2\right\} \\
&= \frac{1}{k^2}[(k-1)^2 + k - 1]\sigma_0^2 \\
&= \frac{k-1}{k}\sigma_0^2
\end{aligned}
\tag{2.5.3-7}
$$

Thus the sample variance (2.5.3-5) is *biased*, even though as $k \to \infty$ it becomes unbiased (i.e., it is *asymptotically unbiased*). In order to be unbiased, the denominator in (2.5.3-5) should be $k - 1$ rather than k:

$$
[\hat{\sigma}(k)]^2 = \frac{1}{k-1}\sum_{i=1}^{k}\left[z(j) - \frac{1}{k}\sum_{i=1}^{k} z(i)\right]^2
\tag{2.5.3-8}
$$

Expression (2.5.3-8) is the more common sample variance used. However, for reasonably large k this is not going to make a significant difference. (See also problem 2-4.)

2.6 THE VARIANCE OF AN ESTIMATOR

2.6.1 Definitions of Estimator Variances

Non-Bayesian Case

For a non-Bayesian estimator, $\hat{x}(Z)$, (LS or ML) the **variance of the estimator** is

$$\text{var}[\hat{x}(Z)] \triangleq E\left[\{\hat{x}(Z) - E[\hat{x}(Z)]\}^2\right] \qquad (2.6.1\text{-}1)$$

where the averaging is over the observation set Z.

If this estimator is *unbiased*, that is,

$$E[\hat{x}(Z)] = x_0 \qquad (2.6.1\text{-}2)$$

where x_0 is the true value, then

$$\text{var}[\hat{x}(Z)] = E\left[[\hat{x}(Z) - x_0]^2\right] \qquad (2.6.1\text{-}3)$$

If this estimator is *biased* then (2.6.1-3) is its **mean square error (MSE)** [2]

$$\text{MSE}[\hat{x}(Z)] = E\left[[\hat{x}(Z) - x_0]^2\right] \qquad (2.6.1\text{-}4)$$

Bayesian Case

For a Bayesian estimator, the **unconditional MSE** is

$$\text{MSE}[\hat{x}(Z)] \triangleq E\left[[\hat{x}(Z) - x]^2\right] \qquad (2.6.1\text{-}5)$$

where the averaging is with respect to the joint pdf of the observations Z and the random parameter x. The above can be rewritten, using the smoothing property of expectations (see Subsection 1.4.12), as follows:

$$\begin{aligned}
\text{MSE}[\hat{x}(Z)] &= E\left[E\{[\hat{x}(Z) - x]^2|Z\}\right] \\
&= E\left[\text{MSE}[\hat{x}(Z)|Z]\right] \qquad (2.6.1\text{-}6)
\end{aligned}$$

where the last expression inside the braces is the **conditional MSE** (i.e., for a given realization (value) of the observations Z).

---

[2] Mean square is a personality type from the official list compiled by psycho-statisticians.

For the MMSE estimator, the conditional MSE is

$$E\left[[\hat{x}^{\mathrm{MMSE}}(Z) - x]^2 | Z\right] = E\left[[x - E(x|Z)]^2 | Z\right]$$
$$= \mathrm{var}(x|Z) \qquad (2.6.1\text{-}7)$$

that is, the **conditional variance** of x *given* Z. Note that the expectations in (2.6.1-7) are with respect to $p(x|Z)$.

Averaging over Z yields

$$E[\mathrm{var}(x|Z)] = E\left[[x - E(x|Z)]^2\right] \qquad (2.6.1\text{-}8)$$

which is the **unconditional variance** of the estimate $\hat{x}^{\mathrm{MMSE}}$. This is the "average squared error over all the possible observations."

General Definition

With the definition of the estimation error

$$\tilde{x} \overset{\Delta}{=} x - \hat{x} \qquad (2.6.1\text{-}9)$$

one can say in a unified manner that the expected value of the square of the estimation error is the estimator's variance or MSE:

$$E[\tilde{x}^2] = \begin{cases} \mathrm{var}(\hat{x}) & \text{if } \hat{x} \text{ is unbiased} \\ \mathrm{MSE}(\hat{x}) & \text{if } \hat{x} \text{ is biased or unbiased} \end{cases} \qquad (2.6.1\text{-}10)$$

where the expectations are to be taken according to the discussion above.

The square root of the variance (or MSE) of an estimator

$$\sigma_{\hat{x}} \overset{\Delta}{=} \sqrt{\mathrm{var}(\hat{x})} \qquad (2.6.1\text{-}11)$$

is its **standard error**, also called the **standard deviation associated with the estimator** or the **standard deviation of the estimation error**.

The standard error provides a measure of the accuracy of the estimator: assuming the estimation error to be Gaussian, the difference between the estimate and the true value will be up to 2 standard errors with 95% probability.

2.6.2   Comparison of Variances of an ML and a MAP Estimator

The "qualities" of the ML and MAP estimators discussed in Subsection 2.3.2 (from a single observation), as measured by their variances, will be compared.

For the MLE given by (2.3.2-4) one has

$$
\begin{aligned}
\mathrm{var}(\hat{x}^{\mathrm{ML}}) &= E[(\hat{x}^{\mathrm{ML}} - x_0)^2] \\
&= E[(z - x_0)^2] \\
&\stackrel{\triangle}{=} \sigma^2
\end{aligned}
\tag{2.6.2-1}
$$

For the MAP estimate given by (2.3.2-11), which has a Gaussian prior in this case, one has

$$
\begin{aligned}
\mathrm{var}(\hat{x}^{\mathrm{MAP}}) &= E[(\hat{x}^{\mathrm{MAP}} - x)^2] \\
&= E\left\{\left[\frac{\sigma^2}{\sigma_0^2 + \sigma^2}\bar{x} + \frac{\sigma_0^2}{\sigma_0^2 + \sigma^2}(x + w) - x\right]^2\right\} \\
&= E\left[\left[\frac{\sigma^2}{\sigma_0^2 + \sigma^2}(x - \bar{x}) + \frac{\sigma_0^2}{\sigma_0^2 + \sigma^2}w\right]^2\right] \\
&= \frac{\sigma_0^2 \sigma^2}{\sigma_0^2 + \sigma^2} \\
&< \sigma^2 \\
&= \mathrm{var}(\hat{x}^{\mathrm{ML}})
\end{aligned}
\tag{2.6.2-2}
$$

Thus it can be seen that the variance of the MAP estimator (given by the "parallel resistors formula") is *smaller* than that of the MLE — this is due to the availability of *prior information.*

Note that in (2.6.2-1) the averaging is only over z (or, equivalently, w) while in (2.6.2-2) the averaging is over w *and* x, which is assumed random. (See also problem 2-1.)

2.6.3   The Variances of the Sample Mean and Sample Variance

The **variance of the sample mean** (2.4.2-5) — the square of its **standard error** — is obtained as

$$E\left[[\hat{x}^{ML}(k) - x_0]^2\right] = E\left\{\left[\frac{1}{k}\sum_{j=1}^{k}[z(j) - x_0]\right]^2\right\} = \frac{\sigma^2}{k} \tag{2.6.3-1}$$

which, as $k \to \infty$, converges to zero, that is, this estimator is consistent (the definition is given in the next subsection).

The **variance of the sample variance** (2.5.3-5) is computed next. For simplicity, it is assumed that the mean is zero and known. The estimator of the variance is in this case

$$(\hat{\sigma}^{ML})^2 = \frac{1}{k}\sum_{j=1}^{k} z(j)^2 \tag{2.6.3-2}$$

and it can be easily shown that it is unbiased.

The variance of this estimator is, with the true value denoted by σ^2,

$$E\left[[(\hat{\sigma}^{ML})^2 - \sigma^2]^2\right] = E\left\{\left[\frac{1}{k}\sum_{j=1}^{k} z(j)^2 - \sigma^2\right]^2\right\}$$

$$= \frac{1}{k^2}\sum_{j=1}^{k}\sum_{i=1}^{k} E[w(j)^2 w(i)^2] - 2\sigma^2\frac{1}{k}\sum_{j=1}^{k} E[w(j)^2] + \sigma^4$$

$$= \frac{1}{k^2}[k(k-1)\sigma^4 + k3\sigma^4] - \frac{2}{k}k\sigma^4 + \sigma^4 = \frac{2\sigma^4}{k} \tag{2.6.3-3}$$

which also converges to zero as $k \to \infty$.

The following relationship has been used in (2.6.3-3)

$$E[w(i)^2 w(j)^2] = \begin{cases} \sigma^4 & \text{if } i \neq j \\ 3\sigma^4 & \text{if } i = j \end{cases} \tag{2.6.3-4}$$

The fourth moment of w is needed here and, assuming it to be Gaussian, use was made of (1.4.15-7).

The **standard error of the sample variance** from k samples is therefore, from (2.6.3-3), given by

$$\sigma_{(\hat{\sigma}^{ML})^2} = \sigma^2\sqrt{2/k} \tag{2.6.3-5}$$

In the above the notation σ_ξ has been used to denote the **standard error of the estimate** ξ.

Application — The Number of Samples Needed to Estimate a Variance with a Given Accuracy

Based on this result, the number of samples needed to obtain the sample variance *within 10% of the true value with probability of 95%* can be obtained as follows.

Assuming for convenience that the sample variance is normally distributed about its mean (equal to the true variance) with standard error as above, one has the 95% confidence region

$$P\left\{|(\hat{\sigma}^{\mathrm{ML}})^2 - \sigma^2| \leq 1.96\sigma^2\sqrt{2/k}\right\} = 0.95 \qquad (2.6.3\text{-}6)$$

The requirement of at most 10% error in the variance estimate, that is,

$$\frac{|(\hat{\sigma}^{\mathrm{ML}})^2 - \sigma^2|}{\sigma^2} = 0.1 \qquad (2.6.3\text{-}7)$$

leads to setting

$$1.96\sqrt{2/k} = 0.1 \qquad (2.6.3\text{-}8)$$

which yields

$$k \approx 800 \qquad (2.6.3\text{-}9)$$

The resulting very large number of samples necessary for the required accuracy justifies the use of the CLT in (2.6.3-6). (See also problem 1-2.)

2.7 CONSISTENCY AND EFFICIENCY OF ESTIMATORS

2.7.1 Consistency

An estimator of a *nonrandom parameter* is said to be a **consistent estimator** if the estimate (which is a random variable) converges to the true value in some stochastic sense. Using the **convergence in mean square** criterion, then

$$\lim_{k \to \infty} E\left[[\hat{x}(k, Z^k) - x_0]^2\right] = 0 \qquad (2.7.1\text{-}1)$$

is the condition for **consistency in the mean square sense**. The expectation is taken over Z^k, as in (2.5.1-1).

For a *random parameter*, convergence of its estimator in the mean square sense requires

$$\lim_{k \to \infty} E\left[[\hat{x}(k, Z^k) - x]^2\right] = 0 \qquad (2.7.1\text{-}2)$$

where the expectation is over Z^k *and* x, as in (2.5.1-2).

Similarly to the unbiasedness case, consistency can be expressed as the requirement that the estimation error converges to zero, that is,

$$\lim_{k \to \infty} \tilde{x}(k, Z^k) = 0 \qquad (2.7.1\text{-}3)$$

in some stochastic (e.g., mean square) sense.

Remark

The consistency defined above is an *asymptotic* property; that is, it is defined for the case when the sample size k tends to infinity. Later, in the context of state estimation, there will be another definition of consistency as a *finite sample size* property.

2.7.2 The Cramer-Rao Lower Bound and the Fisher Information Matrix

According to the **Cramer-Rao lower bound (CRLB)**, the mean square error corresponding to the estimator of a parameter *cannot be smaller* than a certain quantity related to the likelihood function.

Scalar Case

In the estimation of a scalar *nonrandom* parameter x with an *unbiased* estimator $\hat{x}(Z)$, the variance is bounded from below as follows:

$$\boxed{E\left[[\hat{x}(Z) - x_0]^2\right] \geq J^{-1}} \qquad (2.7.2\text{-}1)$$

where

$$
\begin{aligned}
J &\triangleq -E\left[\frac{\partial^2 \ln \Lambda(x)}{\partial x^2}\right]\Bigg|_{x=x_0} \\
&= E\left\{\left[\frac{\partial \ln \Lambda(x)}{\partial x}\right]^2\right\}\Bigg|_{x=x_0}
\end{aligned}
\qquad (2.7.2\text{-}2)
$$

is the **Fisher information**, $\Lambda(x) = p(Z|x)$ is the likelihood function (2.2.2-2) denoted for simplicity without subscript, and x_0 is the true value of the unknown constant x.

For a scalar *random* parameter x estimated by an unbiased estimator $\hat{x}(k)$, the variance is bounded from below by a similar expression

$$E\left[[\hat{x}(Z) - x]^2\right] \geq J^{-1} \qquad (2.7.2\text{-}3)$$

where

$$J \triangleq -E\left[\frac{\partial^2 \ln p(Z,x)}{\partial x^2}\right] = E\left\{\left[\frac{\partial \ln p(Z,x)}{\partial x}\right]^2\right\} \qquad (2.7.2\text{-}4)$$

The expectations in (2.7.2-2) and (2.7.2-4) are taken as in (2.7.1-1) and (2.7.1-2), respectively.

Note that the Fisher information has two forms in (2.7.2-2) as well as in (2.7.2-4). The proof of (2.7.2-1) and the equivalence of the two forms in (2.7.2-2) — one with first partial derivatives, the other one with second partials — is given later.

If an estimator's variance is equal to the CRLB, then such an estimator is called **efficient**.

Multidimensional Case

For *nonrandom vector parameters*, the CRLB states that the covariance matrix of an unbiased estimator is bounded from below as follows:

$$\boxed{E\left[[\hat{x}(Z) - x_0][\hat{x}(Z) - x_0]'\right] \geq J^{-1}} \tag{2.7.2-5}$$

where the **Fisher information matrix (FIM)** is

$$\boxed{J \triangleq -E[\nabla_x \nabla_x' \ln \Lambda(x)]\Big|_{x=x_0} = E\left[[\nabla_x \ln \Lambda(x)][\nabla_x \ln \Lambda(x)]'\right]\Big|_{x=x_0}} \tag{2.7.2-6}$$

and x_0 is the true value of the vector parameter x.

As in the scalar case, note the two forms of the FIM: one with the Hessian of the log-likelihood function, the other with the dyad of its gradient.

The matrix inequality in (2.7.2-5) is to be interpreted as follows:

$$A \geq B \qquad \Longleftrightarrow \qquad C \triangleq A - B \geq 0 \tag{2.7.2-7}$$

that is, the difference C of the two matrices is positive semidefinite.

A similar expression holds for the case of a multidimensional random parameter.

Remark

A necessary condition for an estimator to be consistent in the mean square sense is that there must be an increasing amount of information (in the sense of Fisher) about the parameter in the measurements — the Fisher information has to tend to infinity as $k \to \infty$. Then the CRLB converges to zero as $k \to \infty$ and thus the variance can also converge to zero.

Note

For estimators that are *biased*, there is a modified version of the CRLB (e.g., [Van68]).

2.7.3   Proof of the Cramer-Rao Lower Bound

Let $\hat{x}(z)$ be an unbiased estimate of the nonrandom real-valued parameter x based on the observation (or set of observations) denoted now as z. The likelihood function of x is

$$\Lambda(x) = p(z|x) \qquad (2.7.3\text{-}1)$$

It will be assumed that the first and second derivatives of (2.7.3-1) with respect to x exist and are absolutely integrable.

From the unbiasedness condition on the estimate $\hat{x}(z)$ one has (the true value is denoted now also as x)

$$E[\hat{x}(z) - x] = \int_{-\infty}^{\infty} [\hat{x}(z) - x]p(z|x)dz = 0 \qquad (2.7.3\text{-}2)$$

The derivative of the above with respect to x is

$$\frac{d}{dx}\int_{-\infty}^{\infty}[\hat{x}(z) - x]p(z|x)dz = \int_{-\infty}^{\infty}\frac{\partial}{\partial x}\Big\{[\hat{x}(z) - x]p(z|x)\Big\}dz$$

$$= -\int_{-\infty}^{\infty} p(z|x)dz + \int_{-\infty}^{\infty}[\hat{x}(z) - x]\frac{\partial p(z|x)}{\partial x}dz = 0 \qquad (2.7.3\text{-}3)$$

Using the fact that the first integral in the last line above is equal to unity and the identity

$$\frac{\partial p(z|x)}{\partial x} = \frac{\partial \ln p(z|x)}{\partial x}p(z|x) \qquad (2.7.3\text{-}4)$$

yields from (2.7.3-3)

$$\int_{-\infty}^{\infty}[\hat{x}(z) - x]\frac{\partial \ln p(z|x)}{\partial x}p(z|x)dz = 1 \qquad (2.7.3\text{-}5)$$

Equation (2.7.3-5) can be rewritten as

$$\int_{-\infty}^{\infty}\Big\{[\hat{x}(z) - x]\sqrt{p(z|x)}\Big\}\Big\{\frac{\partial \ln p(z|x)}{\partial x}\sqrt{p(z|x)}\Big\}dz = 1 \qquad (2.7.3\text{-}6)$$

The **Schwarz inequality** for (real-valued) functions, which is a generalized version of (1.3.4-2), is

$$\langle f_1, f_2\rangle \triangleq \int_{-\infty}^{\infty} f_1(z)f_2(z)dz \le \|f_1\|\,\|f_2\| \qquad (2.7.3\text{-}7)$$

where

$$\|f_i\| \triangleq \{\langle f_i, f_i\rangle\}^{1/2} = \Big\{\int_{-\infty}^{\infty} f_i(z)^2 dz\Big\}^{1/2} \qquad (2.7.3\text{-}8)$$

The equality in (2.7.3-7) holds if and only if

$$f_1(z) = cf_2(z) \qquad \forall z \qquad (2.7.3\text{-}9)$$

113

Note that the left hand side of (2.7.3-6) is an inner product of two functions as in (2.7.3-7). Using (2.7.3-7) to majorize the left hand side of (2.7.3-6) yields

$$\left\{ \int_{-\infty}^{\infty} [\hat{x}(z) - x]^2 p(z|x) dz \right\}^{1/2} \left\{ \int_{-\infty}^{\infty} \left[\frac{\partial \ln p(z|x)}{\partial x} \right]^2 p(z|x) dz \right\}^{1/2} \geq 1 \qquad (2.7.3\text{-}10)$$

which can be rewritten as

$$E\{[\hat{x}(z) - x]^2\} \geq \left\{ E \left[\frac{\partial \ln p(z|x)}{\partial x} \right]^2 \right\}^{-1} \qquad (2.7.3\text{-}11)$$

with equality holding if and only if

$$\frac{\partial \ln p(z|x)}{\partial x} = c(x)[\hat{x}(z) - x] \qquad \forall z \qquad (2.7.3\text{-}12)$$

Equation (2.7.3-11) is equivalent to (2.7.2-1), which completes the proof of the CRLB for a nonrandom scalar parameter. In view of (2.7.3-2), which holds at the true value of the parameter, all the partial derivatives are to be evaluated at the true value of the parameter, which is indicated explicitly only in (2.7.2-2).

Equivalence of the Two Forms of the Fisher Information

To prove the equivalence of the two forms of the Fisher information in (2.7.2-2), consider the identity

$$\int_{-\infty}^{\infty} p(z|x) dz = 1 \qquad (2.7.3\text{-}13)$$

Taking the derivative of the above with respect to x yields

$$\int_{-\infty}^{\infty} \frac{\partial p(z|x)}{\partial x} dz = 0 \qquad (2.7.3\text{-}14)$$

Using identity (2.7.3-4), the above can be rewritten as

$$\int_{-\infty}^{\infty} \frac{\partial \ln p(z|x)}{\partial x} p(z|x) dz = 0 \qquad (2.7.3\text{-}15)$$

Taking now the derivative of (2.7.3-15) with respect to x leads to

$$\int_{-\infty}^{\infty} \frac{\partial^2 \ln p(z|x)}{\partial x^2} p(z|x) dz + \int_{-\infty}^{\infty} \left[\frac{\partial \ln p(z|x)}{\partial x} \right]^2 p(z|x) dz = 0 \qquad (2.7.3\text{-}16)$$

which proves the equivalence of the expression of the Fisher information with the second partial derivative of the log-likelihood function with the one that has the square of the first partial derivative, as in (2.7.2-2).

2.7.4   An Example of Efficient Estimator

Consider the likelihood function (2.4.2-4) for the estimation of the mean x from a set of k independent and identically distributed measurements.

The Fisher information is in this case the (scalar) quantity

$$J = -E \left[\frac{\partial^2 \ln \Lambda_k(x)}{\partial x^2} \right] \bigg|_{x=x_0} = \frac{k}{\sigma^2} \qquad (2.7.4\text{-}1)$$

Thus

$$E \left[[\hat{x}^{\mathrm{ML}}(k) - x_0]^2 \right] \geq J^{-1} = \frac{\sigma^2}{k} \qquad (2.7.4\text{-}2)$$

Comparing this to (2.6.3-1) it is seen that the CRLB is met, that is, the ML estimator (which is in this case the sample mean) is efficient.

Since the variance (2.7.4-2) converges to zero as $k \to \infty$, this estimator is also consistent. (See also problem 2-2.)

Evaluation of the CRLB

In the simple case considered above, the expression of J is independent of x. In general this is not the case and there is need to evaluate J at the *true value* of x.

If the true value of the parameter is not available, then the **evaluation of the CRLB**, which amounts to a linearization, is done *at the estimate*. Caution has to be exercised in this case, since the unavoidable estimation errors can lead to a *possibly incorrect value* of the resulting Fisher information J, which is in general a matrix.

2.7.5   Large Sample Properties of the ML Estimator

The following are the **large-sample properties of the ML estimator**:

1. It is **asymptotically unbiased**.
2. It is **asymptotically efficient**.

Thus, if there is "enough information" in the measurements, in which case the CRLB will tend to zero, the variance of the ML estimate will also converge to zero. Therefore, the ML estimate will converge to the true value — it will be consistent.

Another property of the ML estimator is the following:

3. It is **asymptotically Gaussian**.

Combining all the above, *the ML estimate is asymptotically Gaussian with the mean equal to the true value of the parameter to be estimated and variance given by the CRLB.*

This can be summarized for a vector parameter as

$$\hat{x}^{\text{ML}}(k) \sim \mathcal{N}(x, J^{-1}) \qquad \text{for large } k \qquad (2.7.5\text{-}1)$$

where J is the Fisher information matrix.

Comparing the (non-Bayesian) MLE (2.7.5-1) with the (Bayesian) MMSE estimate — the conditional mean

$$\hat{x}^{\text{MMSE}} = E[x|z] \qquad (2.7.5\text{-}2)$$

points out the contrast between these two philosophies:

1. In (2.7.5-1), given x, the estimate $\hat{x}^{\text{ML}}$ is a random variable, function of z;
2. In (2.7.5-2), given z, the true value x is a random variable.

116

2.8 SUMMARY

2.8.1 Summary of Estimators

Estimator of a parameter — a function of the measurements that yields a "best approximation" of the value of a parameter.

Estimate of a parameter — the value taken by the estimator function for given values of the measurements.

Models for the parameter to be estimated:

1. *Unknown constant (nonrandom).*
2. *Random*: realization of a random variable according to a certain prior pdf.

Model 2 yields the **Bayesian approach** while model 1 leads to what is called the **non-Bayesian approach**.

Likelihood function of a (nonrandom) parameter — pdf of the measurements conditioned on the parameter.

Bayes' formula — given a prior pdf of a (random) parameter, it yields its posterior pdf conditioned on the measurements.

ML estimate (of a nonrandom parameter) — the value of the parameter that maximizes its likelihood function.

MAP estimate (of a random parameter) — the value of the parameter that maximizes its posterior pdf.

The MAP estimate of a parameter with a *diffuse (noninformative)* prior pdf coincides with its MLE.

LS estimate (of a nonrandom parameter) — minimizes the square of the error between the measurements and the observed function of the parameter.

MMSE estimate (of a random parameter) — minimizes the expected value (mean) of the square of the parameter estimation error conditioned on the measurements. This estimate is the *conditional mean* of the parameter given the measurements.

If in a given set of measurements the errors are additive, zero mean, *Gaussian*, and independent, then the *LS* estimate coincides with the *ML* estimate.

The *MAP* estimate of a *Gaussian* random variable coincides with its *MMSE* estimate (conditional mean).

2.8.2 Summary of Estimator Properties

Unbiased estimator — if the corresponding error has mean zero.

Variance/MSE of an estimator — the expected value of the square of the estimation error of an unbiased/biased estimator. The variance of the estimator of a parameter modeled as random (with some prior) is *smaller* than when it is modeled as an unknown constant.

Consistent estimator — if the corresponding error converges to zero in some stochastic sense (most common: in mean square).

CRLB — lower bound on the achievable variance in the estimation of a parameter. It is given by the inverse of the **Fisher information matrix** (for an unbiased estimator).

Efficient estimator — if its variance meets the CRLB.

On the Terminology

In (most of) the literature there is little or no distinction between the terms LS and MMSE estimation. The MMSE estimation is sometimes called LS, which is incorrect according to our definition, or **least mean square**, which is a valid alternate designation. Another term used is **minimum variance (MV)**.

2.9 NOTES AND PROBLEMS

2.9.1 Bibliographical Notes

The basic concepts in estimation are discussed, for example, in [Van68, SM71, MC78]. The proof of the CRLB for vector-valued parameters can be found in [Van68] and [Lju87, (p.206)].

Another model of uncertainty in parameter estimation is the "unknown but bounded" approach discussed in [Sch73].

2.9.2 Problems

2-1 **Estimators for a discrete-valued parameter.** A discrete-valued parameter with the prior pdf

$$p(x) = \sum_{i=1}^{2} p_i \delta(x - i)$$

is measured with the additive noise $w \sim \mathcal{N}(0, \sigma^2)$

$$z = x + w$$

1. Find the posterior pdf of the parameter.

2. Find its MAP estimate and the associated MSE conditioned on z.

3. Find its MMSE estimate and the associated variance.

4. Evaluate these estimates and MSE for

Case	p_1	σ	z
A	0.5	1	1.5
B	0.5	1	3
C	0.3	1	1.5
D	0.5	0.1	1.8

5. Comment on the meaningfulness of the two estimates in the above 4 cases.

2-2 **Estimation with correlated noises.** A parameter x is measured with correlated rather than independent additive Gaussian noises

$$z_k = x + w_k \qquad k = 1, \ldots, n$$

with

$$E w_k = 0 \qquad E[w_k w_j] = \begin{cases} 1 & k = j \\ \rho & |k - j| = 1 \\ 0 & |k - j| > 1 \end{cases}$$

For $n = 2$:

1. Write the likelihood function of the parameter x.

2. Find the MLE of x. What happens if $\rho = 1$?

3. Find the CRLB for the estimation of x. Show the effect of $\rho > 0$ versus $\rho < 0$. Explain what happens at $\rho = 1$.

4. Is the MLE efficient? Can one have a perfect (zero-variance) estimate?

(The remaining items are more challenging.) For general n, let

$$z \overset{\Delta}{=} [z_1 \; \cdots \; z_n]' \qquad 1 \overset{\Delta}{=} [1 \; \cdots \; 1]' \qquad w \overset{\Delta}{=} [w_1 \; \cdots \; w_n]' \qquad P \overset{\Delta}{=} E[ww']$$

5. Using the above notations, write the likelihood function of x.

6. Find the MLE of x.

2-3 Estimation criteria that lead to the conditional mean. Show that, in estimating a random vector x with the following criteria

1. $\min\limits_{\hat{x}} E[(x - \hat{x})' A (x - \hat{x}) | z]$, $\forall A > 0$ (positive definite).

2. $\min\limits_{\hat{x}} \operatorname{tr}[P]$ with $P \overset{\Delta}{=} E[(x - \hat{x})(x - \hat{x})' | z]$.

3. $\min\limits_{\hat{x}} \operatorname{tr}[AP]$ with A and P as above all yield the same result $\hat{x} = E[x|z]$.

2-4 Estimate of the variance with the smallest MSE. Consider the problem of estimating the mean and the variance of a set of independent and identically distributed random variables $z(j)$, $j = 1, \ldots, k$, as in Subsection 2.5.3, with the true mean x_0 and true variance σ_0^2.

1. Show that the value of n in

$$[\hat{\sigma}(k)]^2 = \frac{1}{n} \sum_{i=1}^{k} \left[z(j) - \frac{1}{k} \sum_{i=1}^{k} z(i) \right]^2$$

that minimizes the MSE of the above (defined according to (2.6.1-4) with respect to σ_0^2) is $n = k + 1$.

2. Can a biased estimator have a smaller MSE than an unbiased one?

2-5 MAP estimate with two-sided exponential (Laplacian) prior pdf. Consider the same problem as in Subsection 2.3.2 but with a two-sided exponential prior

$$p(x) = \frac{a}{2} e^{-a|x|}$$

1. Write the posterior pdf of x.

2. Find $\hat{x}^{\mathrm{MAP}}$.

2-6 Two-sided exponential prior made diffuse.

1. Specify the limiting process that will make the prior from problem 2-5 into a diffuse one.

2. Show that the resulting MAP estimate coincides with the MLE.

2-7 Minimum magnitude error estimate. Show that the Bayesian estimation that minimizes the expected value of the cost function

$$C(x - \hat{x}) \triangleq |x - \hat{x}|$$

yields $\hat{x} = x_m$, the **median** of x, defined as

$$\int_{-\infty}^{x_m} p(x)dx = \frac{1}{2}$$

2-8 MAP with Gaussian prior — vector version. Given $z = x + w$, where all the variables are n-vectors, with

$$w \sim \mathcal{N}(0, P) \qquad\qquad x \sim \mathcal{N}(\bar{x}, P_0)$$

and x independent of w. Find the MAP estimator of x in terms of w and the covariance of this estimator.

2-9 Conditional variance versus unconditional variance. Let

$$\bar{x} \triangleq E[x] \qquad\qquad \mathrm{var}(x) \triangleq E[(x - \bar{x})^2]$$

$$\hat{x} \triangleq E[x|z] \qquad\qquad \mathrm{var}(x|z) \triangleq E[(x - \hat{x})^2|z]$$

Prove that

$$\mathrm{var}(x) \geq E[\mathrm{var}(x|z)]$$

Chapter 3

LINEAR ESTIMATION IN STATIC SYSTEMS

3.1 INTRODUCTION

3.1.1 Outline

This chapter presents the minimum mean square error (MMSE) estimation of Gaussian random vectors (Section 3.2) and the *linear* MMSE estimator for arbitrarily distributed random vectors (Section 3.3). The latter is the estimator constrained to have a linear form. The estimation of unknown constant vectors according to the least squares (LS) criterion is then discussed in Section 3.4, where both the batch and recursive versions are derived. These results are then applied to polynomial fitting in Section 3.5.

Section 3.6 presents the statistical tools for deciding what is the appropriate order of the polynomial when fitting a set of data points. This latter method is especially important because in practice the models, for instance, for the motion of targets, are not known a priori and have to be inferred from the data.

A realistic example that deals with the localization of a target based on bearings-only measurements from a platform is presented in Section 3.7. This illustrates the use of nonlinear LS to a practical problem.

3.1.2 Linear Estimation in Static Systems — Summary of Objectives

Present

- MMSE estimation of Gaussian random vectors
- Linear MMSE estimator for arbitrarily distributed random vectors
- LS estimation of unknown constant vectors from *linear* observations

    - batch form
    - recursive form.

Apply the LS technique to

- Polynomial fitting
- Choice of order of the polynomial when fitting a set of data points.

Illustrate the use of *nonlinear* LS to a practical problem — target localization from bearings-only measurements.

3.2 ESTIMATION OF GAUSSIAN RANDOM VECTORS

Consider two random vectors x and z that are *jointly* normally (Gaussian) distributed.

Define the **stacked vector**

$$y \triangleq \begin{bmatrix} x \\ z \end{bmatrix}$$

(3.2.0-1)

The notation

$$y \sim \mathcal{N}[\bar{y}, P_{yy}]$$

(3.2.0-2)

will indicate that the variable y is **normally (Gaussian) distributed** with mean

$$\bar{y} = \begin{bmatrix} \bar{x} \\ \bar{z} \end{bmatrix}$$

(3.2.0-3)

and covariance matrix (assumed nonsingular)

$$P_{yy} = \begin{bmatrix} P_{xx} & P_{xz} \\ P_{zx} & P_{zz} \end{bmatrix}$$

(3.2.0-4)

written in partitioned form, where $\bar{x}$ is the mean of x,

$$P_{xx} = E[(x - \bar{x})(x - \bar{x})']$$

(3.2.0-5)

is its covariance, and

$$P_{xz} = E[(x - \bar{x})(z - \bar{z})'] = P'_{zx}$$

(3.2.0-6)

is the covariance between x and z, and so on.

In the above, x is the random variable to be estimated and z is the **measurement** or the **observation**.

As shown in (2.4.1-7), the **estimate** of the random variable x in terms of z according to the **minimum mean square error (MMSE)** criterion — the **MMSE estimator** — is the conditional mean of x given z.

For x and z *jointly Gaussian*, as assumed in (3.2.0-2), the **conditional mean** is

$$\boxed{\hat{x} \triangleq E[x|z] = \bar{x} + P_{xz}P_{zz}^{-1}(z - \bar{z})}$$

(3.2.0-7)

and the corresponding **conditional covariance matrix** is

$$\boxed{P_{xx|z} \triangleq E[(x - \hat{x})(x - \hat{x})'|z] = P_{xx} - P_{xz}P_{zz}^{-1}P_{zx}}$$

(3.2.0-8)

(see Subsection 1.4.14 for proof).

This follows from the fact that the conditional pdf of x given z is Gaussian with mean (3.2.0-7) and covariance (3.2.0-8).

Note that the optimal estimator (in the MMSE sense) of x in terms of z is a *linear function* of z. This is a consequence of the Gaussian assumption.

Another important property specific to this Gaussian problem is that the *conditional covariance* (3.2.0-8), which measures the "quality" of the estimate, is *independent of the observation z.*

Summary

The *MMSE estimate* — the *conditional mean* — of a *Gaussian* random vector in terms of another *Gaussian* random vector (the measurement) is a *linear combination* of

- The prior (unconditional) mean of the variable to be estimated;
- The difference between the measurement and its prior mean.

The *conditional covariance* of one *Gaussian* random vector given another *Gaussian* random vector (the measurement) is *independent* of the measurement.

Both of the above properties hinge strictly on the assumption that the two random vectors under consideration are *jointly Gaussian.*

3.3 LINEAR MINIMUM MEAN SQUARE ERROR ESTIMATION

3.3.1 The Principle of Orthogonality

The *minimum mean square error (MMSE) estimate* of a random variable x in terms of another random variable z is, according to (2.4.1-7), the conditional mean $E[x|z]$.

In many problems the distributional information needed for the evaluation of the conditional mean is not available. Furthermore, even if it were available, the evaluation of the conditional mean can be prohibitively complicated.

In view of this, a method that: (1) is simple — yields the estimate as a linear function of the observable(s), and (2) requires little information — only first and second moments, is highly desirable. Such a method, called **linear MMSE** estimation relies on the **principle of orthogonality**:

The best linear estimate (in the sense of MMSE) of a random variable in terms of another random variable — the observable(s) — is such that

1. The estimate is unbiased — the estimation error has mean zero, and
2. The estimation error is uncorrelated from the observable(s),

that is, they are **orthogonal**.

Linear MMSE Estimation for Zero-Mean Random Variables

The linear MMSE estimation can be formulated in terms of a (normed linear) space of random variables as follows.

The set of real-valued (scalar) **zero-mean random variables** z_i, $i = 1, \ldots, n$, can be considered as **vectors in an abstract vector space** or **linear space**. Such a space is closed under addition of its elements and multiplication by scalars — the linear combination of two random variables is another element in this space.

A vector space in which one defines an **inner product** is a Hilbert space. The inner product that can be defined is

$$\langle z_i, z_k \rangle = E[z_i z_k] \qquad (3.3.1\text{-}1)$$

Since the random variables under consideration are zero mean, it is clear that

$$\langle z_i, z_i \rangle = E[z_i^2] = \|z_i\|^2 \qquad (3.3.1\text{-}2)$$

satisfies the properties of a **norm** and can be taken as such.

3.3.1 The Principle of Orthogonality

With this definition of the norm, **linear dependence** is defined by stating that the norm of a linear combination of vectors is zero

$$E\left[\left(\sum_{i=1}^{m}\alpha_i z_i\right)^2\right] = 0 \tag{3.3.1-3}$$

If $\alpha_1 \neq 0$, then z_1 is a linear combination of $z_2, \ldots, z_m$

$$z_1 = -\frac{1}{\alpha_1}\sum_{i=2}^{m}\alpha_i z_i \tag{3.3.1-4}$$

that is, it is an element of the **subspace** spanned by $z_2, \ldots, z_m$.

Two vectors are **orthogonal**, denoted as $z_i \perp z_k$, if and only if

$$\langle z_i, z_k \rangle = 0 \tag{3.3.1-5}$$

which is equivalent to these (zero-mean) random variables being *uncorrelated*.

The **linear MMSE estimator** of a zero-mean random variable x in terms of z_i, $i = 1$, $\ldots, n$, is given by

$$\hat{x} = \sum_{i=1}^{n}\beta_i z_i \tag{3.3.1-6}$$

and has to be such that the norm of the **estimation error**

$$\tilde{x} \stackrel{\Delta}{=} x - \hat{x} \tag{3.3.1-7}$$

is minimum. The linear MMSE estimate is denoted also by a circumflex ("hat"), even though it not the conditional mean as in (3.2.0-7).

Thus the norm of the estimation error

$$\|\tilde{x}\|^2 = E[(x - \hat{x})^2] = E\left[(x - \sum_{i=1}^{n}\beta_i z_i)^2\right] \tag{3.3.1-8}$$

will have to be minimized with respect to β_i, $i = 1, \ldots, n$.

Setting to zero the derivative of (3.3.1-8) with respect to β_k

$$-\frac{1}{2}\frac{\partial}{\partial \beta_k}\|\tilde{x}\|^2 = E\left[(x - \sum_{i=1}^{n}\beta_i z_i)z_k\right] = E[\tilde{x}z_k] = \langle \tilde{x}, z_k \rangle = 0 \qquad k = 1, \ldots, n \tag{3.3.1-9}$$

is seen to be equivalent to requiring the following *orthogonality* property

$$\tilde{x} \perp z_k \qquad \forall k \tag{3.3.1-10}$$

128

This is the **principle of orthogonality**: in order for the error to have minimum norm, it has to be *orthogonal to the observations*. This is equivalent to stating that the estimate $\hat{x}$ has to be the **orthogonal projection** of x into the space spanned by the observables, as illustrated in Figure 3.3.1-1.

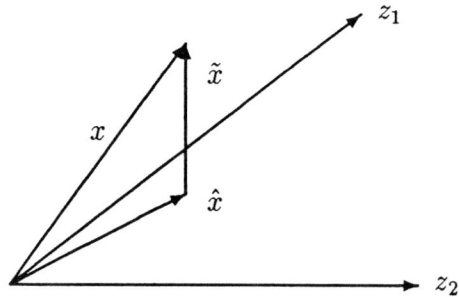

Figure 3.3.1-1: Orthogonal projection of the random variable x into the subspace of $\{z_1, z_2\}$.

Linear MMSE Estimation for Nonzero Mean Random Variables

For a random variable x with nonzero mean $\bar{x}$, the best linear estimator is of the form (an affine function)

$$\hat{x} = \beta_0 + \sum_{i=1}^{n} \beta_i z_i \qquad (3.3.1\text{-}11)$$

Since the MSE is the sum of the square of the mean and the variance

$$E[\tilde{x}^2] = \left(E[\tilde{x}]\right)^2 + \text{var}(\tilde{x}) \qquad (3.3.1\text{-}12)$$

in order to minimize it, the estimate should have the **unbiasedness property**:

$$E[\tilde{x}] = 0 \qquad (3.3.1\text{-}13)$$

It can be easily shown that this follows from the fact that β_0 enters only into the mean of the error (3.3.1-13), whose minimum norm is zero. Equation (3.3.1-13) yields

$$\beta_0 = \bar{x} - \sum_{i=1}^{n} \beta_i \bar{z}_i \qquad (3.3.1\text{-}14)$$

where

$$\bar{z}_i = E[z_i] \qquad (3.3.1\text{-}15)$$

Inserting (3.3.1-14) into (3.3.1-11) leads to

$$\hat{x} = \bar{x} + \sum_{i=1}^{n} \beta_i (z_i - \bar{z}_i) \qquad (3.3.1\text{-}16)$$

The error corresponding to the estimate (3.3.1-16) is

$$\begin{aligned}
\tilde{x} \;&\triangleq\; x - \hat{x} \\
&=\; x - \bar{x} - \sum_{i=1}^{n} \beta_i (z_i - \bar{z}_i)
\end{aligned} \qquad (3.3.1\text{-}17)$$

This has transformed the nonzero-mean case into the zero-mean case.

The orthogonality principle (3.3.1-10) then yields the coefficients β_i from the following equations:

$$\langle \tilde{x}, z_k \rangle = E[\tilde{x} z_k] = E\left[\left[x - \bar{x} - \sum_{i=1}^{n} \beta_i (z_i - \bar{z}_i)\right] z_k\right] = 0 \qquad k = 1, \ldots, n \qquad (3.3.1\text{-}18)$$

The estimator (3.3.1-16) is also known as the **best linear unbiased estimator**.

3.3.2 Linear MMSE Estimation for Vector Random Variables

Consider the vector-valued random variables x and z, which are not necessarily Gaussian or zero mean. The "best linear" estimate of x in terms of z is obtained as follows. The criterion for "best" is the MMSE, that is, find the estimator

$$\hat{x} = Az + b \qquad (3.3.2\text{-}1)$$

that minimizes the *scalar MSE criterion*, which in the multidimensional case is the *expected value of the squared norm* of the estimation error,

$$J \triangleq E[(x - \hat{x})'(x - \hat{x})] \qquad (3.3.2\text{-}2)$$

According to the previous discussion, the *linear MMSE estimator* is such that the estimation error

$$\tilde{x} = x - \hat{x} \qquad (3.3.2\text{-}3)$$

is *unbiased* and *orthogonal to the observation* z. In other words, the estimate $\hat{x}$ is the orthogonal projection of the vector x into the space spanned by the (random components of the) observation vector z.

The unbiasedness requirement (3.3.1-13) is

$$E[\tilde{x}] = \bar{x} - (A\bar{z} + b) = 0 \qquad (3.3.2\text{-}4)$$

and it yields

$$b = \bar{x} - A\bar{z} \qquad (3.3.2\text{-}5)$$

The estimation error is then

$$\tilde{x} = x - \bar{x} - A(z - \bar{z}) \qquad (3.3.2\text{-}6)$$

The orthogonality requirement is, in the multidimensional case, that *each component* of $\tilde{x}$ be orthogonal to *each component* of z.

The orthogonality requirement can thus be written as

$$E[\tilde{x}z'] = E\{[x - \bar{x} - A(z - \bar{z})]z'\} \qquad (3.3.2\text{-}7)$$
$$= E\{[x - \bar{x} - A(z - \bar{z})](z - \bar{z})'\} = P_{xz} - AP_{zz} = 0 \qquad (3.3.2\text{-}8)$$

The subtraction of $\bar{z}$ from z in the transition from (3.3.2-7) to (3.3.2-8) could be done in view of the property (3.3.2-4) that $\tilde{x}$ is zero mean.

131

The solution for the **weighting matrix** A is thus

$$A = P_{xz}P_{zz}^{-1} \tag{3.3.2-9}$$

Combining (3.3.2-5) and (3.3.2-9) yields the expression of the **linear MMSE estimator for the multidimensional case** as

$$\boxed{\hat{x} = \bar{x} + P_{xz}P_{zz}^{-1}(z - \bar{z})} \tag{3.3.2-10}$$

which is *identical* to the conditional mean (3.2.0-7) from the Gaussian case.

The **matrix MSE** corresponding to (3.3.2-10) is given by

$$E[\tilde{x}\tilde{x}'] = E\left[[x - \bar{x} - P_{xz}P_{zz}^{-1}(z - \bar{z})][x - \bar{x} - P_{xz}P_{zz}^{-1}(z - \bar{z})]'\right] \tag{3.3.2-11}$$

This becomes, after simple manipulations,

$$\boxed{E[\tilde{x}\tilde{x}'] = P_{xx} - P_{xz}P_{zz}^{-1}P_{zx} = P_{xx|z}} \tag{3.3.2-12}$$

that is, *identical* expression to the conditional covariance (3.2.0-8) in the Gaussian case. Note, however, that strictly speaking the matrix MSE (3.3.2-12) is not a covariance matrix since (3.3.2-10) is not the conditional mean.

Equations (3.3.2-10) and (3.3.2-12) are the **fundamental equations of linear estimation**.

Remarks

Note the distinction between the scalar MSE criterion (3.3.2-2), an inner product, and the matrix MSE (3.3.2-12), an outer product. The **matrix MSE** is sometimes called, by abuse of language, a **covariance matrix**.

From the above derivations it follows that

- the *best estimator* (in the MMSE sense) for *Gaussian random variables*

is identical to

- the *best linear estimator* for *arbitrarily distributed* random variables with the *same first and second order moments.*

The linear estimator (3.3.2-10) is the overall best if the random variables are Gaussian; otherwise, it is only the *best within the class of linear estimators.*

Gaussian Assumption as "Worst Case" in MMSE Estimation

The following statement can be made: From the point of view of MMSE estimation, one can view the **Gaussian assumption as the worst case**:

- If the random variables are Gaussian, the minimum achievable matrix MSE is (3.2.0-8);
- If they are not Gaussian, but with the same first two moments, one can achieve (with the linear estimator) the matrix MSE (3.3.2-12), which is the same as (3.2.0-8).

However, in the non-Gaussian case,

> The conditional mean (if one can compute it), being the *absolute best* as opposed to the *best within the class of linear estimators*, would give a matrix MSE *less or equal* to (3.3.2-12).

On the Terminology

The LMMSE estimator is also called in the literature as the **least mean square (LMS)** or **minimum variance (MV)** or **least squares (LS)**.

(See also problem 3-1.)

3.3.3 Linear MMSE Estimation — Summary

The **linear MMSE estimator** of one random vector in terms of another random vector (the measurement) is such that the estimation error is

1. Unbiased
2. Uncorrelated from the measurements.

These two properties imply that the error is orthogonal to the measurements. This is the **principle of orthogonality**.

The expression of the *linear MMSE estimator* is identical to the expression of the *conditional mean* of *Gaussian* random vectors if they have the same first two moments.

Similarly, the *matrix MSE* associated with the LMMSE estimator has the same expression as the *conditional covariance* in the Gaussian case.

The *linear MMSE estimator* is

1. The *overall best* if the random variables are *Gaussian*
2. The *best* within the class of *linear* estimators otherwise.

3.4 LEAST SQUARES ESTIMATION

3.4.1 The Batch LS Estimation

In the **linear least squares (LS)** problem it is desired to estimate the n_x-vector x modeled as an *unknown constant* from the linear observations (n_z-vectors)

$$z(i) = H(i)x + w(i) \qquad i = 1, \ldots, k \tag{3.4.1-1}$$

such as to minimize the quadratic error

$$J(k) = \frac{1}{2} \sum_{i=1}^{k} [z(i) - H(i)x]' R(i)^{-1} [z(i) - H(i)x] \tag{3.4.1-2}$$

weighted with the (inverses of the) positive definite matrices $R(i)$.

The above can be rewritten in a compact form as

$$J(k) = \frac{1}{2} [z^k - H^k x]' (R^k)^{-1} [z^k - H^k x] \tag{3.4.1-3}$$

where

$$z^k = \begin{bmatrix} z(1) \\ \vdots \\ z(k) \end{bmatrix} \tag{3.4.1-4}$$

is the **stacked vector** of measurements (of dimension $k n_z \times 1$),

$$H^k = \begin{bmatrix} H(1) \\ \vdots \\ H(k) \end{bmatrix} \tag{3.4.1-5}$$

is the **stacked measurement matrix** (of dimension $k n_z \times n_x$),

$$w^k = \begin{bmatrix} w(1) \\ \vdots \\ w(k) \end{bmatrix} \tag{3.4.1-6}$$

is the stacked vector of the measurement errors, and

$$R^k = \begin{bmatrix} R(1) & \cdots & 0 \\ \vdots & \ddots & \vdots \\ 0 & \cdots & R(k) \end{bmatrix} = \text{diag}[R(k)] \tag{3.4.1-7}$$

Note that the matrix (3.4.1-7) is a block-diagonal positive definite $kn_z \times kn_z$ matrix.

If in (3.4.1-1) there is a nonlinear function of the unknown vector x, then one has a **nonlinear LS** problem.

The **LS estimator** that minimizes (3.4.1-3) is obtained by setting its gradient with respect to x to zero

$$\nabla_x J(k) = -H^{k'}(R^k)^{-1}[z^k - H^k x] = 0 \qquad (3.4.1\text{-}8)$$

which yields

$$\boxed{\hat{x}(k) = [H^{k'}(R^k)^{-1}H^k]^{-1}H^{k'}(R^k)^{-1}z^k} \qquad (3.4.1\text{-}9)$$

assuming the required inverse exists.

It can be easily shown that since R^k, defined in (3.4.1-7), is positive definite, the Hessian of (3.4.1-3) with respect to x is positive definite, and consequently the extremum point (3.4.1-9) is a minimum.

Note that (3.4.1-9) is a **batch estimator** — the entire data have to be processed simultaneously for every k.

Remark

In this approach x is an unknown constant. The estimate $\hat{x}(k)$ is a random variable if the disturbances (measurement errors) $w(i)$ are modeled as random.

Relationship to the ML Estimator

If the measurement errors $w(i)$ are *independent Gaussian* random variables with mean zero and covariance $R(i)$, then minimizing the LS criterion (3.4.1-2) is equivalent to maximizing the likelihood function

$$
\begin{aligned}
\Lambda_k(x) &= p[z^k|x] \\
&= \prod_{i=1}^{k} p[z(i)|x] \\
&= c\, e^{-\frac{1}{2}\sum_{i=1}^{k}[z(i)-H(i)x]'R(i)^{-1}[z(i)-H(i)x]}
\end{aligned}
\qquad (3.4.1\text{-}10)
$$

that is, the LS and ML estimators coincide in this case.

The LS criterion (3.4.1-2) implicitly assumes that $w(i)$ are independent, zero mean, and with covariance $R(i)$, and leads to the minimization of the sum of their weighted norms. Since this is equivalent to the maximization of the likelihood function under the additional Gaussian assumption, the LS is clearly a "disguised" ML technique.

Properties of the LS Estimator

With the assumption that $w(i)$ are independent, zero-mean random variables with covariance $R(i)$, but without any further distributional assumptions, the LS estimator (3.4.1-9) is *unbiased*, that is,

$$
\begin{aligned}
E[\hat{x}(k)] &= [H^{k'}(R^k)^{-1}H^k]^{-1}H^{k'}(R^k)^{-1}E[H^k x + w^k] \\
&= x
\end{aligned}
\tag{3.4.1-11}
$$

The estimation error is

$$
\begin{aligned}
\tilde{x}(k) &= x - \hat{x}(k) \\
&= -[H^{k'}(R^k)^{-1}H^k]^{-1}H^{k'}(R^k)^{-1}w^k
\end{aligned}
\tag{3.4.1-12}
$$

Thus, the **covariance matrix of the LS estimator** is

$$
\begin{aligned}
P(k) &\triangleq E\Big[\{\hat{x}(k) - E[\hat{x}(k)]\}\{\hat{x}(k) - E[\hat{x}(k)]\}'\Big] \\
&= E\Big[[\hat{x}(k) - x][\hat{x}(k) - x]'\Big] \\
&= E[\tilde{x}(k)\tilde{x}(k)'] \\
&= [H^{k'}(R^k)^{-1}H^k]^{-1}H^{k'}(R^k)^{-1}R^k(R^k)^{-1}H^k[H^{k'}(R^k)^{-1}H^k]^{-1}
\end{aligned}
\tag{3.4.1-13}
$$

where in the last line above use was made of (3.4.1-12) and the fact that based on (3.4.1-7) one has

$$
E[w^k w^{k'}] = R^k
\tag{3.4.1-14}
$$

Equation (3.4.1-13) yields, after cancellations,

$$
\boxed{P(k) = [H^{k'}(R^k)^{-1}H^k]^{-1}}
\tag{3.4.1-15}
$$

Note that when carrying out the expectation in (3.4.1-11) and (3.4.1-13) it is over w^k.

Existence of the Solution — Parameter Observability

The *existence of the inverse* required in (3.4.1-9) is equivalent to having the covariance of the error (3.4.1-15) *finite*. This amounts to requiring the parameter x to be **observable**, that is, that it can be estimated from the observations. (See also problem 3-3.)

3.4.2 The Recursive LS Estimator

A useful feature of the LS estimator (3.4.1-9) is that it can be rewritten in recursive form (i.e., suitable for sequential rather than batch processing). In this case k is interpreted as "discrete time."

When $z(k+1)$ is obtained one can write the following partitioned forms

$$z^{k+1} = \begin{bmatrix} z^k \\ z(k+1) \end{bmatrix} \tag{3.4.2-1}$$

$$H^{k+1} = \begin{bmatrix} H^k \\ H(k+1) \end{bmatrix} \tag{3.4.2-2}$$

$$w^{k+1} = \begin{bmatrix} w^k \\ w(k+1) \end{bmatrix} \tag{3.4.2-3}$$

$$R^{k+1} = \begin{bmatrix} R^k & 0 \\ 0 & R(k+1) \end{bmatrix} \tag{3.4.2-4}$$

The Recursion for the Inverse Covariance

Expression (3.4.1-15) at $k+1$ can be expressed recursively as

$$
\begin{aligned}
P(k+1)^{-1} &= H^{k+1'}(R^{k+1})^{-1}H^{k+1} \\
&= \begin{bmatrix} H^{k'} & H(k+1)' \end{bmatrix} \begin{bmatrix} R^k & 0 \\ 0 & R(k+1) \end{bmatrix}^{-1} \begin{bmatrix} H^k \\ H(k+1) \end{bmatrix} \\
&= H^{k'}(R^k)^{-1}H^k + H(k+1)'R(k+1)^{-1}H(k+1) \tag{3.4.2-5}
\end{aligned}
$$

or,

$$\boxed{P(k+1)^{-1} = P(k)^{-1} + H(k+1)'R(k+1)^{-1}H(k+1)} \tag{3.4.2-6}$$

This can be interpreted as follows: the **information** (in the sense of Fisher, that is, the inverse covariance) at $k+1$ equals the sum of the information at k and the new information about x obtained from $z(k+1)$. The information is *additive* here because

1. the problem is a static one — the parameter is fixed, and
2. the observations are modeled as *independent*.

Using the matrix inversion lemma (1.3.3-11), (3.4.2-6) can be rewritten as

$$
\begin{aligned}
P(k+1) &= [P(k)^{-1} + H(k+1)'R(k+1)^{-1}H(k+1)]^{-1} \\
&= P(k) - P(k)H(k+1)'[H(k+1)P(k)H(k+1)' + R(k+1)]^{-1}H(k+1)P(k)
\end{aligned}
$$

$$\tag{3.4.2-7}$$

The Residual Covariance and the Update Gain

Denote the matrices

$$\boxed{S(k+1) \triangleq H(k+1)P(k)H(k+1)' + R(k+1)}$$

(3.4.2-8)

$$\boxed{W(k+1) \triangleq P(k)H(k+1)'S(k+1)^{-1}}$$

(3.4.2-9)

which, as will be seen later, have the interpretations of **covariance of the residual** and parameter **update gain**, respectively.

The Recursion for the Covariance

With (3.4.2-8) and (3.4.2-9), recursion (3.4.2-7) can be rewritten more compactly as

$$\boxed{P(k+1) = [I - W(k+1)H(k+1)]P(k) = P(k) - W(k+1)S(k+1)W(k+1)'}$$

(3.4.2-10)

which is an alternative to (3.4.2-6).

Alternative Expression for the Gain

Using (3.4.2-7), one has the following identity

$$
\begin{aligned}
P(k+1)H(k+1)'R(k+1)^{-1} &= \{P(k)H(k+1)' - P(k)H(k+1)' \\
&\quad \cdot [H(k+1)P(k)H(k+1)' + R(k+1)]^{-1} \\
&\quad \cdot H(k+1)P(k)H(k+1)'\}R(k+1)^{-1} \\
&= P(k)H(k+1)'[H(k+1)P(k)H(k+1)' + R(k+1)]^{-1} \\
&\quad \cdot \{H(k+1)P(k)H(k+1)' + R(k+1) \\
&\quad - H(k+1)P(k)H(k+1)'\}R(k+1)^{-1} \\
&= P(k)H(k+1)'S(k+1)^{-1} \\
&= W(k+1)
\end{aligned}
$$

(3.4.2-11)

This gives an alternative expression for the update gain (3.4.2-9) as

$$\boxed{W(k+1) = P(k+1)H(k+1)'R(k+1)^{-1}}$$

(3.4.2-12)

The Recursion for the Estimate

The batch estimation equation (3.4.1-9) for $k + 1$ is rewritten as

$$
\begin{aligned}
\hat{x}(k+1) &= P(k+1)H^{k+1'}(R^{k+1})^{-1}z^{k+1} \\
&= P(k+1)\left[H^{k'}\ \ H(k+1)'\right]\begin{bmatrix} R^k & 0 \\ 0 & R(k+1) \end{bmatrix}^{-1}\begin{bmatrix} z^k \\ z(k+1) \end{bmatrix} \\
&= P(k+1)H^{k'}(R^k)^{-1}z^k + P(k+1)H(k+1)'R(k+1)^{-1}z(k+1) \\
&= [I - W(k+1)H(k+1)]P(k)H^{k'}(R^k)^{-1}z^k + W(k+1)z(k+1) \\
&= [I - W(k+1)H(k+1)]\hat{x}(k) + W(k+1)z(k+1) \qquad (3.4.2\text{-}13)
\end{aligned}
$$

where (3.4.2-10) and (3.4.2-12) were used.

The above is the **recursive parameter estimate updating equation** — the **recursive LS estimator**, written as

$$
\boxed{\hat{x}(k+1) = \hat{x}(k) + W(k+1)[z(k+1) - H(k+1)\hat{x}(k)]} \qquad (3.4.2\text{-}14)
$$

The new (updated) estimate $\hat{x}(k+1)$ is therefore equal to the previous one plus a **correction term**. This correction term consists of the **gain** $W(k+1)$ multiplying the **residual** — the difference between the observation $z(k+1)$ and the **predicted value** of this observation from the previous k measurements.

The Residual Covariance

It can be easily shown that $S(k+1)$ defined in (3.4.2-8) is the **covariance of the residual** from (3.4.2-14), that is,

$$
E\Big[[z(k+1) - H(k+1)\hat{x}(k)][z(k+1) - H(k+1)\hat{x}(k)]'\Big] = S(k+1) \qquad (3.4.2\text{-}15)
$$

3.4.3 Examples and Incorporation of Prior Information

The Sample Mean

Consider noisy observations on a constant scalar x

$$z(i) = x + w(i) \qquad i = 1, \ldots, k \tag{3.4.3-1}$$

For the batch LS formulation, one has

$$H^k = \begin{bmatrix} 1 \\ \vdots \\ 1 \end{bmatrix} \tag{3.4.3-2}$$

a k-dimensional vector, and let

$$R^k = I\sigma^2 \tag{3.4.3-3}$$

where I is the $k \times k$ identity matrix.

Then, using (3.4.1-9) with (3.4.3-2) and (3.4.3-3) from above, one has the LS estimate

$$
\begin{aligned}
\hat{x}(k) &= [H^{k'}(R^k)^{-1}H^k]^{-1}H^{k'}(R^k)^{-1}z^k \\
&= \left\{ [1 \, \cdots \, 1](I\sigma^2)^{-1} \begin{bmatrix} 1 \\ \vdots \\ 1 \end{bmatrix} \right\}^{-1} [1 \, \cdots \, 1](I\sigma^2)^{-1} \begin{bmatrix} z(1) \\ \vdots \\ z(k) \end{bmatrix} \\
&= \frac{1}{k}\sum_{i=1}^{k} z(i)
\end{aligned}
\tag{3.4.3-4}
$$

that is, the **sample mean**. The variance of this estimate, assuming $w(i)$ to be a sequence of independent and identically distributed random variables that are zero mean and with variance σ^2, follows from (3.4.1-13) as

$$
\begin{aligned}
P(k) &= [H^{k'}(R^k)^{-1}H^k]^{-1} \\
&= \frac{\sigma^2}{k}
\end{aligned}
\tag{3.4.3-5}
$$

Note that this is the same result as obtained in (2.6.3-1).

To obtain the recursive form of the LS estimation, using (3.4.2-8) and (3.4.2-9) one has

$$S(k+1) = H(k+1)P(k)H(k+1)' + R(k+1) = \frac{\sigma^2}{k} + \sigma^2 = \frac{k+1}{k}\sigma^2 \qquad (3.4.3\text{-}6)$$

$$W(k+1) = P(k)H(k+1)'S(k+1)^{-1} = \frac{\sigma^2}{k}\left(\frac{k+1}{k}\sigma^2\right)^{-1} = \frac{1}{k+1} \qquad (3.4.3\text{-}7)$$

and thus, using (3.4.2-14)

$$\boxed{\hat{x}(k+1) = \hat{x}(k) + \frac{1}{k+1}[z(k+1) - \hat{x}(k)]} \qquad (3.4.3\text{-}8)$$

The above recursion could have also been obtained directly from the batch expression by the following simple algebraic manipulation

$$
\begin{aligned}
\hat{x}(k+1) &= \frac{1}{k+1}\sum_{i=1}^{k+1} z(i) \\
&= \frac{1}{k+1}\left[\sum_{i=1}^{k} z(i) + z(k+1)\right] \\
&= \frac{1}{k+1}[k\hat{x}(k) + z(k+1) - \hat{x}(k) + \hat{x}(k)] \\
&= \hat{x}(k) + \frac{1}{k+1}[z(k+1) - \hat{x}(k)] \qquad k = 1,\dots \qquad (3.4.3\text{-}9)
\end{aligned}
$$

with the initial condition $\hat{x}(1) = z(1)$.

Estimation of the Mean with Prior Information

The previous example will be reconsidered with x now assumed a *random variable* with **prior information** consisting of the mean $\bar{x}$ and the variance

$$P_{xx} = \sigma_0^2 \qquad (3.4.3\text{-}10)$$

The estimation will be cast in the framework of the LMMSE estimation discussed in Section 3.3.

The measurements are as in (3.4.3-1) and $w(i)$ are independent and identically distributed $\mathcal{N}(0, \sigma^2)$ and independent of x. Thus

$$z = H^k x + w^k \qquad (3.4.3\text{-}11)$$

142

Averaging over x and w, one has

$$
\begin{aligned}
P_{zz} &= E[(z - \bar{z})(z - \bar{z})'] = E\left[[H^k(x - \bar{x}) + w^k)][H^k(x - \bar{x}) + w^k]'\right] \\
&= H^k \sigma_0^2 H^{k'} + I\sigma^2
\end{aligned}
\tag{3.4.3-12}
$$

and, similarly,

$$
P_{xz} = E[(x - \bar{x})(z - \bar{z})'] = E\left[(x - \bar{x})[H^k(x - \bar{x}) + w^k]'\right] = \sigma_0^2 H^{k'}
\tag{3.4.3-13}
$$

The inverse of (3.4.3-12) is, using the matrix inversion lemma (1.3.3-12),

$$
\begin{aligned}
P_{zz}^{-1} &= I\sigma^{-2} - \sigma^{-2} H^k (H^{k'} \sigma^{-2} H^k + \sigma_0^{-2})^{-1} H^{k'} \sigma^{-2} \\
&= I\sigma^{-2} - \frac{\sigma^{-4}}{k\sigma^{-2} + \sigma_0^{-2}} H^k H^{k'}
\end{aligned}
\tag{3.4.3-14}
$$

The MMSE estimate (3.3.2-10) of x is then

$$
\begin{aligned}
\hat{x}(k) &= \bar{x} + P_{xz} P_{zz}^{-1} (z - H^k \bar{x}) \\
&= \bar{x} + \sigma_0^2 H^{k'} \left(I\sigma^{-2} - \frac{\sigma^{-4}}{k\sigma^{-2} + \sigma_0^{-2}} H^k H^{k'}\right)(z - H^k \bar{x}) \\
&= \bar{x} + \sigma_0^2 \left(\sigma^{-2} - \frac{\sigma^{-4}}{k\sigma^{-2} + \sigma_0^{-2}} H^{k'} H^k\right) H^{k'} (z - H^k \bar{x}) \\
&= \bar{x} + \sigma_0^2 \left(\sigma^{-2} - \frac{k\sigma^{-4}}{k\sigma^{-2} + \sigma_0^{-2}}\right) \sum_{i=1}^{k} [z(i) - \bar{x}] \\
&= \bar{x} + \frac{\sigma^{-2}}{k\sigma^{-2} + \sigma_0^{-2}} \sum_{i=1}^{k} [z(i) - \bar{x}]
\end{aligned}
\tag{3.4.3-15}
$$

or

$$
\hat{x}(k) = \frac{\sigma_0^{-2}}{k\sigma^{-2} + \sigma_0^{-2}} \bar{x} + \frac{\sigma^{-2}}{k\sigma^{-2} + \sigma_0^{-2}} \sum_{i=1}^{k} z(i)
\tag{3.4.3-16}
$$

Effect of Diffuse Prior

Note that if $\sigma_0 \to \infty$, the prior information becomes **diffuse**, which amounts to lack of prior information — this is the motivation of the term **noninformative**. In this case (3.4.3-16) becomes

$$
\hat{x}(k) = \frac{1}{k} \sum_{i=1}^{k} z(i)
\tag{3.4.3-17}
$$

which is the LS or ML estimate, as expected. (See also problem 3-2.)

3.4.4 LS Estimation — Summary

The LS estimator based on a set (or a sequence) of linear measurements of an *unknown constant parameter* is a linear function of the *stacked measurement vector* — this is the *batch* form of the LS.

The LS criterion is really a disguised ML criterion under suitable Gaussian assumptions on the observation noises (zero mean, uncorrelated, and with the covariance given by the inverses of the LS criterion weighting matrices).

The LS estimator can be rewritten in a *recursive* form:

The estimate based on a given set of measurements is a *linear combination* of

1. the *previous estimate* (available prior to the latest measurement), and
2. the *latest measurement.*

The inverse of the covariance of the parameter estimate (the Fisher information matrix) is also obtained *recursively* as the sum of:

1. the inverse covariance of the estimate prior to the latest measurement, and
2. the information about the parameter in the latest measurement.

3.5 POLYNOMIAL FITTING

3.5.1 Fitting a First Order Polynomial to Noisy Measurements

Assume that one measures, in the presence of additive noise, the position of an object moving in one dimension with constant velocity, that is,

$$z(i) = x_0 + \dot{x}_0 t_i + w(i) \qquad i = 1, \ldots \qquad (3.5.1\text{-}1)$$

This motion is characterized by the (unknown) parameter

$$x = [x_0 \ \dot{x}_0]' \qquad (3.5.1\text{-}2)$$

consisting of the object's *initial position and velocity*.

Equation (3.5.1-1), which is known in statistics as a first order **regression**, can be written as

$$z(i) = H(i)x + w(i) \qquad (3.5.1\text{-}3)$$

where

$$H(i) = [1 \ \ t_i] \qquad (3.5.1\text{-}4)$$

The problem of estimating the parameter x amounts to **polynomial fitting**: in this case fitting a first order polynomial — a straight line — to a set of noisy measurements.

If the noises $w(i)$ are independent, identically distributed, zero mean, and with variances σ^2, then according to (3.4.1-7), the covariance matrix of the stacked measurement noise vector is

$$R^k = I\sigma^2 \qquad (3.5.1\text{-}5)$$

where I denotes the identity matrix of dimension k, not indicated for simplicity.

The batch solution is, noting that σ^2 cancels,

$$
\begin{aligned}
\hat{x}(k) &= [H^{k'}(R^k)^{-1}H^k]^{-1}H^{k'}(R^k)^{-1}z^k = [H^{k'}(I\sigma^2)^{-1}H^k]^{-1}H^{k'}(I\sigma^2)^{-1}z^k \\
&= (H^{k'}H^k)^{-1}H^{k'}z^k \qquad (3.5.1\text{-}6)
\end{aligned}
$$

Because of the special form of the noise covariance (3.5.1-5), the parameter estimate covariance can be written as

$$P(k) = [H^{k'}(R^k)^{-1}H^k]^{-1} = [H^{k'}(I\sigma^2)^{-1}H^k]^{-1} = (H^{k'}H^k)^{-1}\sigma^2 \qquad (3.5.1\text{-}7)$$

The recursive solution is

$$\hat{x}(k+1) = \hat{x}(k) + W(k+1)[z(k+1) - H(k+1)\hat{x}(k)] \qquad (3.5.1\text{-}8)$$

The **initialization** of recursion (3.5.1-8) is to be done using a batch estimate. This requires a minimum of two observations in (3.5.1-6), because otherwise the required inversion cannot be done. For $k = 2$, one has

$$P(2) = (H^{2'}H^2)^{-1}\sigma^2 = \left\{\begin{bmatrix} 1 & 1 \\ t_1 & t_2 \end{bmatrix}\begin{bmatrix} 1 & t_1 \\ 1 & t_2 \end{bmatrix}\right\}^{-1}\sigma^2$$

$$= \frac{1}{(t_2 - t_1)^2}\begin{bmatrix} t_1^2 + t_2^2 & -t_1 - t_2 \\ -t_1 - t_2 & 2 \end{bmatrix}\sigma^2 \tag{3.5.1-9}$$

and,

$$\hat{x}(2) = P(2)H^{2'}\sigma^{-2}z^2 = \frac{1}{(t_2 - t_1)^2}\begin{bmatrix} t_1^2 + t_2^2 & -t_1 - t_2 \\ -t_1 - t_2 & 2 \end{bmatrix}\begin{bmatrix} 1 & 1 \\ t_1 & t_2 \end{bmatrix}\begin{bmatrix} z(1) \\ z(2) \end{bmatrix}$$

$$= \frac{1}{t_2 - t_1}\begin{bmatrix} z(1)t_2 - z(2)t_1 \\ z(2) - z(1) \end{bmatrix} \tag{3.5.1-10}$$

The covariance matrix of the estimation error of the parameter vector (3.5.1-2) is, from $k + 1$ measurements, given by

$$P(k + 1) = (H^{k+1'}H^{k+1})^{-1}\sigma^2 = \left\{\begin{bmatrix} 1 & \cdots & 1 \\ t_1 & \cdots & t_{k+1} \end{bmatrix}\begin{bmatrix} 1 & t_1 \\ \vdots & \vdots \\ 1 & t_{k+1} \end{bmatrix}\right\}^{-1}\sigma^2$$

$$= \begin{bmatrix} s_0 & s_1 \\ s_1 & s_2 \end{bmatrix}^{-1} = \frac{1}{s_0 s_2 - s_1^2}\begin{bmatrix} s_2 & -s_1 \\ -s_1 & s_0 \end{bmatrix}\sigma^2 \tag{3.5.1-11}$$

where

$$s_j \triangleq \sum_{i=1}^{k+1}(t_i)^j \qquad j = 0, 1, 2 \tag{3.5.1-12}$$

denote compactly the three functions of the sampling times needed in (3.5.1-11).

Then, the gain (a 2-dimensional vector) is obtained from (3.4.2-11) as

$$W(k + 1) = P(k + 1)H(k + 1)'R(k + 1)^{-1} = \frac{1}{s_0 s_2 - s_1^2}\begin{bmatrix} s_2 - s_1 t_{k+1} \\ -s_1 + s_0 t_{k+1} \end{bmatrix} \tag{3.5.1-13}$$

These expressions are general — the sampling times are arbitrary. They can be simplified for uniformly spaced samples, as will be seen next.

3.5.1 Fitting a First Order Polynomial to Noisy Measurements

If the samples are uniformly spaced with sampling interval (period) T, that is,

$$t_i = iT \qquad i = 1, 2, \ldots \qquad (3.5.1\text{-}14)$$

then the three functions in (3.5.1-12) have the following closed form expressions

$$s_0 = k + 1 \qquad (3.5.1\text{-}15)$$

$$s_1 = \frac{(k+1)(k+2)}{2} T \qquad (3.5.1\text{-}16)$$

$$s_2 = \frac{(k+1)(k+2)(2k+3)}{6} T^2 \qquad (3.5.1\text{-}17)$$

In this case the gain (3.5.1-13) is

$$W(k+1) = \begin{bmatrix} -\dfrac{2}{(k+1)} \\ \\ \dfrac{6}{(k+1)(k+2)T} \end{bmatrix} \qquad (3.5.1\text{-}18)$$

The explicit expression of the covariance (3.5.1-11) is

$$P(k+1) = \frac{\sigma^2}{s_0 s_2 - s_1^2} \begin{bmatrix} s_2 & -s_1 \\ -s_1 & s_0 \end{bmatrix} = \frac{\sigma^2}{k(k+1)} \begin{bmatrix} 2(2k+3) & -\dfrac{6}{T} \\ \\ -\dfrac{6}{T} & \dfrac{12}{(k+2)T^2} \end{bmatrix} \qquad (3.5.1\text{-}19)$$

Remark

The gain (3.5.1-18) tends to zero as $k \to \infty$ since the covariance $P(k)$ tends to zero as more observations are made. The reason for the gain tending to zero is that *zero variance* implies *perfect estimate*, in which case *there is no more need to update the estimate.*

Note

In statistics, regression is used to relate one variable z, to a set of variables y_j

$$z(i) = \sum_{j=1}^{n} a_j y_j(i) + w(i) \qquad (3.5.1\text{-}20)$$

with the variance of $w(i)$ assumed, in general, *unknown*, to be estimated together with the coefficients a_j. In the case discussed above we had $n = 2$, $y_1(i) = 1$, $y_2(i) = t_i$, $a_1 = x_0$, $a_2 = \dot{x}_0$, and the variance of $w(i)$ was assumed to be known. This last assumption is reasonable if we know that the measurements are made with a sensor with known accuracy, which is the case in (most of the) engineering problems.

147

3.5.2 Fitting a General Polynomial to a Set of Noisy Measurements

Assume that the evolution of the position of an object is modeled as a polynomial in time, that is,

$$\xi(t) = \sum_{j=0}^{n} a_j \frac{t^j}{j!} \tag{3.5.2-1}$$

with the parameters being the polynomial coefficients a_j, $j = 0, 1, \ldots, n$, to be estimated.

The coefficient a_j is the j-th derivative of the position at the **reference time** $t = 0$.

The LS technique from Section 3.4.1 will be used to estimate these parameters via **polynomial fitting** — of order n in this case.

The noisy measurements of the position (3.5.2-1) can be written as

$$z(i) = h(i)'a + w(i) \qquad i = 1, \ldots, k \tag{3.5.2-2}$$

where

$$a = [a_0 \ a_1 \ \cdots \ a_n]' \tag{3.5.2-3}$$

is the $(n + 1)$-dimensional parameter vector to be estimated and the row vector

$$h(i)' = \left[1 \ t_i \ \cdots \ \frac{t_i^n}{n!}\right] \tag{3.5.2-4}$$

plays the role of $H(i)$ from (3.4.1-1).

The measurement disturbances $w(i)$ are assumed to be a zero-mean white sequence with variance σ^2, assumed to be known.

The stacked measurement matrix (3.4.1-5) is

$$H^k = \begin{bmatrix} h(1)' \\ \vdots \\ h(k)' \end{bmatrix} \tag{3.5.2-5}$$

Then, since

$$R^k = \sigma^2 I \tag{3.5.2-6}$$

one has

$$H^{k'}(R^k)^{-1}H^k = \sigma^{-2}H^{k'}H^k = \sigma^{-2}[h(1) \ \cdots \ h(k)]\begin{bmatrix} h(1)' \\ \vdots \\ h(k)' \end{bmatrix} = \sigma^{-2}\sum_{i=1}^{k}h(i)h(i)' \tag{3.5.2-7}$$

148

3.5.2 Fitting a General Polynomial to a Set of Noisy Measurements

The estimate of the parameter vector a is then, using (3.4.1-9),

$$\hat{a}(k) = \left\{ \sum_{i=1}^{k} h(i)h(i)' \right\}^{-1} \sum_{i=1}^{k} h(i)z(i) \qquad (3.5.2\text{-}8)$$

with the covariance matrix

$$P(k) = \sigma^2 \left\{ \sum_{i=1}^{k} h(i)h(i)' \right\}^{-1} \qquad (3.5.2\text{-}9)$$

Note that the term to be inverted in (3.5.2-8) and in (3.5.2-9), being an $(n+1) \times (n+1)$ matrix that is the sum of k dyads, one needs $k \geq n+1$ in order for the inverse to exist, that is, *at least as many measurements as the number of the parameters are needed.*

Using (3.5.2-4) one can write the i-th dyad in (3.5.2-8) or (3.5.2-9) as

$$h(i)h(i)' = \begin{bmatrix} 1 & t_i & \cdots & t_i^n/n! \\ t_i & t_i^2 & \cdots & \\ \vdots & \vdots & \ddots & \vdots \\ t_i^n/n! & & \cdots & (t_i^n/n!)^2 \end{bmatrix} \qquad (3.5.2\text{-}10)$$

With this, explicit expressions can be obtained for (3.5.2-8) and (3.5.2-9) if the samples are uniformly spaced.

Let

$$t_i = \frac{2i - k - 1}{2}T \qquad i = 1, \dots, k \qquad (3.5.2\text{-}11)$$

where T is the sampling period and the sampling times are centered around $t = 0$ for convenience.

Then the parameter a_j is the j-th derivative of the position *at the center of the batch,* which is the **reference time.**

The mapping of these estimates to an *arbitrary time* is presented in Subsection 3.5.3.

The explicit expressions of the parameter estimates and their covariances for polynomials of order $n = 1, 2, 3$ are given next.

First Order Polynomial

For $n = 1$, which corresponds to a **constant velocity motion** (i.e., **straight line fit**), one has

$$P(k) = \frac{\sigma^2}{k} \begin{bmatrix} 1 & 0 \\ 0 & \dfrac{12}{(k-1)(k+1)T^2} \end{bmatrix} \qquad (3.5.2\text{-}12)$$

and

$$\begin{bmatrix} \hat{a}_0(k) \\ \hat{a}_1(k) \end{bmatrix} = \sigma^{-2} P(k) \begin{bmatrix} \sum_{i=1}^{k} z(i) \\ \sum_{i=1}^{k} z(i)t_i \end{bmatrix} \qquad (3.5.2\text{-}13)$$

In the above $\hat{a}_0$ and $\hat{a}_1$ are the position and velocity estimates, respectively, at the *center of the batch*, which corresponds to the reference time $t = 0$.

Note that (3.5.2-12) and (3.5.2-13) are equivalent to the results of the example from Subsection 3.5.1. The only difference is that in the latter the parameters were the initial position and velocity.

Second Order Polynomial

For $n = 2$, which corresponds to a **constant acceleration motion (parabolic fit)**, one has

$$P(k) = \sigma^2 \begin{bmatrix} \dfrac{3(3k^2-7)}{4k(k^2-4)} & 0 & \dfrac{-30}{k(k^2-4)T^2} \\ 0 & \dfrac{12}{k(k^2-1)T^2} & 0 \\ \dfrac{-30}{k(k^2-4)T^2} & 0 & \dfrac{720}{k(k^2-1)(k^2-4)T^4} \end{bmatrix} \qquad (3.5.2\text{-}14)$$

and

$$\hat{a}(k) = \begin{bmatrix} \hat{a}_0(k) \\ \hat{a}_1(k) \\ \hat{a}_2(k) \end{bmatrix} = \sigma^{-2} P(k) \begin{bmatrix} \sum_{i=1}^{k} z(i) \\ \sum_{i=1}^{k} z(i)t_i \\ \sum_{i=1}^{k} z(i)t_i^2/2 \end{bmatrix} \qquad (3.5.2\text{-}15)$$

In the above $\hat{a}_0$, $\hat{a}_1$, and $\hat{a}_2$ are the position, velocity, and acceleration estimates, respectively, for the center of the batch ($t = 0$).

Third Order Polynomial

For $n = 3$, which corresponds to a **constant jerk motion (cubic fit)**, one has the following parameter estimate covariance

$$P(k) = \sigma^2 \begin{bmatrix} \frac{3(3k^2-7)}{4k(k^2-4)} & 0 & \frac{-30}{k(k^2-4)T^2} & 0 \\ 0 & \frac{25(3k^4-18k^2+31)}{k(k^2-1)(k^2-4)(k^2-9)T^2} & 0 & \frac{-840(3k^2-7)}{k(k^2-1)(k^2-4)(k^2-9)T^4} \\ \frac{-30}{k(k^2-4)T^2} & 0 & \frac{720}{k(k^2-1)(k^2-4)T^4} & 0 \\ 0 & \frac{-840(3k^2-7)}{k(k^2-1)(k^2-4)(k^2-9)T^4} & 0 & \frac{100{,}800}{k(k^2-1)(k^2-4)(k^2-9)T^6} \end{bmatrix}$$

(3.5.2-16)

The parameter estimates are given by

$$\hat{a}(k) = \begin{bmatrix} \hat{a}_0(k) \\ \hat{a}_1(k) \\ \hat{a}_2(k) \\ \hat{a}_3(k) \end{bmatrix}$$

$$= \sigma^{-2} P(k) \begin{bmatrix} \sum_{i=1}^k z(i) \\ \sum_{i=1}^k z(i)t_i \\ \sum_{i=1}^k z(i)t_i^2/2 \\ \sum_{i=1}^k z(i)t_i^3/6 \end{bmatrix}$$

(3.5.2-17)

Remark

Comparing (3.5.2-12), (3.5.2-14), and (3.5.2-16) it can be seen that as the order of the polynomial fit increases, the parameter variances increase. This is because there is *less information per parameter* when more parameters are fitted to the same number of data points.

This increase takes place (for the present choice of the parameters — position and its derivatives at the center of the batch) for every other parameter. For example, comparing (3.5.2-14) with (3.5.2-16) indicates that

$$P_{11}(k, n = 3) = P_{11}(k, n = 2) \tag{3.5.2-18}$$

$$P_{22}(k, n = 3) > P_{22}(k, n = 2) \tag{3.5.2-19}$$

A second argument, indicating the order of the polynomial fit, has been used in the above equations to distinguish between the various covariance matrices.

3.5.3 Mapping of the Estimates to an Arbitrary Time

Assume a constant acceleration motion characterized by the 3-dimensional vector consisting of position, velocity, and acceleration at the reference time (center of the batch) estimated in (3.5.2-15). The estimate of the corresponding position-velocity-acceleration vector $x(t)$ at an arbitrary time t, that is, the **prediction** (or **extrapolation**) based on k measurements uniformly spaced according to (3.5.2-11), is

$$\hat{x}(t|k) = \Phi(t)\hat{a}(k) \tag{3.5.3-1}$$

where $\hat{a}(k)$ is given in (3.5.2-15), and

$$\Phi(t) = \begin{bmatrix} 1 & t & t^2/2 \\ 0 & 1 & t \\ 0 & 0 & 1 \end{bmatrix} \tag{3.5.3-2}$$

The corresponding covariance is

$$P(t|k) = \Phi(t)P(k)\Phi(t)' \tag{3.5.3-3}$$

where $P(k)$ is given in (3.5.2-14).

Similar transformations are used for lower or higher dimensional parameter vectors.

As an example, for the 2-dimensional case (straight line fitting, i.e., constant velocity assumption) it can be shown that the closed-form expression of the covariance matrix for the **one-step prediction** $(t = t_{k+1} = (k+1)T/2)$ is

$$P(t_{k+1}|k) \triangleq P(k+1|k) = \frac{\sigma^2}{(k-1)k} \begin{bmatrix} 2(2k+1) & 6/T \\ 6/T & \dfrac{12}{(k+1)T^2} \end{bmatrix} \tag{3.5.3-4}$$

Based on (3.5.3-1), Table 3.5.3-1 shows the values of the normalized one-step position prediction variances for a linear extrapolation based on k uniformly spaced observations and the corresponding velocity estimate variances.

k	2	3	4	5	6
$P_{11}(k+1|k)/\sigma^2$	5	2.33	1.5	1.1	0.867
$T^2 P_{22}(k+1|k)/\sigma^2$	2	0.5	0.2	0.1	0.057

Table 3.5.3-1: One-step prediction variances for linear extrapolation from k measurements.

In tracking, when the motion of the object of interest is, say, with constant velocity, and one has k measurements, a *prediction of the location* of the next measurement $z(k+1)$ at $t = t_{k+1} = (k+1)T/2$ is made according to the expression

$$\hat{x}_1(t_{k+1}|k) \triangleq \hat{x}_1(k+1|k) = \hat{a}_1 + \hat{a}_2 t_{k+1} \qquad (3.5.3\text{-}5)$$

The measurement $z(k+1)$ will be, with a certain probability, in a *region around this predicted location*, called **gate**, determined by the variance associated with the prediction, $P_{11}(k+1|k)$, given above.

In practice one can use a "3σ gate." Then the measurement will be in this gate with a probability of 99.8% under the Gaussian assumption. Such a gate is the interval

$$\left[\hat{x}_1(k+1|k) - 3\sqrt{P_{11}(k+1|k)}, \ \hat{x}_1(k+1|k) - 3\sqrt{P_{11}(k+1|k)}\right] \qquad (3.5.3\text{-}6)$$

Figure 3.5.3-1 illustrates the 1σ region corresponding to the one-step position prediction based on two observations.

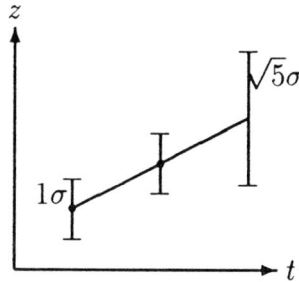

Figure 3.5.3-1: Uncertainty for a one-step position prediction (straight-line motion).

Similarly, the position variance for an n-step prediction is

$$P_{11}(k+n|k) = 2\sigma^2 \frac{(k-1)(2k-1) + 6n(k-1) + 6n^2}{(k-1)(k+1)k} \qquad (3.5.3\text{-}7)$$

For example,

$$P_{11}(4|2) = 13\sigma^2 - P_{11}(5|3) = \frac{29}{6}\sigma^2 \qquad (3.5.3\text{-}8)$$

These values are useful in track initiation when the motion of the target is assumed to be described by a straight line. For example, if the target was not detected at t_{k+1}, then one has to wait until t_{k+2}. Note the increase of the two-step prediction variances compared to the one-step prediction variances from Table 3.5.3-1.

3.5.4 Polynomial Fitting — Summary

Polynomials in time can be fitted to a set of arbitrarily spaced data points (noisy position measurements) via the least squares method.

Explicit expressions have been presented for the coefficients of polynomials up to third order fitted to a set of uniformly spaced data points.

The coefficients, as presented, yield estimates of the position and its derivatives at the center of the batch of measurements.

The covariances associated with estimates also have explicit expressions.

A simple linear mapping transforms these coefficients to the corresponding estimates to an arbitrary time — this yields the extrapolation or prediction of the motion. The same transformation matrix can be used to obtain the covariance of the position and its derivatives at an arbitrary time.

The predicted position can be used as the center of the region in which the next measurement will be with a high probability, called gate. The size of the gate is determined from the variance of this prediction.

3.6 GOODNESS OF FIT AND STATISTICAL SIGNIFICANCE OF PARAMETER ESTIMATES

3.6.1 Hypothesis Testing Formulation of the Problem

When a set of parameters is estimated to fit a polynomial to a number of data points, there is always the following question:

What is the appropriate order of the polynomial?

The following fundamental result of estimation/polynomial fitting is relevant to this question:

Theorem. Through *any three points* on a sheet of paper one can pass a straight line. [1]

Given a set of data points (scalar measurements), when fitting a polynomial to these points one can encounter the following situations:

- If the order of the polynomial is too low, that is, **underfitting**, then the fit will be poor,

or, at the other extreme;

- If the order of the polynomial is too high, that is, **overfitting**, the estimates of some of the parameters are not "statistically significant," that is, they are "noise."

As shown in the next subsection, the sum of the squares of the residuals in an LS estimation problem, that is, the minimized value of the LS criterion (3.4.1-2), also called the **goodness of fit** or **fitting error**,

$$J^*(k) = [z^k - H^k \hat{x}(k)]'(R^k)^{-1}[z^k - H^k \hat{x}(k)] \qquad (3.6.1\text{-}1)$$

has, if the noises are Gaussian, a chi-square distribution with $kn_z - n_x$ degrees of freedom. In the above, the notations (3.4.1-4) to (3.4.1-7) and (3.4.1-9) have been used and k is the number of measurements of dimension n_z, while n_x is the dimension of the parameter vector.

The matrix R^k, consisting of the noise covariances, is assumed to be known — a similar result is available for the situation where this matrix is unknown and estimated together with the "regression coefficients" x.

---

[1] Proof: Left to the reader. (Hint: Use a pencil that is thick enough.)

155

Test for Underfitting

The order of the polynomial fit to a set of data points is *too low* if the fit is not "good enough," that is,

$$J^* > c = \chi^2_{kn_z - n_x}(1 - \alpha) \qquad (3.6.1\text{-}2)$$

where c is obtained from Table 1.5.4-1 such that the probability of a $kn_z - n_x$ degrees of freedom chi-square random variable exceeding it is α (usually 5% or 1%).

If the first choice of the polynomial is too low, one can increase it until an "acceptable" fit is obtained, that is, the resulting J^* falls below the maximum allowed.

Test for Overfitting

If the order of the polynomial is *too high* then the estimate (usually of the highest power coefficient) will be **statistically insignificant**.

Assuming the noises to be normal with zero mean and known variances, the estimate of the i-th component of the parameter vector is

$$\hat{x}_i(k) \sim \mathcal{N}[x_i, P_{ii}(k)] \qquad (3.6.1\text{-}3)$$

that is, it is normal with mean equal to the unknown true value x_i, and with variance $P_{ii}(k)$.

The **parameter estimate significance test** is the test between the following hypotheses:

$$H_0 : x_i = 0 \qquad (3.6.1\text{-}4)$$

$$H_1 : x_i \neq 0 \qquad (3.6.1\text{-}5)$$

subject to

$$P\{\text{accept } H_1 | H_0 \text{ true}\} = \alpha \qquad (3.6.1\text{-}6)$$

Then one accepts H_1 (i.e., that the parameter is nonzero) if and only if

$$\frac{|\hat{x}_i(k)|}{[P_{ii}(k)]^{1/2}} > c' = \mathcal{G}(1 - \frac{\alpha}{2}) \qquad (3.6.1\text{-}7)$$

The above implies a **two-sided probability region**. The threshold c' is obtained from the tables of Subsection 1.5.4 such that the probability of a standard Gaussian random variable exceeding it is $\alpha/2$. For example, for $\alpha = 5\%$, one has $c' = 1.96$.

If (3.6.1-7) does not hold, then the estimate of the parameter is *statistically insignificant* and it is better to accept H_0. Then the problem is solved again for a lower dimension parameter obtained by deleting the component found insignificant.

3.6.2 The Fitting Error in a Least Squares Estimation Problem

Consider the LS estimation of the n_x dimensional vector x based on k measurements $z(i)$ of dimension n_z described in (3.4.1-1). The stacked vector of measurements, of dimension kn_z, is

$$z^k = H^k x + w^k \qquad (3.6.2\text{-}1)$$

and the estimate was obtained as (the superscripts will be dropped for simplicity in the sequel)

$$\hat{x} = (H'R^{-1}H)^{-1}H'R^{-1}z \qquad (3.6.2\text{-}2)$$

We want to evaluate the minimized value of the criterion (3.4.1-2), that is, the **fitting error**, or the **goodness of fit**, or the **norm of the residual**

$$J^* \triangleq (z - H\hat{x})'R^{-1}(z - H\hat{x}) \qquad (3.6.2\text{-}3)$$

Note that the above, which is the sum of the squares of the normalized residuals, is a scalar and a (physically) *dimensionless quantity*.

The vector residual is

$$
\begin{aligned}
z - H\hat{x} &= Hx + w - H(H'R^{-1}H)^{-1}H'R^{-1}(Hx + w) \\
&= [I - H(H'R^{-1}H)^{-1}H'R^{-1}]w \qquad (3.6.2\text{-}4)
\end{aligned}
$$

where I in (3.6.2-4) denotes the $kn_z \times kn_z$ identity matrix. This follows from the fact that the stacked vectors z and w are of dimension kn_z.

Using (3.6.2-4) in (3.6.2-3) yields

$$
\begin{aligned}
J^* &= w'[I - H(H'R^{-1}H)^{-1}H'R^{-1}]'R^{-1}[I - H(H'R^{-1}H)^{-1}H'R^{-1}]w \\
&= w'[R^{-1} - R^{-1}H(H'R^{-1}H)^{-1}H'R^{-1}]w \\
&= w'R^{-1/2}[I - R^{-1/2}H(H'R^{-1}H)^{-1}H'R^{-1/2}]R^{-1/2}w \\
&\triangleq \omega' A \omega \qquad (3.6.2\text{-}5)
\end{aligned}
$$

where $R^{-1/2}$ denotes a *square root* of R^{-1} (see (1.3.2-15)),

$$\omega \triangleq R^{-1/2}w \qquad (3.6.2\text{-}6)$$

is a vector of dimension kn_z and

$$A \triangleq I - R^{-1/2}H(H'R^{-1}H)^{-1}H'R^{-1/2} \qquad (3.6.2\text{-}7)$$

157

3.6.2 The Fitting Error in a Least Squares Estimation Problem

Assuming

$$w \sim \mathcal{N}(0, R) \qquad (3.6.2\text{-}8)$$

it follows that

$$\omega \sim \mathcal{N}(0, I) \qquad (3.6.2\text{-}9)$$

that is, the components of ω are *independent standardized Gaussians*.

It can be easily verified that the symmetric matrix A defined in (3.6.2-7) is *idempotent* (see (1.3.2-14)), that is,

$$AA = A \qquad (3.6.2\text{-}10)$$

Such a matrix can have eigenvalues equal to 0 or 1 only. Furthermore,

$$
\begin{aligned}
\text{tr}(A) &= \text{tr}[I - R^{-1/2}H(H'R^{-1}H)^{-1}H'R^{-1/2}] \\
&= \text{tr}(I) - \text{tr}[R^{-1/2}H(H'R^{-1}H)^{-1}H'R^{-1/2}]
\end{aligned}
\qquad (3.6.2\text{-}11)
$$

Note that, since the dimension of I is $kn_z \times kn_z$, one has

$$\text{tr}(I) = kn_z \qquad (3.6.2\text{-}12)$$

Using circular permutations for matrices multiplying each other under the trace operator (1.3.2-13) yields

$$\text{tr}[R^{-1/2}H(H'R^{-1}H)^{-1}H'R^{-1/2}] = \text{tr}[(H'R^{-1}H)^{-1}H'R^{-1/2}R^{-1/2}H] = \text{tr}[I_{n_x}] = n_x$$
$$(3.6.2\text{-}13)$$

where I_{n_x} is the $n_x \times n_x$ identity matrix. This follows from the fact that the dimension of $H'R^{-1}H$ is $n_x \times n_x$.

Thus, combining (3.6.2-12) and (3.6.2-13) into (3.6.2-11) results in

$$\text{tr}(A) = kn_z - n_x \qquad (3.6.2\text{-}14)$$

that is, the matrix A in (3.6.2-5), which has dimension $kn_z \times kn_z$, has $kn_z - n_x$ unity eigenvalues and n_x zero eigenvalues.

Now, since ω, defined in (3.6.2-6), is zero mean normal with identity covariance matrix (of dimension $kn_z \times kn_z$) and A has $kn_z - n_x$ unity eigenvalues (and the rest zero), the **fitting error**

$$J^* = \omega' A \omega \qquad (3.6.2\text{-}15)$$

is, as shown next, the sum of the squares of $kn_z - n_x$ independent scalar random variables that are normal with mean zero and unity variance.

Proof

The spectral representation (1.3.6-12) of the symmetric matrix A is

$$A = \sum_{i=1}^{kn_z} \lambda_i u_i u_i' \tag{3.6.2-16}$$

where λ_i are the eigenvalues of A, and u_i are its normalized eigenvectors that are orthogonal to each other, that is,

$$u_i' u_j = \delta_{ij} \tag{3.6.2-17}$$

Such vectors are called **orthonormal**.

Using (3.6.2-16) in (3.6.2-5) one has

$$
\begin{aligned}
\omega' A \omega &= \omega' \sum_{i=1}^{kn_z} \lambda_i u_i u_i' \omega = \sum_{i=1}^{kn_z} \lambda_i \omega' u_i u_i' \omega \\
&= \sum_{i=1}^{kn_z} \lambda_i u_i' \omega u_i' \omega = \sum_{i=1}^{kn_z} \lambda_i (u_i' \omega)^2 \\
&\triangleq \sum_{i=1}^{kn_z} \lambda_i \xi_i^2
\end{aligned} \tag{3.6.2-18}
$$

where

$$\xi \triangleq \mathrm{col}(\xi_i) \sim \mathcal{N}(0, I) \tag{3.6.2-19}$$

since ξ_i is a linear combination of the components of ω and

$$
\begin{aligned}
E[\xi_i \xi_j] &= E[\omega' u_i \omega' u_j] = E[u_i' \omega \omega' u_j] \\
&= u_i' E[\omega \omega'] u_j = u_i' I u_j = u_i' u_j = \delta_{ij}
\end{aligned} \tag{3.6.2-20}
$$

that is, ξ_i, $i = 1, \ldots, kn_z$, are *independent standardized Gaussians*.

Therefore, since in (3.6.2-18) there are $kn_z - n_x$ unity eigenvalues λ_i and the rest are zero, it follows that J^* is the sum of $kn_z - n_x$ independent standardized Gaussians squared, that is, it has a chi-square distribution with $kn_z - n_x$ degrees of freedom. Therefore,

$$\omega' A \omega = \sum_{i=1}^{kn_z} \lambda_i \xi_i^2 = \sum_{i=1}^{kn_z - n_x} \xi_i^2 \sim \chi^2_{kn_z - n_x} \tag{3.6.2-21}$$

Remark

The meaning of **degrees of freedom** can be seen as being the number of observations, kn_z, minus the number of parameters estimated, n_x.

3.6.3   A Polynomial Fitting Example

Consider the following numerical example. The true parameter vector is assumed to be of dimension 3, that is, a constant acceleration motion (polynomial of order 2) models the truth.

The measurements are given as in (3.5.2-2) by

$$z(i) = x_1 + x_2 t_i + x_3 \frac{t_i^2}{2} + w(i) \qquad (3.6.3\text{-}1)$$

where

$$x_1 = 10 \qquad (3.6.3\text{-}2)$$

$$x_2 = 1 \qquad (3.6.3\text{-}3)$$

$$x_3 = 0.2 \qquad (3.6.3\text{-}4)$$

$$t_i = \frac{1}{2}(2i - k - 1) \qquad i = 1, \ldots, k; \qquad k = 15 \qquad (3.6.3\text{-}5)$$

with the noise sequence white and

$$w(i) \sim \mathcal{N}(0, 1) \qquad (3.6.3\text{-}6)$$

Table 3.6.3 presents the results of the fitting of a sequence of measurements generated according to the above model with polynomials of order 1, 2, and 3.

For the linear model ($n_x = 2$) the fitting error J^* is too large — well above the threshold, chosen as 95% point from the chi-square tables.

For the quadratic model ($n_x = 3$), J^* is below the threshold (i.e., acceptable) and all the parameter estimates are significant.

For the cubic model ($n_x = 4$) the last parameter (third derivative of the position — jerk) estimate is statistically insignificant: $0.843 < \mathcal{G}(97.5\%) = 1.96$; the fit has only slightly improved (the so-called F test would indicate that this improvement is statistically insignificant — this is equivalent to the parameter estimate significance test).

Thus $n_x = 4$ is clearly an "overparametrization." It can also be seen that for $n_x = 4$ the velocity standard deviation $\sqrt{P_{22}}$ is *much larger* than for $n_x = 3$ because of the additional parameter to be estimated.

Figure 3.6.3-1 illustrates the resulting fit with these three polynomials. Also shown are the "uncertainty tubes" — the 2σ confidence region (95%) around the predicted position. Note how much more rapidly this widens for the third order polynomial fit ($n_x = 4$) in comparison with the second order ($n_x = 3$) case.

Assumed model	Linear	Quadratic	Cubic
Order n_x	2	3	4
$\hat{x}_1$	11.97	10.34	10.34
$\sqrt{P_{11}}$	0.258	0.388	0.388
$|\hat{x}_1|/\sqrt{P_{11}}$	46.38	26.60	26.60
$\hat{x}_2$	0.996	0.997	1.114
$\sqrt{P_{22}}$	0.059	0.059	0.151
$|\hat{x}_2|/\sqrt{P_{22}}$	16.67	16.67	7.341
$\hat{x}_3$		0.174	0.174
$\sqrt{P_{33}}$		0.031	0.031
$|\hat{x}_3|/\sqrt{P_{33}}$		5.611	5.611
$\hat{x}_4$			0.021
$\sqrt{P_{44}}$			0.025
$|\hat{x}_4|/\sqrt{P_{44}}$			0.843
J^*	44.67	13.18	12.47
$\chi^2_{k-n_x}(95\%)$	22.4	21.0	19.7

Table 3.6.3-1: Fitting of various order polynomial models.

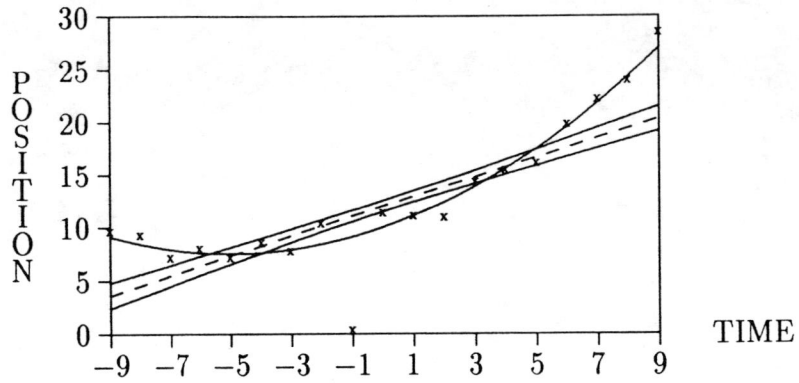

(a) $n_x = 2$

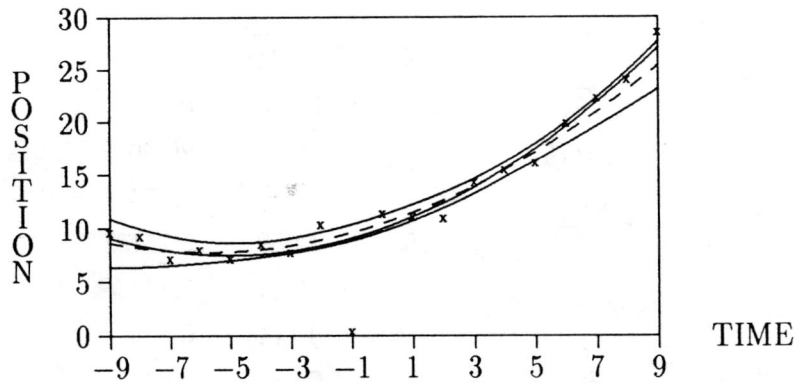

(b) $n_x = 3$

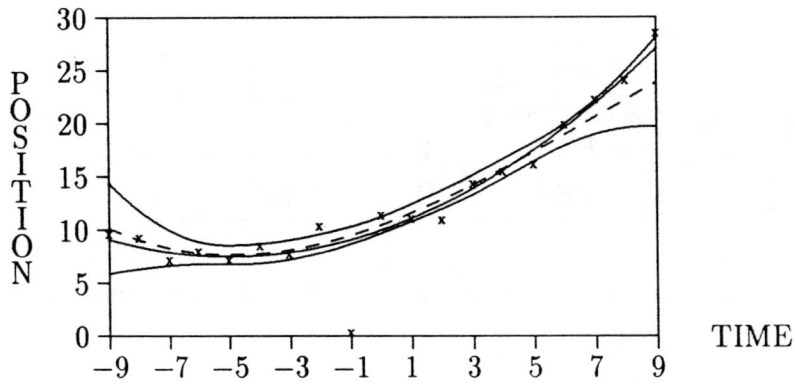

(c) $n_x = 4$

$- \cdot -$ true trajectory; $\times$ measurements;
$--$ estimated trajectory; $-$ 95% confidence region

Figure 3.6.3-1: Fitting of various order polynomials to a constant acceleration motion.

3.6.4 Order Selection in Polynomial Fitting — Summary

Fitting of polynomials to noisy measurements, which is a particular case of regression, consists of LS estimation of its coefficients.

The *order* of the polynomial chosen for fitting is

1. *too low* if the **fitting error** is poor (sum of squares of the residuals is too large);
2. *too high* if some estimates of coefficients are **statistically insignificant** ("buried in noise").

The fitting error is, under the Gaussian assumption, chi-square distributed with number of degrees of freedom equal to the number of measurements minus the number of estimated parameters (this is the origin of the term degrees of freedom). The fitting error ("goodness of fit") has to be *below a threshold in order to be acceptable.*

The statistical significance is the ratio of the *magnitude of a parameter estimate* to its standard deviation and this ratio has to be *above a threshold* for the estimate to be significant.

Fitting a polynomial (or, in general, a model) of *unnecessarily high order* decreases the accuracy of some of the estimated coefficients (model parameters) — this *wastes information.*

3.7 USE OF LS FOR A NONLINEAR PROBLEM: BEARINGS-ONLY TARGET MOTION ANALYSIS

3.7.1 The Problem

The **nonlinear least squares** problem, defined in Subsection 2.4.1, will be used to estimate the motion parameters of a constant velocity target — **target motion analysis** — based on observations from a passive sensor that measures only the direction of arrival of a signal emitted by the target. This problem is also called **passive localization** or **passive ranging**.

The target, which is moving in a plane, is observed from a platform with a known position $[\xi_p(k), \eta_p(k)]$ in the same plane.

The target "localization parameter" is the vector of dimension $n_x = 4$ consisting of its initial position and velocity in Cartesian coordinates

$$
\begin{aligned}
x &\triangleq [x_1 \ x_2 \ x_3 \ x_4]' \\
&\triangleq [\xi(0) \ \eta(0) \ \dot\xi \ \dot\eta]'
\end{aligned}
\tag{3.7.1-1}
$$

The position of the target at time t_k is

$$
\begin{aligned}
\xi(k) &\triangleq \xi(t_k) = \xi(0) + \dot\xi t_k \\
&= x_1 + x_3 t_k
\end{aligned}
\tag{3.7.1-2}
$$

$$
\begin{aligned}
\eta(k) &\triangleq \eta(t_k) = \eta(0) + \dot\eta t_k \\
&= x_2 + x_4 t_k
\end{aligned}
\tag{3.7.1-3}
$$

It is assumed that the available measurements are bearings ("line of sight" angles with respect to some reference direction) only, given by

$$
z(k) \triangleq z(t_k) = h(k, x) + w(k) \qquad k = 1, \dots, n
\tag{3.7.1-4}
$$

where

$$
h(k, x) \triangleq \tan^{-1} \frac{\eta(k) - \eta_p(k)}{\xi(k) - \xi_p(k)}
\tag{3.7.1-5}
$$

and $w(k)$ is the measurement noise, assumed to be a zero-mean Gaussian white sequence with known variance r, that is,

$$
E[w(k)w(j)] = r\delta_{ij}
\tag{3.7.1-6}
$$

164

3.7.2 Observability of the Target Parameter in Passive Localization

A question of interest is the effect of the motion of the platform on the ability to estimate the target parameters. It will be shown that if the platform moves with a constant velocity, then the target motion parameter cannot be estimated — it is **unobservable**.

Denoting the platform velocity components as $\dot{\xi}_p$ and $\dot{\eta}_p$, then the true bearing to the target is at time t

$$
\begin{aligned}
\tan^{-1} \frac{\eta(t) - \eta_p(t)}{\xi(t) - \xi_p(t)} &= \tan^{-1} \frac{\eta(0) + \dot{\eta}t - \eta_p(0) - \dot{\eta}_p t}{\xi(0) + \dot{\xi}t - \xi_p(0) - \dot{\xi}_p t} \\
&= \tan^{-1} \frac{\eta(0) - \eta_p(0) + (\dot{\eta} - \dot{\eta}_p)t}{\xi(0) - \xi_p(0) + (\dot{\xi} - \dot{\xi}_p)t} \\
&= \tan^{-1} \frac{[\eta(0) - \eta_p(0)]\alpha + (\dot{\eta} - \dot{\eta}_p)\alpha t}{[\xi(0) - \xi_p(0)]\alpha + (\dot{\xi} - \dot{\xi}_p)\alpha t} \\
& \qquad \forall \alpha \neq 0
\end{aligned}
\tag{3.7.2-1}
$$

The above is seen to hold for all α, that is, one will obtain the same sequence of true bearings if the relative position and relative velocity are multiplied by an arbitrary constant α.

Thus, an infinity of values of the target parameter vector can yield the same observations and, consequently, its parameter vector cannot be estimated in this case due to the above described nonuniqueness — the observability requirement of full and unique recovery of the initial state is not satisfied.

Therefore, in order to estimate the target parameter vector, the platform has to have an acceleration. A constant speed platform with a change of course (heading) satisfies this requirement. In general, the platform trajectory has to have *one more nonzero derivative than the target trajectory*.

Figure 3.7.2-1 illustrates two cases where the target localization parameter is not observable from angle-only measurements: (a) moving target and fixed sensor platform, (b) constant velocity target and constant velocity platform.

Figure 3.7.2-2 illustrates two observable cases: (a) fixed target and moving platform, (b) constant velocity target and platform with acceleration.

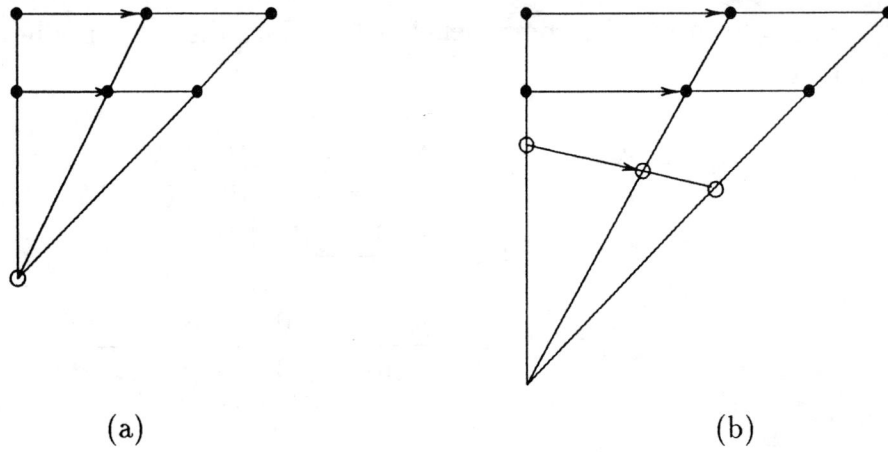

(a)   (b)

Figure 3.7.2-1: Unobservable target: (• target; ○ platform).

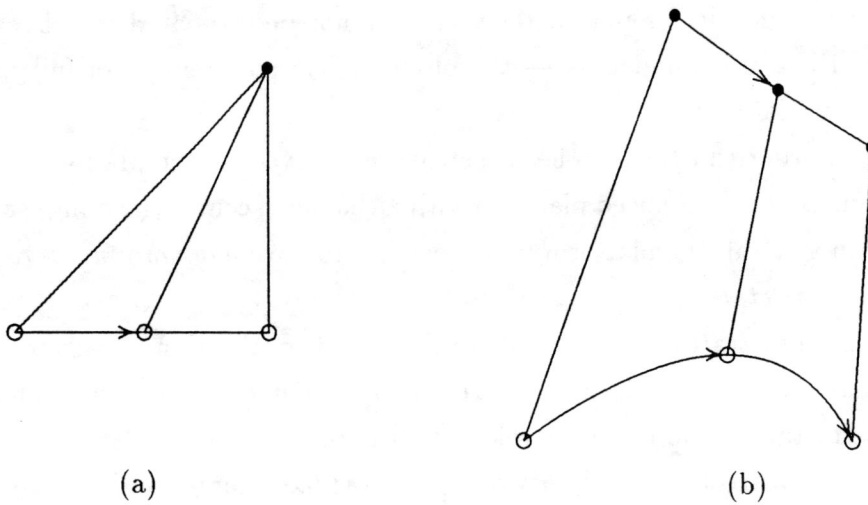

(a)   (b)

Figure 3.7.2-2: Observable target (• target; ○ platform).

3.7.3   The Likelihood Function for Target Parameter Estimation

The likelihood function of the target parameter vector (3.7.1-1) is

$$
\begin{aligned}
\Lambda(x) &= p(Z^n|x) \\
&= p[z(1),\ldots,z(n)|x] \\
&= \prod_{k=1}^{n} p[z(k)|x]
\end{aligned}
\tag{3.7.3-1}
$$

where, in view of (3.7.1-4),

$$
\begin{aligned}
p[z(k)|x] &= \mathcal{N}[z(k); h(k,x), r] \\
&= ce^{-\frac{1}{2r}[z(k)-h(k,x)]^2}
\end{aligned}
\tag{3.7.3-2}
$$

and $h(k,x)$, given in (3.7.1-5), is the expected value (average) of the observation $z(k)$ for a given target parameter vector x, and r is its variance. This follows from the assumption (3.7.1-6) that the measurement noises are white, zero mean, and with variance r.

In view of (3.7.3-2), the maximization of (3.7.3-1), which yields the maximum likelihood estimate (MLE), is equivalent to the following *nonlinear* LS problem:

$$
\begin{aligned}
\hat{x} &= \arg\max_{x} \Lambda(x) \\
&= \arg\min_{x} \lambda(x)
\end{aligned}
\tag{3.7.3-3}
$$

where

$$
\lambda(x) \triangleq \frac{1}{2r} \sum_{k=1}^{n} [z(k) - h(k,x)]^2
\tag{3.7.3-4}
$$

is the *negative **log-likelihood function*** with the irrelevant additive constants omitted.

The minimization of the log-likelihood function (3.7.3-4) can be carried out via one of the many existing numerical optimization algorithms. The Newton-Raphson or quasi-Newton techniques are the most effective in this case.

3.7.4   The Fisher Information Matrix for the Target Parameter

The **Cramer-Rao lower bound (CRLB)** on the covariance matrix of the target parameter estimate $\hat{x}$ is (assuming this estimate to be unbiased)

$$E[(\hat{x} - x)(\hat{x} - x)'] \geq J^{-1} \tag{3.7.4-1}$$

where J is the **Fisher information matrix (FIM)**

$$
\begin{aligned}
J &= E\left\{[\nabla_x \ln \Lambda(x)][\nabla_x \ln \Lambda(x)]'\right\}\Big|_{x=x_0} \\
&= E\left\{[\nabla_x \lambda(x)][\nabla_x \lambda(x)]'\right\}\Big|_{x=x_0}
\end{aligned}
\tag{3.7.4-2}
$$

The FIM is to be evaluated at the true value of the parameter x_0; in practice, when this is not known the evaluation is done at the estimate.

Parameter Observability and the FIM

In order to have **parameter observability** (i.e., to allow its estimation without the ambiguity discussed in Subsection 3.7.2), the *FIM must be invertible*. If the FIM is not invertible, then the lower bound (3.7.4-1) will not exist (actually, it will have one or more infinite eigenvalues, which means total uncertainty in a subspace of the parameter space, i.e., ambiguity).

The gradient of the log-likelihood function (3.7.3-4) is

$$\nabla_x \lambda(x) = -r^{-1} \sum_{k=1}^{n} [\nabla_x h(k,x)'][z(k) - h(k,x)] \tag{3.7.4-3}$$

which, when inserted into (3.7.4-2) yields (see problem 3-4) the sum of dyads

$$J = r^{-1} \sum_{k=1}^{n} h_x(k,x) h_x(k,x)'\Big|_{x=x_0} \tag{3.7.4-4}$$

where

$$h_x(k,x) \triangleq \nabla_x h(k,x) \tag{3.7.4-5}$$

168

Remarks

The Cramer-Rao lower bound quantifies in this case the **stochastic observability**, which is not a binary property as in the deterministic case:

1. Lack of invertibility of J (in practice, ill-conditioning) indicates that the parameter is **unobservable**. This happens if the condition number (1.3.6-13) of J is too large.
2. If J is invertible but the position confidence region (3.7.4-14) is "large"[2], one has **marginal observability** . This occurs if the gradient vectors in (3.7.4-4) are "nearly collinear."
3. A "small" confidence region — **good observability** — is obtained if the n gradient vectors in (3.7.4-4) span "well" the n_x-dimensional space, that is, they are "far" from being collinear. The actual measure of this is the condition number of J.

Expressions of the Gradient Vector Components

From (3.7.1-5) one has the following expressions for the components of the gradient vector entering into the FIM

$$h_{x_1}(k, x) = -\frac{\eta(k) - \eta_p(k)}{[\xi(k) - \xi_p(k)]^2 + [\eta(k) - \eta_p(k)]^2} \qquad (3.7.4\text{-}15)$$

$$h_{x_2}(k, x) = \frac{\xi(k) - \xi_p(k)}{[\xi(k) - \xi_p(k)]^2 + [\eta(k) - \eta_p(k)]^2} \qquad (3.7.4\text{-}16)$$

$$h_{x_3}(k, x) = t_k h_{x_1}(k, x) \qquad (3.7.4\text{-}17)$$

$$h_{x_4}(k, x) = t_k h_{x_2}(k, x) \qquad (3.7.4\text{-}18)$$

---

[2]In the eye of the engineer, like beauty in the eye of the beholder.

3.7.5 The Goodness of Fit Test

A test based on data from a *single run*, which can be used with real data is presented next. This test does not require knowledge of the true parameter.

Similarly to the linear LS, the minimized value of the log-likelihood function (3.7.3-4), multiplied by 2 for convenience, is

$$
\begin{aligned}
\lambda^* \; &\triangleq \; \lambda(\hat{x}) \\
&= \; \frac{1}{r} \sum_{k=1}^{n} [z(k) - h(k, \hat{x})]^2
\end{aligned}
\tag{3.7.5-1}
$$

This is the **normalized sum of the squares of the residuals** or the **fitting error**, and can be used as a measure of the goodness of fit. Note that (3.7.5-1) is a physically dimensionless quantity.

In the *linear LS* case, under the *Gaussian noise assumptions*, the fitting error was shown to be chi-square distributed in Subsection 3.6.2. In the present *nonlinear LS* problem, the same result can be assumed to hold approximately. Then, with n being the number of (scalar) measurements, one has

$$
\lambda^* \sim \chi^2_{n-n_x}
\tag{3.7.5-2}
$$

and a suitable probability region check can be made to ascertain that the model used for the problem is valid. Namely, λ^* should be, with 95% probability, below the threshold $\chi^2_{n-n_x}(0.95)$, with the notation (1.5.4-3).

This test can also be used with the results of **Monte Carlo runs**. In this case, by summing up the fitting error from N runs *with independent random variables* one obtains a total error that is chi-square distributed with $N(n - n_x)$ degrees of freedom and has to be below a threshold obtained similarly to the one discussed above.

3.7.6   Testing for Efficiency with Monte Carlo Runs

The practical procedure to check the estimator efficiency is using **Monte Carlo simulations** as follows. Let ϵ_x^i be the NEES (3.7.4-8) in run i, $i = 1, \ldots, N$, and the sample average NEES from N independent such runs be

$$\bar{\epsilon}_x = \frac{1}{N} \sum_{i=1}^{N} \epsilon_x^i \qquad (3.7.6\text{-}1)$$

The quantity $N\bar{\epsilon}_x$ is chi-square distributed with Nn_x degrees of freedom. Let the $1 - Q$ **two-sided probability region** for $N\bar{\epsilon}_x$ be the interval $[\epsilon_1', \epsilon_2']$, that is, using the notation from (1.5.4-3),

$$\epsilon_1' = \chi_{Nn_x}^2 \left(\frac{Q}{2} \right) \qquad (3.7.6\text{-}2)$$

$$\epsilon_2' = \chi_{Nn_x}^2 \left(1 - \frac{Q}{2} \right) \qquad (3.7.6\text{-}3)$$

For example, for $n_x = 4$, $N = 25$, $Q = 5\%$, one has $\epsilon_1' = 74$, $\epsilon_2' = 130$. The $1 - Q$ ($= 95\%$) two-sided probability region for $\bar{\epsilon}_x$ is

$$[\epsilon_1, \epsilon_2] = [3, 5.2] \qquad (3.7.6\text{-}4)$$

where, in view of the division by N in (3.7.6-1), one has

$$\epsilon_i = \frac{\epsilon_i'}{N} \qquad i = 1, 2 \qquad (3.7.6\text{-}5)$$

Thus, if the estimator is *efficient*, one has to have

$$P\left\{ \bar{\epsilon}_x \in [\epsilon_1, \epsilon_2] \right\} = 1 - Q \qquad (3.7.6\text{-}6)$$

Remarks

The division by N in (3.7.6-1) is a convenience: it yields results in the neighborhood of n_x regardless of N. Note that the interval (3.7.6-4) is not symmetric about the mean $n_x = 4$. For a single run ($N = 1$) the corresponding interval is $[0.5, 11.1]$; also note how much smaller is (3.7.6-4), which corresponds to $N = 25$. This illustrates how Monte Carlo runs *reduce the variability*.

Another option is to use a *one-sided probability region* and check only for the upper limit (the lower limit is zero).

3.7.7 A Localization Example

The following example, using the software $BearDAT^{TM}$, illustrates the problem of estimating the localization parameter of a constant velocity target based on noisy bearings-only observations. The target parameter vector is

$$x = [1000m \quad 3000m \quad -10m/s \quad 0m/s]' \tag{3.7.7-1}$$

Bearing measurements with standard deviation of $1°$ are made from a sensor moving on a platform, every $4s$ over a total period of $900s$. The measurement noises are independent, identically distributed Gaussian with mean zero.

Two scenarios are depicted in Figure 3.7.7-1 — they differ in the trajectory of the sensor platform. The target moves over the period of $900s$ from its initial location, designated as "$\times$". At the same time, the sensor platform moves in a two-leg constant speed trajectory starting from point "o". This sensor motion makes the target localization parameter observable.

The 95% probability mass ellipses around the target's initial and final position based on the Cramer-Rao lower bound, also shown, quantify the stochastic observability of the target's localization parameter. The target's final position uncertainty ellipse has been obtained using the transformation technique discussed in Section 3.4.

Figure 3.7.7-1(b) illustrates the sequences of bearing measurements obtained in one realization of each scenario.

In scenario 2, the resulting target localization uncertainty is larger than in scenario 1: the change of course in the platform trajectory leads to less information from the measurements.

Monte Carlo simulations can be used to verify estimator efficiency, that is, the *validity of the Cramer-Rao lower bound as the actual parameter estimate's covariance matrix*. This makes it possible to obtain a *confidence region for the location of the target* at any time (under the constant velocity motion assumption).

Figure 3.7.7-2 shows the scattering of the initial and final position estimates for 100 runs of scenario 1. As can be seen, the 90% ellipses for the position indeed contain all but 3 of the estimated positions.

The sample average of the NEES (3.7.4-8) was obtained from these 100 runs as 3.87, which is very close to the theoretical value of $n_x = 4$. This confirms the validity of the **CRLB as the actual covariance** for the present problem.

3.7.7 A Localization Example

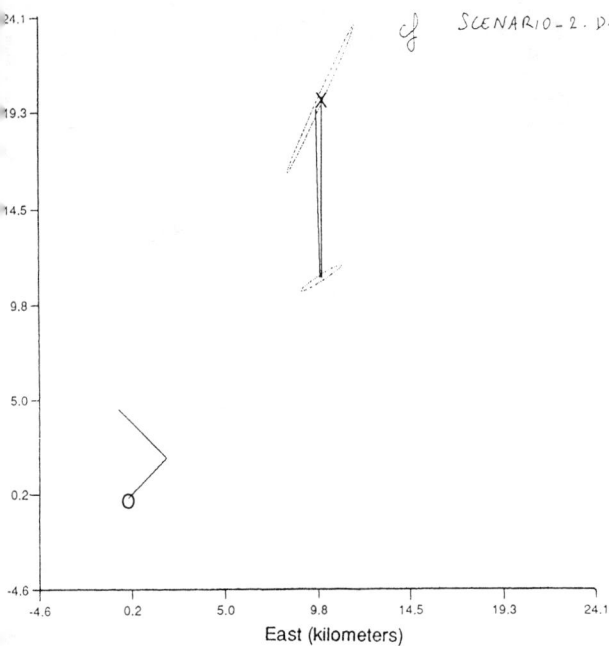

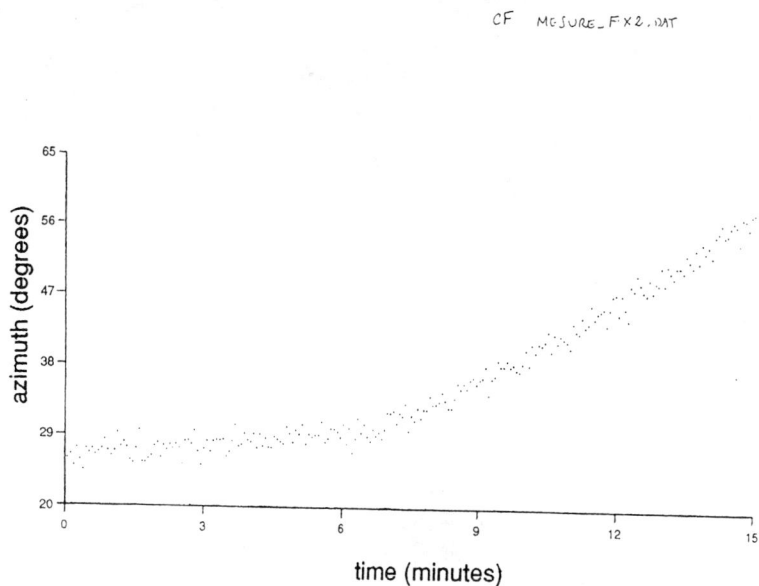

(a)   Scenario 1   (b)

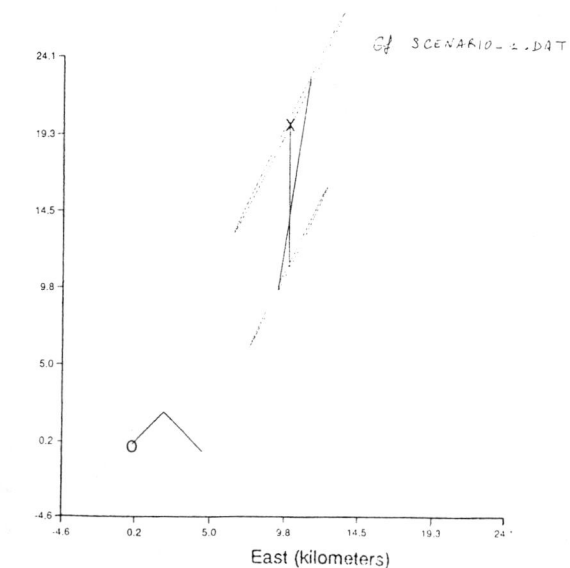

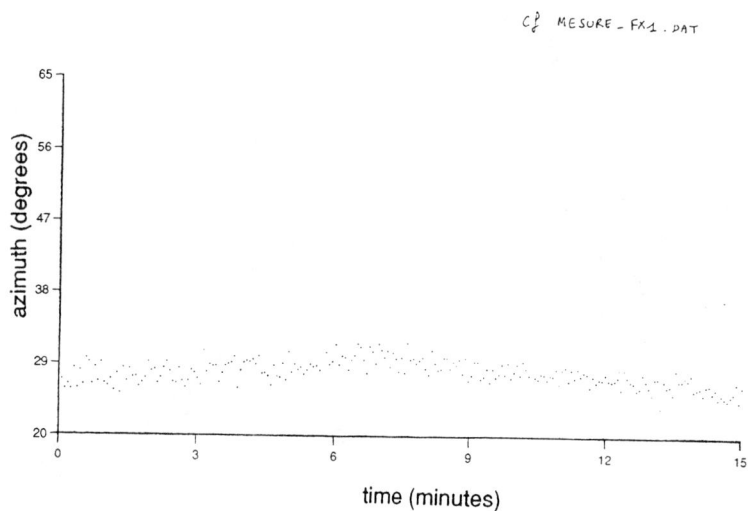

(a)   Scenario 2   (b)

o Initial platform location;   × Initial target location

Figure 3.7.7-1: (a) Target and platform trajectories with target localization uncertainty and its estimated trajectory (b) Sequence of bearing measurements (one realization).

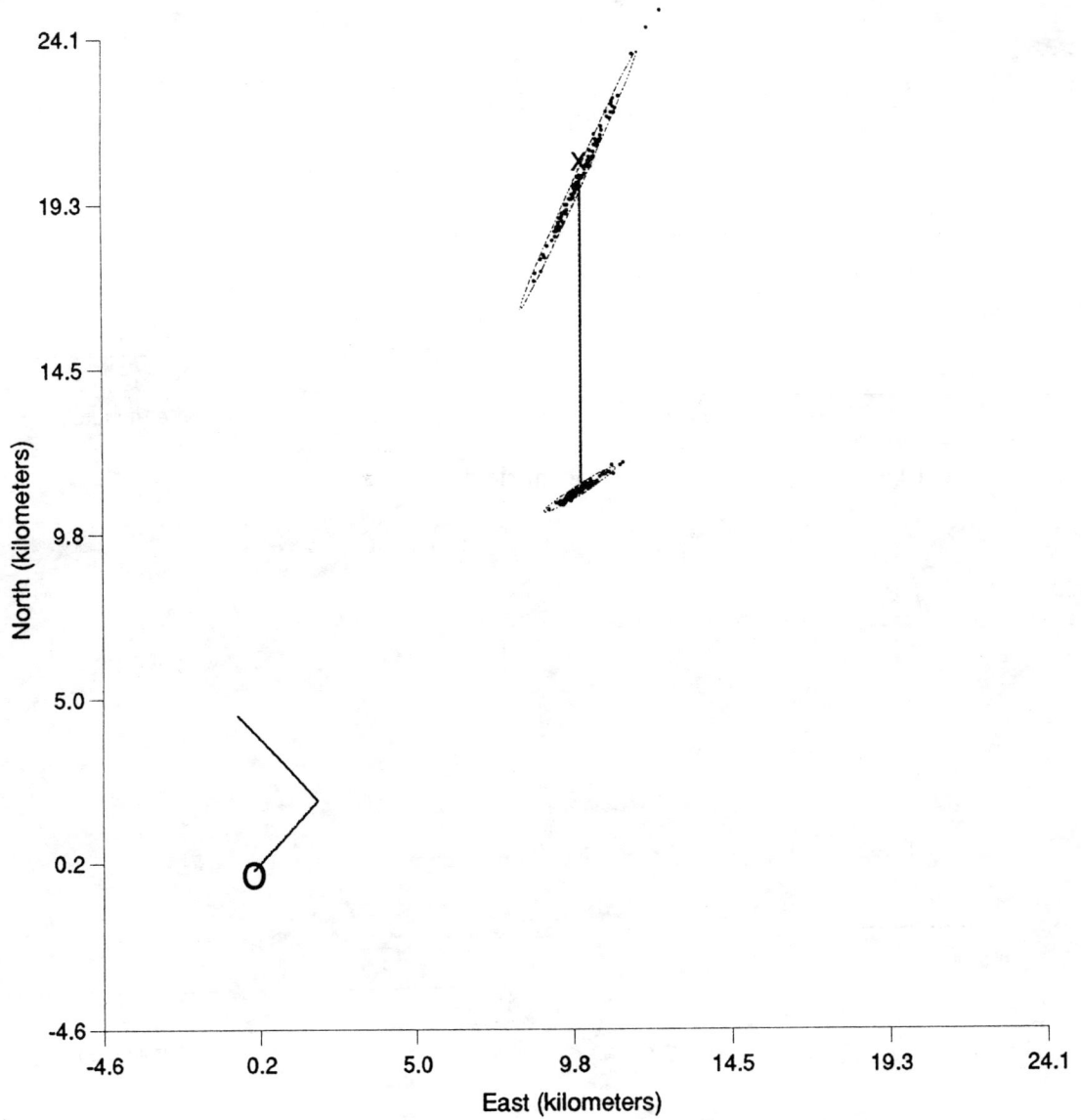

Figure 3.7.7-2: Estimated target positions in 100 runs (scenario 1).

3.7.8   Passive Localization — Summary

The technique of least squares has been illustrated for the problem of estimating the localization parameter of a constant velocity target based on noisy bearings-only observations. The noises were assumed independent and identically distributed Gaussian with *zero mean and known variance.*

In order for the target parameter vector to be observable, the platform from which the measurements are made has to undergo an acceleration. A constant-speed platform motion with a change of course satisfies this requirement.

The likelihood function of the target parameter is a nonlinear function of the parameter and therefore a numerical search technique (Newton-Raphson or quasi-Newton) is needed to find the MLE.

The FIM for this problem has been obtained, which allows the evaluation of the Cramer-Rao lower bound for the parameter estimate's covariance.

The target's position estimate at an *arbitrary time* and the corresponding covariance matrix have been obtained.

The goodness of fit test can be used to ascertain the acceptability of the parameter estimate — the fitting error has to be below a certain threshold.

Monte Carlo simulations can be used to verify estimator efficiency, that is, the *validity of the CRLB as the actual parameter estimate's covariance matrix.* This makes it possible to obtain a *confidence region for the location of the target* at any time (under the constant velocity motion assumption).

3.8 NOTES AND PROBLEMS

3.8.1 Bibliographical Notes

The static linear estimation material can be also found in [SM71] and, in more detail — with simultaneous estimation of the parameters and the noise variance — in statistics or econometrics texts, for example, [Joh72]. The reason the noise variance was assumed known here is that in engineering systems the measurements are obtained from a sensor and signal processor whose accuracy is (usually) known. In contrast to this, in statistics the noises reflect modeling errors and their variances are unknown. The concepts of goodness of fit and statistical significance are treated mainly in statistics and econometrics texts, for example, [Joh72]. This text (and most other statistics texts) present in detail the general regression theory where the variance is estimated together with the regression coefficients, as well as the t test and the F test used for significance testing in this case.

The software $BearDAT^{\text{TM}}$ is available in [Bar91] and is based on [JB90].

3.8.2 Problems

3-1 **Estimation with correlated noises.** Given the prior information $x \sim \mathcal{N}(\bar{x}, \sigma_0^2)$ and the measurements

$$z(j) = x + w(j) \qquad j = 1, 2$$

with the jointly Gaussian measurement noises $w(j) \sim \mathcal{N}(0, \sigma^2)$ independent of x but correlated among themselves, with

$$E[w(1)w(2)] = \rho\sigma^2$$

1. Find the variance of the MMSE estimator of x conditioned on these measurements.

2. What is the "effective number" of measurements (the number of measurements with independent noises with the same variance σ^2, which yield the same variance $P_{xx|z}$ for the MMSE estimator)?

3. If $\bar{x} = 10$, $\sigma_0 = 1$, and $\rho = 0.5$, how accurate should the measurements be (i.e., find σ) if we want the estimate to be within 10% of the true value with 99% probability?

4. Repeat (3) if $\rho = 0$.

3-2 **Effect of incorrect prior variance.** The random variable x with prior mean $\bar{x}$ and variance σ_0^2 is measured via

$$z = x + w$$

where w is zero mean, with variance σ^2, and independent of x.

1. Write the *linear* MMSE estimator $\hat{x}$ of x in terms of z and the MSE σ_1^2 associated with this estimator.

178

2. Write the estimate x^* of x as above but under the *incorrect* assumption that the prior variance is σ_p^2.

3. Find the actual MSE, σ_a^2, associated with (2), and the MSE σ_c^2 computed by the estimator in (2).

4. Let $\sigma_p^2 = s\sigma_0^2$. Assuming $\sigma_0^2 = \sigma^2 = 2$, evaluate the computed and the actual MSE for $s = 0.5, 1, 2, 10, 20$ and compare with the optimal one.

5. Indicate the limits as $s \to 0$ and $s \to \infty$ and interpret the results.

3-3 Fisher information matrix in the LS problem. Derive the FIM for (3.4.1-10).

3-4 Fisher information matrix in bearings-only target localization. Prove (3.7.4-4).

3-5 Passive localization with direction cosine measurements. Rework the problem of Section 3.7 with measurements

$$h(k, x) = \frac{\xi - \xi_p}{(\xi - \xi_p)^2 + (\eta - \eta_p)^2}$$

3-6 PROJECT: An interactive program for bearings-only target localization.

Set up an interactive program BEAR.EXE that accepts the following user-specified inputs for the problem of Section 3.7:

A1. True value of the target parameter x corresponding to $t = 0$.

A2. Sampling period T.

A3. Number of samples n.

A4. Measurement noise variance r.

A5. Platform motion — initial position, initial velocity.

A6. Platform maneuver — as accelerations in each coordinate with starting and ending time (k_1 and k_2, with the accelerations being in effect from $k_1 T$ to $k_2 T$).

A7. An initial estimate of the parameter to start the minimization. A set of default input values should be readable in from file BEAR.DEF.

The program should carry out the following:

B1. Evaluation of the Fisher information matrix corresponding to the true target parameter vector. Based on this, indicate the observability by stating whether the FIM is invertible.

B2. If the inverse can be obtained, it should be displayed.

B3. The target position uncertainty for each sampling time should be computed and the corresponding covariance matrices displayed.

B4. Generate noisy bearing measurements for the configuration defined.

B5. The MLE of the target parameter should be obtained based on the noisy measurements using a minimization algorithm (Newton-Raphson or quasi-Newton recommended) starting from the initial estimate specified as part of the input and the result will be displayed.

B6. Repeat B1-B3 for the estimated parameter.

B7. The fitting error (sum of the squares of the residuals) should be evaluated and displayed together with its number of degrees of freedom.

B8. The "normalized estimation error squared" (NEES) for the parameter vector of interest, $\tilde{x}'J\tilde{x}$, should be calculated and it should be indicated whether it is within its probability limit (e.g., 95%).

Another program, DBEAR.EXE, should display graphically the position uncertainty ellipses for the target, at user-specified times, centered at the estimate and their size according to the covariance matrices from B6. The ellipses will be of user-specified size (number of sigmas).

Document the algorithm that takes a 2×2 symmetric positive definite matrix

$$P = \begin{bmatrix} P_{11} & P_{12} \\ P_{12} & P_{22} \end{bmatrix}$$

and yields the "g-sigma" ellipse corresponding to it

$$x'P^{-1}x = g^2$$

Chapter 4

LINEAR DYNAMIC SYSTEMS WITH RANDOM INPUTS

4.1 INTRODUCTION

4.1.1 Outline

This chapter deals with the modeling of linear dynamic systems excited by random inputs, called **noise**. [1] Continuous time systems are discussed in Section 4.2. Section 4.3 deals with discrete time systems.

The state space models for continuous time and discrete time are presented and it is shown how the latter can be derived from the former by discretization. The state space model directly defined in discrete time is also discussed.

The Markov property of the state of a linear system driven by white noise is discussed and used to obtain the propagation equations for the mean and covariance of the state.

The power spectral density (the Fourier transform of the autocorrelation function) of the output of a linear system is related to the state space representation via the transfer function and it is shown how its factorization makes it possible to prewhiten an autocorrelated random process or sequence.

---

[1]According to a former program manager at a major Federal research agency, noise is beneficial — it lubricates the system. One can add to this that it also provides opportunities for research. More importantly, it was found via psychological experiments that people will go crazy without noise, which is particularly true for experts on stochastic systems (especially those dealing with estimation and tracking).

4.1.2 Linear Stochastic Systems — Summary of Objectives

Present the state space models for

- continuous time linear stochastic systems
- discrete time linear stochastic systems

and the connection between them.

Discuss the implications of the Markov property.

Derive the propagation equations for the mean and covariance of the state of a linear system driven by white noise.

Frequency domain approach — connect the power spectral density with the state space representation.

Show how spectral factorization can be used to prewhiten an autocorrelated random process.

4.2 CONTINUOUS TIME LINEAR STOCHASTIC DYNAMIC SYSTEMS

4.2.1 The Continuous Time State Space Model

The *state space representation* of *continuous time linear stochastic systems* can be written as

$$\dot{x}(t) = A(t)x(t) + B(t)u(t) + D(t)\tilde{v}(t) \tag{4.2.1-1}$$

where

> x is the **state vector** of dimension n_x,
>
> u is the **input vector** (control) of dimension n_u,
>
> $\tilde{v}$ is the (continuous time) input disturbance or **process noise**, also called **plant noise**, a vector of dimension n_v,
>
> A, B, and D are known matrices of dimensions $n_x \times n_x$, $n_x \times n_u$, and $n_x \times n_v$, respectively;
>
>> A is called the **system matrix**,
>>
>> B is the (continuous time) **input gain**,
>>
>> D is the (continuous time) **noise gain**.

Equation (4.2.1-1) is known as the **dynamic equation** or the **plant equation**. The output of the system is, in general, a vector of dimension n_z

$$z(t) = C(t)x(t) + \tilde{w}(t) \tag{4.2.1-2}$$

where

> $\tilde{w}$ is the output disturbance or **measurement noise**, and
>
> C is a known $n_z \times n_x$ matrix, called the **measurement matrix**.

Equation (4.2.1-2) is known as the **output equation** or the **measurement equation**.

In the absence of the disturbances $\tilde{v}$ and $\tilde{w}$, that is, in the deterministic case, given the initial condition $x(t_0)$ and the input *function in the interval* $[t_0, t]$ denoted as

$$u_{[t_0,t]} \triangleq \{u(\tau), t_0 \leq \tau \leq t\} \qquad (4.2.1\text{-}3)$$

one can compute the future output at any time $t > t_0$

$$z(t) = z[x(t_0), u_{[t_0,t]}, t, t_0] \qquad (4.2.1\text{-}4)$$

The **state** (of a deterministic system) is defined as the smallest vector that *summarizes the past of the system.*

Any linear differential equation that describes an input-output relationship can be put in the form of a first order vector differential equation as in (4.2.1-1). For example, an n-th order scalar differential equation can be rewritten as a first order differential equation for an n-vector, that is, n first order equations, by a suitable definition of state variables.

The initial conditions of the n-th order differential equation can be taken as state variables or any invertible linear transformation of them.

In the stochastic case, as will be discussed in detail in Chapter 10, the pdf of the (deterministic) state vector of the system *summarizes the past in a probabilistic sense.* This requires that the process noise be *white.* Then the pdf of the state vector is called the **information state**.

In the stochastic case the noises are usually assumed to be

1. zero-mean,
2. white, and
3. mutually independent

stochastic processes. If the noise is not zero mean, its mean (if known) can be taken as a known input.

4.2.2 Solution of the Continuous Time State Equation

The state equation (4.2.1-1) has the following solution:

$$x(t) = F(t, t_0)x(t_0) + \int_{t_0}^{t} F(t, \tau)[B(\tau)u(\tau) + D(\tau)\tilde{v}(\tau)]d\tau \qquad (4.2.2\text{-}1)$$

where $x(t_0)$ is the initial state and $F(t, t_0)$ is the **state transition matrix** from t_0 to t.
The transition matrix has the following properties:

$$\frac{dF(t, t_0)}{dt} = A(t)F(t, t_0) \qquad (4.2.2\text{-}2)$$

$$F(t_2, t_0) = F(t_2, t_1)F(t_1, t_0) \qquad \forall t_1 \qquad (4.2.2\text{-}3)$$

$$F(t, t) = I \qquad (4.2.2\text{-}4)$$

The last two imply that

$$F(t, t_0) = F(t_0, t)^{-1} \qquad (4.2.2\text{-}5)$$

The transition matrix has, in general no explicit form, unless the following *commutativity property* is satisfied:

$$A(t)\int_{t_0}^{t} A(\tau)d\tau = \int_{t_0}^{t} A(\tau)d\tau \, A(t) \qquad (4.2.2\text{-}6)$$

Then (and only then)

$$F(t, t_0) = e^{\int_{t_0}^{t} A(\tau)d\tau} \qquad (4.2.2\text{-}7)$$

Condition (4.2.2-6) is satisfied for time-invariant systems or diagonal $A(t)$.
For a time-invariant system, assuming $t_0 = 0$, one has

$$F(t) \triangleq F(t, 0)$$
$$= e^{At} \qquad (4.2.2\text{-}8)$$

Evaluation of the Transition Matrix

Some of the computational methods for the evaluation of the matrix e^{At} are briefly presented below.

1. Infinite series method:

$$
\begin{aligned}
e^{At} &= \sum_{k=0}^{\infty} \frac{(At)^k}{k!} \\
&= I + At + \frac{A^2 t^2}{2} + \cdots
\end{aligned}
\tag{4.2.2-9}
$$

where I is the identity matrix of the same dimension $n \times n$ as A. This is a numerical method and it requires series truncation (unless a closed-form expression can be found for each term).

2. Laplace transform method:

$$
e^{At} = \mathcal{L}^{-1}\{(sI - A)^{-1}\}
\tag{4.2.2-10}
$$

where $\mathcal{L}^{-1}$ is the inverse Laplace transform. This is practical if one can find a closed-form expression of the required matrix inverse above.

3. **Interpolating polynomial** method. Compute the eigenvalues λ_i of A, $i = 1, \ldots, n_e$, where n_e is the number of *distinct eigenvalues*, with multiplicities m_i, and

$$
\sum_{i=1}^{n_e} m_i = n
\tag{4.2.2-11}
$$

Then find a polynomial of degree $n - 1$

$$
g(\lambda) = \sum_{k=0}^{n-1} g_k \lambda^k
\tag{4.2.2-12}
$$

which is equal to $e^{\lambda t}$ *on the spectrum of A*, that is,

$$
\left. \frac{d^j}{d\lambda^j} g(\lambda) \right|_{\lambda = \lambda_i} = \left. \frac{d^j}{d\lambda^j} e^{\lambda t} \right|_{\lambda = \lambda_i} \qquad i = 1, \ldots, n_e, \quad j = 0, \ldots, m_i - 1
\tag{4.2.2-13}
$$

Then

$$
e^{At} = g(A)
\tag{4.2.2-14}
$$

Example — Coordinated Turn

Consider an object moving with **constant speed** (the *magnitude* of the velocity vector) and turning with a **constant angular rate** (i.e., executing a **coordinated turn** in aviation language).

The equations of motion in the plane (ξ, η) are in this case

$$\ddot{\xi} = -\Omega\dot{\eta} \qquad\qquad \ddot{\eta} = \Omega\dot{\xi} \qquad\qquad (4.2.2\text{-}15)$$

where Ω is the constant angular rate ($\Omega > 0$ implies a counterclockwise turn).

The state space representation of the above with the state vector

$$x \triangleq [\xi \ \ \dot{\xi} \ \ \eta \ \ \dot{\eta}]' \qquad\qquad (4.2.2\text{-}16)$$

is

$$\dot{x} = Ax \qquad\qquad (4.2.2\text{-}17)$$

where

$$A = \begin{bmatrix} 0 & 1 & 0 & 0 \\ 0 & 0 & 0 & -\Omega \\ 0 & 0 & 0 & 1 \\ 0 & \Omega & 0 & 0 \end{bmatrix} \qquad\qquad (4.2.2\text{-}18)$$

It can be easily shown that the eigenvalues of A are 0, 0, and $\pm\Omega j$.

It can be shown, using one of the techniques discussed earlier for evaluating the transition matrix, that for A given above one has

$$e^{At} = \begin{bmatrix} 1 & \dfrac{\sin\Omega t}{\Omega} & 0 & -\dfrac{1-\cos\Omega t}{\Omega} \\[2ex] 0 & \cos\Omega t & 0 & -\sin\Omega t \\[2ex] 0 & \dfrac{1-\cos\Omega t}{\Omega} & 1 & \dfrac{\sin\Omega t}{\Omega} \\[2ex] 0 & \sin\Omega t & 0 & \cos\Omega t \end{bmatrix} \qquad\qquad (4.2.2\text{-}19)$$

This allows, among other things, easy generation of state trajectories for such turns (the position evolves along circular arcs). These turns are common for aircraft as well as other flying objects.

4.2.3   The State as a Markov Process

Assume the process noise entering the state equation (4.2.1-1) to be zero mean and *white*, that is, that $\tilde{v}(t)$ is *independent* of $\tilde{v}(\tau)$ for all $t \neq \tau$. In this case the autocorrelation of $\tilde{v}(t)$ is

$$E[\tilde{v}(t)\tilde{v}(\tau)'] = V(t)\delta(t - \tau) \qquad (4.2.3\text{-}1)$$

The whiteness property of the process noise allows the preservation of the state's property of summarizing the past in the following sense: the pdf of the state at some time t conditioned on its values up to some earlier time t_1 depends only on the last value $x(t_1)$:

$$p[x(t)|x_{[-\infty,t_1]}, u_{[t_1,t]}] = p[x(t)|x(t_1), u_{[t_1,t]}] \qquad (4.2.3\text{-}2)$$

This follows from the complete unpredictability of the process noise due to its whiteness. Were the process noise autocorrelated ("colored"), (4.2.3-2) would not hold because states prior to t_1 could be used to predict the process noise $\tilde{v}_{[t_1,t]}$ and thus $x(t)$ in some fashion.

This can be seen from the solution of the state equation

$$x(t) = F(t,t_1)x(t_1) + \int_{t_1}^{t} F(t,\tau)[B(\tau)u(\tau) + D(\tau)\tilde{v}(\tau)]d\tau \qquad (4.2.3\text{-}3)$$

which indicates that $x(t_1)$ summarizes the past, the input provides the known part of the state's evolution after t_1, and the last term above is the contribution of the process noise, which is *completely unpredictable*.

In other words, the state of a dynamic system driven by white noise is a **Markov process**.

4.2.4   Propagation of the State's Mean and Covariance

Consider (4.2.1-1) with the known input $u(t)$ and **nonstationary white process noise** with nonzero mean

$$E[\tilde{v}(t)] = \bar{v}(t) \tag{4.2.4-1}$$

and autocovariance function

$$E\Big[[\tilde{v}(t) - \bar{v}(t)][\tilde{v}(\tau) - \bar{v}(\tau)]'\Big] = V(t)\delta(t - \tau) \tag{4.2.4-2}$$

The expected value of the state

$$\bar{x}(t) \triangleq E[x(t)] \tag{4.2.4-3}$$

evolves according to the (deterministic) differential equation

$$\boxed{\dot{\bar{x}}(t) = A(t)\bar{x}(t) + B(t)u(t) + D(t)\bar{v}(t)} \tag{4.2.4-4}$$

The above **propagation equation of the mean** follows from taking the expected value of (4.2.1-1) or differentiating the expected value of (4.2.2-1) with Leibniz' rule (see problem 4-1).

The covariance of the state

$$P_{xx}(t) \triangleq E\Big[[x(t) - \bar{x}(t)][x(t) - \bar{x}(t)]'\Big] \tag{4.2.4-5}$$

has the expression

$$P_{xx}(t) = F(t,t_0)P_{xx}(t_0)F(t,t_0)' + \int_{t_0}^{t} F(t,\tau)D(\tau)V(\tau)D(\tau)'F(t,\tau)'d\tau \tag{4.2.4-6}$$

and evolves according to the differential equation

$$\boxed{\dot{P}_{xx}(t) = A(t)P_{xx}(t) + P_{xx}(t)A(t)' + D(t)V(t)D(t)'} \tag{4.2.4-7}$$

This is the **propagation equation of the covariance** and it can be proven by evaluating the covariance of (4.2.2-1) using the whiteness of $\tilde{v}(t)$ and differentiating the result (see problem 4-1).

4.2.5 Frequency Domain Approach

Consider the *time-invariant* system driven by noise only

$$\dot{x}(t) = Ax(t) + D\tilde{v}(t) \tag{4.2.5-1}$$

where the noise is zero mean, *stationary*, and white, with autocorrelation function

$$
\begin{aligned}
R_{\tilde{v}\tilde{v}}(\tau) &= E[\tilde{v}(t+\tau)\tilde{v}(t)'] \\
&= V\delta(\tau)
\end{aligned} \tag{4.2.5-2}
$$

and with output

$$z(t) = Cx(t) \tag{4.2.5-3}$$

If the system is *stable* (i.e., all the eigenvalues of the system matrix A are in the left half plane), then its output becomes a stationary process (when the transient period is over). The autocorrelation of the output is denoted as

$$R_{zz}(\tau) = E[z(t+\tau)z(t)'] \tag{4.2.5-4}$$

The power spectral density — **power spectrum** — of the process noise, which is the Fourier transform of its autocorrelation function, is

$$
\begin{aligned}
S_{\tilde{v}\tilde{v}}(\omega) &= \int_{-\infty}^{\infty} R_{\tilde{v}\tilde{v}}(\tau)e^{-j\omega\tau}d\tau \\
&= V
\end{aligned} \tag{4.2.5-5}
$$

It can be shown that the **power spectral density matrix of the output** — the Fourier transform of (4.2.5-4) — is

$$
\begin{aligned}
S_{zz}(\omega) &= H(j\omega)S_{\tilde{v}\tilde{v}}(\omega)H(j\omega)^* \\
&= H(j\omega)VH(j\omega)^*
\end{aligned} \tag{4.2.5-6}
$$

where the asterisks denote complex conjugate transpose and

$$H(j\omega) \triangleq C(j\omega I - A)^{-1}D \tag{4.2.5-7}$$

is the transfer function matrix of system (4.2.5-1) from the noise $\tilde{v}$ to the output. Note that $H(j\omega)$ is a **rational function** (ratio of polynomials).

190

Spectral Factorization

Equation (4.2.5-6) leads to the following result. Given a **rational spectrum**, one can find a linear time-invariant system which, when driven by a white noise, its output will have that spectrum.

The transfer function of such a system, called **prewhitening system** or **shaping filter** is obtained by **factorization** of the desired spectrum into the product of a function with its complex conjugate. The first factor should correspond to a *causal and stable system.*

Example — Exponentially Decaying Autocorrelation

Consider the scalar stochastic process with **exponentially decaying autocorrelation**

$$R_{zz}(\tau) = \sigma^2 e^{-\alpha|\tau|} \qquad \alpha > 0 \tag{4.2.5-8}$$

The spectrum corresponding to the above is

$$S_{zz}(j\omega) = \sigma^2 \frac{2\alpha}{\alpha^2 + \omega^2} \tag{4.2.5-9}$$

Factorization of this spectrum according to (4.2.5-6) yields

$$H(j\omega) = \frac{1}{\alpha + j\omega} \tag{4.2.5-10}$$

$$V = 2\alpha\sigma^2 \tag{4.2.5-11}$$

The state equation corresponding to (4.2.5-10) is

$$\dot{x}(t) = -\alpha x(t) + \tilde{v}(t) \tag{4.2.5-12}$$

with output

$$z(t) = x(t) \tag{4.2.5-13}$$

and process noise autocorrelation

$$E[\tilde{v}(t+\tau)\tilde{v}(t)] = 2\alpha\sigma^2\delta(\tau) \tag{4.2.5-14}$$

Note that the transfer function in (4.2.5-10) corresponds to a stable system, while its complex conjugate would represent an unstable system.

4.3 DISCRETE TIME LINEAR STOCHASTIC DYNAMIC SYSTEMS

4.3.1 The Discrete Time State Space Model

In the **state space representation** of **discrete time systems** it is assumed that the input is piecewise constant, that is,

$$u(t) = u(t_k) \qquad t_k \le t < t_{k+1} \tag{4.3.1-1}$$

Then the state at sampling time t_{k+1} can be written, from (4.2.2-1), in terms of the state at t_k as

$$x(t_{k+1}) = F(t_{k+1}, t_k)x(t_k) + G(t_{k+1}, t_k)u(t_k) + v(t_k) \tag{4.3.1-2}$$

where F is the (state) **transition matrix** of the system, G is the **discrete time gain** through which the input, assumed to be constant over a sampling period, enters the system and $v(t_k)$ is the **discrete time process noise**.

For a *time-invariant* continuous time system sampled at arbitrary times the transition matrix is

$$F(t_{k+1}, t_k) = F(t_{k+1} - t_k) = e^{(t_{k+1}-t_k)A} \triangleq F(k) \tag{4.3.1-3}$$

the input gain is

$$G(t_{k+1}, t_k) = \int_{t_k}^{t_{k+1}} e^{(t_{k+1}-\tau)A}B d\tau \triangleq G(k) \tag{4.3.1-4}$$

and the discrete time process noise relates to the continuous time noise as

$$v(t_k) = \int_{t_k}^{t_{k+1}} e^{(t_{k+1}-\tau)A}D\tilde{v}(\tau)d\tau \triangleq v(k) \tag{4.3.1-5}$$

Equations (4.3.1-3) to (4.3.1-5) introduce the simplified index-only notation for discrete time systems, to be used (most of the time) in the sequel.

With the zero-mean and white assumption on $\tilde{v}(t)$, as in (4.2.3-1), it follows that

$$\boxed{E[v(k)] = 0} \tag{4.3.1-6}$$

$$\boxed{E[v(k)v(j)'] = Q(k)\delta_{kj}} \tag{4.3.1-7}$$

where δ_{kj} is the Kronecker delta function. The covariance of the discrete time process noise is given by

$$Q(k) = \int_{t_k}^{t_{k+1}} e^{(t_{k+1}-\tau)A}DV(\tau)D'e^{(t_{k+1}-\tau)A'}d\tau \tag{4.3.1-8}$$

The proof of (4.3.1-8) is given at the end of this subsection.

The (dynamic) model for **discrete time linear stochastic systems** can be written with the simplified index-only time notation as

$$\boxed{x(k+1) = F(k)x(k) + G(k)u(k) + v(k)} \qquad (4.3.1\text{-}9)$$

where the input is assumed known along with the matrices $F(k)$ and $G(k)$ and the process noise $v(k)$ is a zero-mean, white random sequence with covariance matrix $Q(k)$. Any (known) nonzero mean of the process noise v can be incorporated into the input.

The **discrete time measurement equation** is, with a similar notation,

$$\boxed{z(k) = H(k)x(k) + w(k)} \qquad (4.3.1\text{-}10)$$

where $H(k)$ is the **measurement matrix**, $w(k)$ is the **measurement noise** — a random sequence with moments

$$\boxed{E[w(k)] = 0} \qquad (4.3.1\text{-}11)$$

$$\boxed{E[w(k)w(j)'] = R(k)\delta_{kj}} \qquad (4.3.1\text{-}12)$$

The measurement given by (4.3.1-10) represents a "short-term" integration, during which the state is assumed to be constant.

Note that (4.3.1-9) and (4.3.1-10) describe a **time-varying discrete time system**.

The process and measurement noise sequences are (usually) assumed uncorrelated, that is,

$$E[v(k)w(j)'] = 0 \qquad \forall k, j \qquad (4.3.1\text{-}13)$$

In some cases it is convenient to define a **direct discrete time model** rather than a discretized version of a continuous time model. In such a case the process noise, also modeled as white, enters through a **noise gain**, denoted as $\Gamma(k)$. Then (4.3.1-9) is replaced by

$$\boxed{x(k+1) = F(k)x(k) + G(k)u(k) + \Gamma(k)v(k)} \qquad (4.3.1\text{-}14)$$

In this case the process noise covariance $Q(k)$ is defined directly. This will be discussed in more detail in Chapter 6.

Derivation of the Covariance of the Discretized Process Noise

$$
\begin{aligned}
E[v(k)v(j)'] &= E\left\{\int_{t_k}^{t_{k+1}} e^{(t_{k+1}-\tau_1)A} D\tilde{v}(\tau_1)d\tau_1 \left[\int_{t_j}^{t_{j+1}} e^{(t_{j+1}-\tau_2)A} D\tilde{v}(\tau_2)d\tau_2\right]'\right\} \\
&= E\left\{\int_{t_k}^{t_{k+1}}\int_{t_j}^{t_{j+1}} e^{(t_{k+1}-\tau_1)A} D\tilde{v}(\tau_1)\tilde{v}(\tau_2)'D'e^{(t_{j+1}-\tau_2)A'}d\tau_1 d\tau_2\right\} \\
&= \int_{t_k}^{t_{k+1}}\int_{t_j}^{t_{j+1}} e^{(t_{k+1}-\tau_1)A} D E[\tilde{v}(\tau_1)\tilde{v}(\tau_2)']D'e^{(t_{j+1}-\tau_2)A'}d\tau_1 d\tau_2 \\
&= \int_{t_k}^{t_{k+1}}\int_{t_j}^{t_{j+1}} e^{(t_{k+1}-\tau_1)A} DV(\tau_1)\delta(\tau_1-\tau_2)D'e^{(t_{j+1}-\tau_2)A'}d\tau_1 d\tau_2 \\
&= \int_{t_k}^{t_{k+1}} e^{(t_{k+1}-\tau_1)A} DV(\tau_1)D'e^{(t_{j+1}-\tau_1)A'}d\tau_1 \qquad (4.3.1\text{-}15)
\end{aligned}
$$

4.3.2 Solution of the Discrete Time State Equation

Using (4.3.1-9) for time k and substituting $x(k-1)$ yields

$$
\begin{aligned}
x(k) &= F(k-1)x(k-1) + G(k-1)u(k-1) + v(k-1) \\
&= F(k-1)[F(k-2)x(k-2) + G(k-2)u(k-2) + v(k-2)] \\
&\quad + G(k-1)u(k-1) + v(k-1) \\
&= F(k-1)F(k-2)x(k-2) + F(k-1)[G(k-2)u(k-2) + v(k-2)] \\
&\quad + G(k-1)u(k-1) + v(k-1)
\end{aligned}
\tag{4.3.2-1}
$$

Repeating the above leads to

$$
x(k) = \left[\prod_{j=0}^{k-1} F(k-1-j)\right]x(0) + \sum_{i=0}^{k-1}\left[\prod_{j=0}^{k-i-2} F(k-1-j)\right][G(i)u(i) + v(i)]
\tag{4.3.2-2}
$$

The notation for the product of matrices used above is

$$
\prod_{j=j_1}^{j_2} F(j) \triangleq F(j_1)F(j_1+1) \ \dots \ F(j_2)
\tag{4.3.2-3}
$$

If the upper index in (4.3.2-3) is smaller than the lower index, then the result is taken as the identity matrix.

Note that

$$
\begin{aligned}
\prod_{j=0}^{k-i-1} F(k-1-j) &= F(k-1)F(k-2) \ \dots \ F(i) \\
&= F(t_k, t_i)
\end{aligned}
\tag{4.3.2-4}
$$

is the transition matrix from sampling time i to sampling time k.

If the discrete time system is **time invariant**, that is,

$$
F(k) = F, \qquad G(k) = G \qquad\qquad \forall k
\tag{4.3.2-5}
$$

then (4.3.2-2) becomes

$$
x(k) = F^k x(0) + \sum_{i=0}^{k-1} F^{k-i-1}[Gu(i) + v(i)]
\tag{4.3.2-6}
$$

4.3.3   The State as a Markov Process

If the discrete time representation is obtained by discretizing a continuous time system with white process noise, the resulting discrete time process noise is a white sequence.

Similarly to the continuous time case, one has, following (4.3.2-1),

$$x(k) = \Big[\prod_{j=0}^{k-l-1} F(k-1-j) \Big] x(l) + \sum_{i=l}^{k-1} \Big[\prod_{j=0}^{k-i-2} F(k-1-j) \Big] [G(i)u(i) + v(i)] \qquad (4.3.3\text{-}1)$$

Thus, since $v(i)$, $i = l, \ldots, k-1$, are independent of

$$X^l \triangleq \{x(j)\}_{j=0}^l \qquad (4.3.3\text{-}2)$$

which depend only on $v(i)$, $i = 0, \ldots, l-1$, one has

$$p[x(k)|X^l, U^{k-1}] = p[x(k)|x(l), U_l^{k-1}] \qquad \forall k > l \qquad (4.3.3\text{-}3)$$

where

$$U_l^{k-1} \triangleq \{u(j)\}_{j=l}^{k-1} \qquad (4.3.3\text{-}4)$$

Thus, the state vector is a **Markov process**, or, more correctly, a **Markov sequence**.

As in the continuous time case, this follows from the complete unpredictability of the process noise due to its whiteness. Were the process noise autocorrelated ("colored"), (4.3.3-3) would not hold because states prior to time l could be used to predict the process noises $v(i)$, $i = l, \ldots, k-1$, and thus $x(k)$ in some fashion.

4.3.4   Propagation of the State's Mean and Covariance

Consider (4.3.1-14), repeated below for convenience

$$x(k+1) = F(k)x(k) + G(k)u(k) + \Gamma(k)v(k) \qquad (4.3.4\text{-}1)$$

with the known input $u(k)$ and the process noise $v(k)$ white, but for the sake of generality, **nonstationary** with nonzero mean:

$$E[v(k)] = \bar{v}(k) \qquad (4.3.4\text{-}2)$$

$$\text{cov}[v(k), v(j)] = E\Big[[v(k) - \bar{v}(k)][v(j) - \bar{v}(j)]'\Big] = Q(k)\delta_{kj} \qquad (4.3.4\text{-}3)$$

Then the expected value of the state

$$\bar{x}(k) \stackrel{\Delta}{=} E[x(k)] \qquad (4.3.4\text{-}4)$$

evolves according to the difference equation

$$\boxed{\bar{x}(k+1) = F(k)\bar{x}(k) + G(k)u(k) + \Gamma(k)\bar{v}(k)} \qquad (4.3.4\text{-}5)$$

The above, which is the **propagation equation of the mean**, follows immediately by applying the expectation operator to (4.3.4-1).

The covariance of the state

$$P_{xx}(k) \stackrel{\Delta}{=} E\Big[[x(k) - \bar{x}(k)][x(k) - \bar{x}(k)]'\Big] \qquad (4.3.4\text{-}6)$$

evolves according to the difference equation — the **covariance propagation equation**

$$\boxed{P_{xx}(k+1) = F(k)P_{xx}(k)F(k)' + \Gamma(k)Q(k)\Gamma(k)'} \qquad (4.3.4\text{-}7)$$

This follows by subtracting (4.3.4-5) from (4.3.4-1), which yields

$$x(k+1) - \bar{x}(k+1) = F(k)[x(k) - \bar{x}(k)] + \Gamma(k)[v(k) - \bar{v}(k)] \qquad (4.3.4\text{-}8)$$

It can be easily shown that multiplying (4.3.4-8) with its transpose and taking the expectation yields (4.3.4-7). The resulting cross-terms on the right hand side vanish when taking the expectation due to the whiteness of the process noise since $x(k)$, being a linear combination of the noises prior to k, is *independent* of $v(k)$.

Example — Fading Memory Average

Consider the scalar system (also called "first order Markov")

$$x(k) = \alpha x(k-1) + v(k) \qquad k = 1, \ldots \qquad (4.3.4\text{-}9)$$

with $x(0) = 0$ and $0 < \alpha < 1$. Note the slight change in the time argument in comparison to (4.3.4-1).

For this system it can be easily shown directly, or, using (4.3.2-2), that its solution is

$$x(k) = \sum_{i=1}^{k} \alpha^{k-i} v(i) \qquad k = 1, \ldots \qquad (4.3.4\text{-}10)$$

Normalizing the above by the sum of the coefficients yields

$$z(k) \triangleq \frac{x(k)}{\sum_{i=1}^{k} \alpha^{k-i}} \qquad (4.3.4\text{-}11)$$

which is the **fading memory average**, or, **exponentially discounted average** of the sequence $v(k)$. The term fading memory average is sometimes used for (4.3.4-10), which is really a **fading memory sum** — without the normalization. Note that for $\alpha = 1$, (4.3.4-11) becomes the sample average.

If the input has constant mean

$$E[v(k)] = \bar{v} \qquad (4.3.4\text{-}12)$$

then

$$\begin{aligned} \bar{x}(k) &\triangleq E[x(k)] \\ &= \bar{v} \sum_{i=1}^{k} \alpha^{k-i} \\ &= \bar{v} \frac{1 - \alpha^k}{1 - \alpha} \end{aligned} \qquad (4.3.4\text{-}13)$$

and

$$\lim_{k \to \infty} \bar{x}(k) = \frac{\bar{v}}{1 - \alpha} \qquad (4.3.4\text{-}14)$$

It can be easily shown that

$$\bar{z}(k) \triangleq E[z(k)] = \bar{v} \qquad \forall k \qquad (4.3.4\text{-}15)$$

The use of the fading memory average is for the case of a "slowly" varying mean $\bar{v}(k)$. The larger weightings on the more recent values of $v(k)$ allow "tracking" of its mean at the expense of the accuracy (larger variance) — see also problem 4-4.

4.3.5 Frequency Domain Approach

Consider the time-invariant system driven by noise only

$$x(k+1) = Fx(k) + \Gamma v(k) \qquad (4.3.5\text{-}1)$$

where the noise is assumed zero mean, stationary, and white, with autocorrelation

$$R_{vv}(k-l) = E[v(k)v(l)'] = Q\delta_{kl} \qquad (4.3.5\text{-}2)$$

The output of the system is

$$z(k) = Hx(k) \qquad (4.3.5\text{-}3)$$

The **power spectral density (spectrum)** of the discrete time process noise is the **discrete Fourier transform (DFT)**

$$S_{vv}(e^{j\omega T}) = \sum_{n=-\infty}^{\infty} e^{-j\omega Tn} R_{vv}(n) = Q \qquad (4.3.5\text{-}4)$$

where ω is the angular frequency and T the sampling period. Note the constant (flat) spectrum that characterizes a white noise.

If system (4.3.5-1) is **stable** (i.e., all the eigenvalues of F are *inside the unit circle*), then the state $x(k)$ will also become a stationary random sequence. The spectrum of the output will be

$$S_{zz}(e^{j\omega T}) = \mathcal{H}(\zeta)S_{vv}(\zeta)\mathcal{H}(\zeta)^*\Big|_{\zeta=e^{j\omega T}} \qquad (4.3.5\text{-}5)$$

where the asterisk denotes complex conjugate transpose, ζ is the (two-sided) z-transform variable, and

$$\mathcal{H}(\zeta) = H(\zeta I - F)^{-1}\Gamma \qquad (4.3.5\text{-}6)$$

is the **discrete time transfer function** of the system given by (4.3.5-1) and (4.3.5-3) from the process noise to the output. Note that the transfer function (4.3.5-6) is a **rational function**.

Using (4.3.5-4) in (4.3.5-5) yields the spectrum of the output as

$$S_{zz}(e^{j\omega T}) = \mathcal{H}(e^{j\omega T})Q\mathcal{H}(e^{j\omega T})^* = \mathcal{H}(e^{j\omega T})Q\mathcal{H}(e^{-j\omega T}) \qquad (4.3.5\text{-}7)$$

which, being the product of two rational functions and a constant matrix, is also a rational function.

Spectral Factorization

In view of (4.3.5-7), given a *rational spectrum*, one can find the linear time-invariant system which, driven by white noise, will have the output with the desired spectrum. Based on (4.3.5-7), the transfer function of such a system is obtained by **factorization** of the spectrum as follows:

$$S_{zz}(\zeta) = \mathcal{H}(\zeta)Q\mathcal{H}(\zeta^{-1}) \qquad (4.3.5\text{-}8)$$

since, for

$$\zeta = e^{j\omega T} \qquad (4.3.5\text{-}9)$$

one has

$$\zeta^* = \zeta^{-1} \qquad (4.3.5\text{-}10)$$

The resulting transfer function $\mathcal{H}(\zeta)$ specifies the **prewhitening system** or **shaping filter** for the sequence $z(k)$.

Remark

The factor in (4.3.5-8) that corresponds to a *causal* and *stable* system is the one that is to be chosen for the transfer function of the prewhitening system.

Example

Given the scalar sequence $z(k)$ with mean zero and autocorrelation

$$R_{zz}(n) = \rho\delta_{n,-1} + \delta_{n,0} + \rho\delta_{n,1} \qquad (4.3.5\text{-}11)$$

Its spectrum — the DFT written as the two-sided z-transform — is

$$S_{zz}(\zeta) = 1 + \rho\zeta^{-1} + \rho\zeta \qquad (4.3.5\text{-}12)$$

This can be factorized as

$$\begin{aligned} S_{zz}(\zeta) &= H(\zeta)H(\zeta^{-1}) \\ &= (\beta_0 + \beta_1\zeta^{-1})(\beta_0 + \beta_1\zeta) \end{aligned} \qquad (4.3.5\text{-}13)$$

where the first factor is the transfer function of a causal system.

The equations for the coefficients β_0 and β_1 are

$$\beta_0^2 + \beta_1^2 = 1 \qquad (4.3.5\text{-}14)$$

$$\beta_0\beta_1 = \rho \qquad (4.3.5\text{-}15)$$

with solution

$$\beta_0 = \frac{1}{2}(\sqrt{1+2\rho} + \sqrt{1-2\rho}) \qquad (4.3.5\text{-}16)$$

$$\beta_1 = \frac{1}{2}(\sqrt{1+2\rho} - \sqrt{1-2\rho}) \qquad (4.3.5\text{-}17)$$

Thus the transfer function of the system is

$$\mathcal{H}(\zeta) = \beta_0 + \beta_1\zeta^{-1} \qquad (4.3.5\text{-}18)$$

which corresponds to the following **moving average (MA)** or **finite impulse response (FIR)** system driven by unity-variance white noise

$$x(k) = \beta_0 v(k) + \beta_1 v(k-1) \qquad (4.3.5\text{-}19)$$

$$z(k) = x(k) \qquad (4.3.5\text{-}20)$$

Equations (4.3.5-19) and (4.3.5-20) specify the prewhitening system corresponding to (4.3.5-11).

4.4 SUMMARY

4.4.1 Summary of State Space Representation

State of a deterministic system — the smallest vector that *summarizes in full* its past.

State equation — a first order differential or difference equation that describes the evolution in time (dynamics) of the state vector.

Markov process — a stochastic process whose current state contains all the information about the probabilistic description of its future evolution.

A *state equation* driven by *white noise* yields a (vector) *Markov process*.

State of a stochastic system described by a Markov process — *summarizes probabilistically* its past.

All the above statements hold for linear as well as nonlinear systems.

A linear stochastic system's *continuous time* representation as a differential equation driven by white noise can be written in *discrete time* as a difference equation driven by a sequence of independent random variables — discrete time white noise, which can be related to the continuous time noise.

Alternatively, one can define directly a discrete time state equation.

Continuous time white noise: a random process with autocorrelation function a Dirac (impulse) delta function. If it is stationary, it has a spectrum (Fourier transform of the autocorrelation) that is constant ("flat").

Discrete time white noise: a random sequence with autocorrelation function a Kronecker delta function. If it is stationary, it has a spectrum (DFT of the autocorrelation) that is constant ("flat").

The *unconditional mean and covariance* of the state of a linear stochastic system driven by white noise have been shown to evolve ("open loop") according to linear differential or difference equations.

4.4.2 Summary of Prewhitening

For a linear time-invariant system, the frequency domain approach relates the power spectral density of the output to the one of the input via the transfer function.

The spectral density of the output of a stable linear time-invariant system is given by:

> The spectrum of the input premultiplied by the system's transfer function (a rational function) and postmultiplied by its complex conjugate.

Given a rational spectrum, one can find a linear time-invariant system which, if driven by a stationary white noise, its output will have that spectrum:

> The transfer function of such a *prewhitening system* is obtained by factoring the desired spectrum into the product of a function with its complex conjugate — this is called *spectral factorization*.

Due to the facts that

- Estimation results for dynamic systems, to be discussed in the sequel, are for Markov systems; and
- (Most) Markov processes (of interest) can be represented by linear time-invariant systems driven by white noise,

it is very important to find the "prewhitening system" for a given random process or sequence. This can be accomplished via spectral factorization.

4.5 NOTES AND PROBLEMS

4.5.1 Bibliographical Notes

The material on state space representation of stochastic systems is also discussed in, for example, [SM71]. More on the continuous time and discrete time representations can be found, for example, in [FH77] and [Che84].

The relationship between the spectra of the output and input of a linear time-invariant system is proven in standard probability texts, for example, in [Pap84]. Spectral factorization is treated in more detail in [AM79].

4.5.2 Problems

4-1 Mean and covariance of the state of a linear stochastic system.

1. Prove (4.2.4-4) by differentiating the expected value of (4.2.2-1) with Leibniz' rule

$$\frac{d}{dt}\int_{a(t)}^{b(t)} f(t,\tau)d\tau = \int_{a(t)}^{b(t)} \frac{\partial f(t,\tau)}{\partial t}d\tau + \frac{db(t)}{dt}f[t,b(t)] - \frac{da(t)}{dt}f[t,a(t)]$$

2. Prove (4.2.4-6).

3. Prove (4.2.4-7).

4-2 Moments of the output of a scalar linear system driven by white noise. Given the system

$$\dot{x}(t) = \alpha x(t) + \tilde{v}(t)$$

with $\tilde{v}(t)$ white, with mean $\bar{v}$ and variance q.

1. Find α such that the mean of the resulting stationary process is $\bar{x} = c$. (Hint: this is the steady state of the differential equation of the mean).

2. Using a time-domain approach, find the autocorrelation of the resulting stationary process.

4-3 Autocovariance of the state of a discrete time system.

1. Find the autocovariance

$$V_{xx}(k,j) \triangleq E\Big[[x(k) - \bar{x}(k)][x(j) - \bar{x}(j)]'\Big]$$

in terms of $P_{xx}(j)$ for system (4.3.1-14). Assume $k > j$.

2. Indicate what happens for a stable system for $k \gg j$. Justify.

4-4 Fading memory average. Consider

$$y(k) = \alpha y(k-1) + (1-\alpha)v(k)$$

204

with $y(0) = 0$ and $0 < \alpha < 1$.

1. Write the solution for $y(k)$.

2. Find the mean and variance of $y(k)$ if $v(k)$ is white with mean $\bar{v}$ and variance σ^2.

3. How does $y(k)$ differ from $z(k)$ in (4.3.4-11)?

4-5 Spectral factorization for prewhitening (shaping). Given the scalar zero-mean random sequence $x(k)$ with autocorrelation

$$R_{xx}(n) = E[x(k)x(k+n)] = \sigma^2 a^{|n|} \qquad 0 < a < 1$$

1. Find its spectrum.

2. Factorize it to find the linear system driven by white noise (prewhitening system or shaping filter) whose output has this autocorrelation.

4-6 Autocovariance of the state of a system driven by nonstationary white noise. Given

$$\dot{x}(t) = Ax(t) + D\tilde{v}(t)$$

$$E[x(t_0)] = \bar{x}(t_0)$$

$$E\Big[[x(t_0) - \bar{x}(t_0)][x(t_0) - \bar{x}(t_0)]'\Big] = P_{xx}(t_0)$$

$$E[\tilde{v}(t)] = \bar{v}(t)$$

$$E\Big[[\tilde{v}(t) - \bar{v}(t)][\tilde{v}(\tau) - \bar{v}(\tau)]'\Big] = Q(t)\delta(t - \tau)$$

Let

$$E[x(t)] \triangleq \bar{x}(t)$$

Find

$$V_{xx}(t, \tau) \triangleq E\Big[[x(t) - \bar{x}(t)][x(\tau) - \bar{x}(\tau)]'\Big]$$

4-7 State prediction in a time-invariant discrete time system.

1. Simplify (4.3.3-1) for a time-invariant system, that is, $F(i) = F$, $G(i) = G$.

2. Find a closed-form solution, similar to the above, for the covariance (4.3.4-7) assuming $F(k) = F$, $\Gamma(k) = \Gamma$, $Q(k) = Q$.

4-8 Coordinated turn transition matrix. Prove (4.2.2-19).

Chapter 5

STATE ESTIMATION IN DISCRETE TIME LINEAR DYNAMIC SYSTEMS

5.1 INTRODUCTION

5.1.1 Outline

This chapter extends the estimation concepts presented previously to the case of dynamic (time-varying) quantities. The estimation of the *state vector* of a *stochastic linear dynamic system* is considered.

The state estimator for discrete time linear dynamic systems driven by white noise — the (discrete time) **Kalman filter** — is introduced in Section 5.2 and its properties are discussed. The continuous-time case is considered later, in Chapter 9. An example that illustrates the discrete time Kalman filter is given in Section 5.3.

The issue of consistency [1] of a dynamic estimator, which is crucial for **evaluation of estimator optimality** in every implementation, is discussed in Section 5.4.

The initialization of estimators and practical ways to make it consistent are presented in Section 5.5.

---

[1]Or, "how to keep the filter honest;" the origin of the preoccupation with this issue is summarized in problem 5-6.

5.1.2 Discrete Time Linear Estimation — Summary of Objectives

Extend the static estimation concepts to dynamic systems.

Derive the state estimator for discrete time linear dynamic systems driven by white noise — the (discrete time) Kalman filter.

Present the properties of the Kalman filter:

- Whiteness of the innovations
- The role of the Riccati equation for the state covariance
- Stability and steady-state
- Connection with the observability and controllability of the system.

Introduce the likelihood function of a filter (to be used later in adaptive filtering). Discuss the issues of

- Consistency of a dynamic estimator and *estimator evaluation*
- Initialization of estimators

and practical ways to make it consistent.

5.2 LINEAR ESTIMATION IN DYNAMIC SYSTEMS — THE KALMAN FILTER

5.2.1 The Dynamic Estimation Problem

Consider a discrete time linear dynamic system described by a vector difference equation with additive white Gaussian noise that models "unpredictable disturbances." The dynamic (plant) equation is

$$x(k+1) = F(k)x(k) + G(k)u(k) + v(k) \qquad k = 0, 1, \ldots \qquad (5.2.1\text{-}1)$$

where $x(k)$ is the n_x-dimensional state vector, $u(k)$ is an n_u-dimensional **known input** vector (e.g., control or sensor platform motion), and $v(k)$, $k = 0, 1, \ldots$, is the sequence of zero-mean white Gaussian **process noise** (also n_x-vectors) with covariance

$$E[v(k)v(k)'] = Q(k) \qquad (5.2.1\text{-}2)$$

The measurement equation is

$$z(k) = H(k)x(k) + w(k) \qquad k = 1, \ldots \qquad (5.2.1\text{-}3)$$

with $w(k)$ the sequence of zero-mean white Gaussian **measurement noise** with covariance

$$E[w(k)w(k)'] = R(k) \qquad (5.2.1\text{-}4)$$

The matrices F, G, H, Q, and R are assumed *known* and possibly time varying. In other words, the system can be *time varying* and the noises *nonstationary.*

The initial state $x(0)$, in general unknown, is modeled as a *random variable*, Gaussian distributed with known mean and covariance. The two noise sequences and the initial state are assumed *mutually independent.*

The above constitutes the **linear-Gaussian (LG) assumption**.

In the dynamic equation (5.2.1-1), the process noise term $v(k)$ is sometimes taken as $\Gamma(k)v(k)$ with $v(k)$ an n_v-vector and $\Gamma(k)$ a known $n_x \times n_v$ matrix. Then the covariance matrix of the disturbance in the state equation, which is $Q(k)$ if $v(k)$ enters directly, is to be replaced by

$$E\Big[[\Gamma(k)v(k)][\Gamma(k)v(k)]'\Big] = \Gamma(k)Q(k)\Gamma(k)' \qquad (5.2.1\text{-}5)$$

The linearity of (5.2.1-1) and (5.2.1-3) leads to the preservation of the Gaussian property of the state and measurements — this is a *Gauss-Markov process.*

The following notation will be used: the conditional mean

$$\hat{x}(j|k) \triangleq E[x(j)|Z^k]$$
(5.2.1-6)

where

$$Z^k \triangleq \{z(j), j \leq k\}$$
(5.2.1-7)

denotes the sequence of observations available at time k, is the

- **Estimate of the state** if $j = k$ (also called filtered value);
- **Smoothed value of the state** if $j < k$;
- **Predicted value of the state** if $j > k$.

The **estimation error** is defined as

$$\tilde{x}(j|k) \triangleq x(j) - \hat{x}(j|k)$$
(5.2.1-8)

The **conditional covariance matrix** of $x(j)$ given the data Z^k or the **covariance associated with the estimate** (5.2.1-6) is

$$P(j|k) \triangleq E\left[[x(j) - \hat{x}(j|k)][x(j) - \hat{x}(j|k)]'|Z^k\right] = E[\tilde{x}(j|k)\tilde{x}(j|k)'|Z^k]$$
(5.2.1-9)

Remarks

The smoothed state has also been called recently the **retrodicted state** [DF93]. The term smoothing is, however, commonly used, even though retrodiction is the correct antonym of prediction. Sometimes the estimated state is called (incorrectly) the smoothed state.

Note that the *covariance of the state* is the same as the *covariance of the estimation error* — this is a consequence of the fact that the estimate is the conditional mean (5.2.1-6). (See also problem 5-3.)

It was shown earlier that the *MMSE criterion* for estimation leads to the *conditional mean* as the *optimal estimate*.

As discussed in Section 3.2, if two vectors are jointly Gaussian then the probability density of one conditioned on the other is also Gaussian. Thus the conditional mean (5.2.1-6) will be evaluated using this previous result.

The Estimation Algorithm

The estimation algorithm starts with the ***initial estimate*** $\hat{x}(0|0)$ of $x(0)$ and the associated ***initial covariance*** $P(0|0)$, assumed to be available. The second (conditioning) index 0 stands for Z^0, the ***initial information***. Practical procedures to obtain the initial estimate and initial covariance will be discussed later.

One cycle of the dynamic estimation algorithm — the ***Kalman filter (KF)*** — will thus consist of mapping the estimate

$$\hat{x}(k|k) \triangleq E[x(k)|Z^k] \tag{5.2.1-10}$$

which is the conditional mean of the state at time k (the "current stage") given the observations up to and including time k, and the associated covariance matrix

$$P(k|k) = E\Big[[x(k) - \hat{x}(k|k)][x(k) - \hat{x}(k|k)]'|Z^k\Big] \tag{5.2.1-11}$$

into the corresponding variables at the next stage, namely, $\hat{x}(k+1|k+1)$ and $P(k+1|k+1)$.

This follows from the fact that a Gaussian random variable is *fully characterized* by its first two moments.

The values of past known inputs are subsumed in the conditioning, but (most of the time) will not be shown explicitly.

5.2.2 Dynamic Estimation as a Recursive Static Estimation

The recursion that yields the state estimate at $k + 1$ and its covariance can be obtained from the static estimation equations (3.2.0-7) and (3.2.0-8)

$$\hat{x} \triangleq E[x|z] = \bar{x} + P_{xz}P_{zz}^{-1}(z - \bar{z}) \tag{5.2.2-1}$$

$$P_{xx|z} \triangleq E[(x - \hat{x})(x - \hat{x})'|z] = P_{xx} - P_{xz}P_{zz}^{-1}P_{zx} \tag{5.2.2-2}$$

by the following substitutions, indicated below by "$\rightarrow$".

The *prior* (unconditional) expectations from the static case become *prior to the availability of the measurement at time* $k + 1$ in the dynamic case, that is, *given the data at* k.

The *posterior* (conditional) expectations become *posterior to obtaining the measurement at time* $k + 1$, that is, *given the data up to and including* $k + 1$.

The variable to be estimated is the state at $k + 1$

$$x \rightarrow x(k + 1) \tag{5.2.2-3}$$

Its mean prior to $k + 1$ — the **(one-step) predicted state** — is

$$\bar{x} \rightarrow \hat{x}(k + 1|k) \triangleq E[x(k + 1)|Z^k] \tag{5.2.2-4}$$

Based on the observation (measurement)

$$z \rightarrow z(k + 1) \tag{5.2.2-5}$$

with prior mean — the **predicted measurement**

$$\bar{z} \rightarrow \hat{z}(k + 1|k) \triangleq E[z(k + 1)|Z^k] \tag{5.2.2-6}$$

one computes the estimate posterior to $k + 1$ — the **updated state estimate** (or, just the **updated state**)

$$\hat{x} \rightarrow \hat{x}(k + 1|k + 1) \triangleq E[x(k + 1)|Z^{k+1}] \tag{5.2.2-7}$$

The prior covariance matrix of the state variable $x(k+1)$ to be estimated — the **state prediction covariance** or **predicted state covariance** — is

$$P_{xx} \rightarrow P(k + 1|k) \triangleq \operatorname{cov}[x(k + 1)|Z^k] = \operatorname{cov}[\tilde{x}(k + 1|k)|Z^k] \tag{5.2.2-8}$$

with the last equality following from (5.2.1-9).

The (prior) covariance of the observation $z(k+1)$ — the **measurement prediction covariance** — is

$$P_{zz} \to S(k+1) \triangleq \text{cov}[z(k+1)|Z^k] = \text{cov}[\tilde{z}(k+1|k)|Z^k] \tag{5.2.2-9}$$

The covariance between the variable to be estimated $x(k+1)$ and the observation $z(k+1)$ is

$$P_{xz} \to \text{cov}[x(k+1), z(k+1)|Z^k] = \text{cov}[\tilde{x}(k+1|k), \tilde{z}(k+1|k)|Z^k] \tag{5.2.2-10}$$

The posterior covariance of the state $x(k+1)$ — the **updated state covariance** — is

$$P_{xx|z} \to P(k+1|k+1) = \text{cov}[x(k+1)|Z^{k+1}] = \text{cov}[\tilde{x}(k+1|k+1)|Z^{k+1}] \tag{5.2.2-11}$$

The weighting matrix from the estimation ("updating") equation (5.2.2-1) becomes the **filter gain**

$$P_{xz}P_{zz}^{-1} \to W(k+1) \triangleq \text{cov}[x(k+1), z(k+1)|Z^k]S(k+1)^{-1} \tag{5.2.2-12}$$

Remark

The reasons for the designation of the above as filter gain are:

1. The recursive estimation algorithm is a filter — it reduces the effect of the various noises on the quantity of interest (the state estimate);
2. The quantity (5.2.2-12) multiplies the observation $z(k+1)$, which is the input to the filter, that is, this quantity is a gain.

213

5.2.3 Derivation of the Dynamic Estimation Algorithm

The **predicted state** (5.2.2-4) follows by applying on the state equation (5.2.1-1) the operator of expectation conditioned on Z^k,

$$E[x(k+1)|Z^k] = E[F(k)x(k) + G(k)u(k) + v(k)|Z^k] \qquad (5.2.3\text{-}1)$$

Since the process noise $v(k)$ is *white and zero mean*, this results in

$$\boxed{\hat{x}(k+1|k) = F(k)\hat{x}(k|k) + G(k)u(k)} \qquad (5.2.3\text{-}2)$$

Subtracting the above from (5.2.1-1) yields the **state prediction error**

$$\tilde{x}(k+1|k) \triangleq x(k+1) - \hat{x}(k+1|k) = F(k)\tilde{x}(k|k) + v(k) \qquad (5.2.3\text{-}3)$$

Note the cancellation of the input $u(k)$ in (5.2.3-3) — it has no effect on the estimation error as long as it is *known*.

The **state prediction covariance** (5.2.2-8) is

$$E[\tilde{x}(k+1|k)\tilde{x}(k+1|k)'|Z^k] = F(k)E[\tilde{x}(k|k)\tilde{x}(k|k)'|Z^k]F(k)' + E[v(k)v(k)'] \quad (5.2.3\text{-}4)$$

which can be rewritten as

$$\boxed{P(k+1|k) = F(k)P(k|k)F(k)' + Q(k)} \qquad (5.2.3\text{-}5)$$

The cross terms in (5.2.3-4) are zero due to the fact that $v(k)$ is zero mean and white and, thus, orthogonal to $\tilde{x}(k|k)$.

The **predicted measurement** (5.2.2-6) follows similarly by taking the expected value of (5.2.1-2) conditioned on Z^k,

$$E[z(k+1)|Z^k] = E[H(k+1)x(k+1) + w(k+1)|Z^k] \qquad (5.2.3\text{-}6)$$

Since the measurement noise $w(k+1)$ is zero mean and white, this becomes

$$\boxed{\hat{z}(k+1|k) = H(k+1)\hat{x}(k+1|k)} \qquad (5.2.3\text{-}7)$$

Subtracting the above from (5.2.1-2) yields the **measurement prediction error**

$$\tilde{z}(k+1|k) \triangleq z(k+1) - \hat{z}(k+1|k) = H(k+1)\tilde{x}(k+1|k) + w(k+1) \qquad (5.2.3\text{-}8)$$

The **measurement prediction covariance** (5.2.2-9) follows from (5.2.3-8), in a manner similar to (5.2.3-5), as

$$\boxed{S(k+1) = H(k+1)P(k+1|k)H(k+1)' + R(k+1)} \qquad (5.2.3\text{-}9)$$

5.2.3 Derivation of the Dynamic Estimation Algorithm

The covariance (5.2.2-10) between the state and measurement is, using (5.2.3-8)

$$
\begin{aligned}
E[\tilde{x}(k+1|k)\tilde{z}(k+1|k)'|Z^k] &= E\left[\tilde{x}(k+1|k)[H(k+1)\tilde{x}(k+1|k) + w(k+1)]'|Z^k\right] \\
&= P(k+1|k)H(k+1)'
\end{aligned}
\tag{5.2.3-10}
$$

The **filter gain** (5.2.2-12) is, using (5.2.3-9) and (5.2.3-10),

$$
\boxed{W(k+1) \triangleq P(k+1|k)H(k+1)'S(k+1)^{-1}}
\tag{5.2.3-11}
$$

Thus the **updated state estimate** (5.2.2-7) can be written according to (5.2.2-1) as

$$
\boxed{\hat{x}(k+1|k+1) = \hat{x}(k+1|k) + W(k+1)\nu(k+1)}
\tag{5.2.3-12}
$$

where

$$
\boxed{\nu(k+1) \triangleq z(k+1) - \hat{z}(k+1|k) = \tilde{z}(k+1|k)}
\tag{5.2.3-13}
$$

is called the **innovation** or **measurement residual**. This is the same as (5.2.3-8) but the notation ν will be used in the sequel. Note that in view of this, S is also the **innovation covariance**.

Finally, the **updated covariance** (5.2.2-11) of the state at $k+1$ is, according to (5.2.2-2)

$$
\begin{aligned}
P(k+1|k+1) &= P(k+1|k) - P(k+1|k)H(k+1)'S(k+1)^{-1}H(k+1)P(k+1|k) \\
&= [I - W(k+1)H(k+1)]P(k+1|k)
\end{aligned}
\tag{5.2.3-14}
$$

or, in symmetric form

$$
\boxed{P(k+1|k+1) = P(k+1|k) - W(k+1)S(k+1)W(k+1)'}
\tag{5.2.3-15}
$$

Equation (5.2.3-12) is called the **state update** equation, since it yields the updated state estimate, and (5.2.3-15) is the **covariance update** equation.

Note the similarity of the state update equation (5.2.3-12) to the recursive LS equation (3.4.2-14). The covariance update equation (5.2.3-15) is analogous to (3.4.2-10). The only difference is that in the LS case the prediction to $k+1$ from k is the same as the updated value at k. This follows from the fact that in the LS formulation one deals with a constant parameter while in a dynamic system the state evolves in time.

215

Alternative Forms for the Covariance Update

Similarly to (3.4.2-6), there is a recursion for the inverse covariance

$$P(k+1|k+1)^{-1} = P(k+1|k)^{-1} + H(k+1)'R(k+1)^{-1}H(k+1) \qquad (5.2.3\text{-}16)$$

It can be easily shown using the matrix inversion lemma that (5.2.3-16) is algebraically equivalent to (5.2.3-15). The filter using (5.2.3-16) instead of (5.2.3-15) is known as the **information matrix filter** (see Chapter 7).

As in (3.4.2-11), the filter gain (5.2.3-11) has the alternate expression

$$W(k+1) = P(k+1|k+1)H(k+1)'R(k+1)^{-1} \qquad (5.2.3\text{-}17)$$

An alternative form for the covariance update equation (5.2.3-15), which holds for an *arbitrary gain W* (see Problem 5-5), called the **Joseph form covariance update**, is

$$\begin{aligned} P(k+1|k+1) \;=\; & [I - W(k+1)H(k+1)]P(k+1|k)[I - W(k+1)H(k+1)]' \\ & + W(k+1)R(k+1)W(k+1)' \end{aligned} \qquad (5.2.3\text{-}18)$$

While this is computationally more expensive than (5.2.3-15), it is less sensitive to round-off errors: it will not lead to negative eigenvalues, as (5.2.3-15) is prone to, due to the subtraction present in it. Numerical techniques that reduce the sensitivity to round-off errors are discussed in Chapter 7.

Equation (5.2.3-18) can be also used for evaluation of the **sensitivity of the filter to an incorrect gain**.

Intuitive Interpretation of the Gain

Note from (5.2.3-17) that the *optimal filter gain* is (taking a simplistic "scalar view" of it)

1. "proportional" to the state prediction variance, and
2. "inverse proportional" to the innovation variance.

Thus, the gain is

- "large" if the state prediction is "inaccurate" (has a large variance) and the measurement is "accurate" (has a relatively small variance);
- "small" if the state prediction is "accurate" (has a small variance) and the measurement is "inaccurate" (has a relatively large variance).

A large gain indicates a "rapid" response to the measurement in updating the state, while a small gain yields a slower response to the measurement. In the frequency domain it can be shown that these properties correspond to a higher/lower **bandwidth of the filter**.

A filter whose **optimal gain** is higher yields less "noise reduction," as one would expect from a filter with a higher bandwidth. This will be seen quantitatively in the next chapter.

Remark

Equations (5.2.3-9) and (5.2.3-15) yield **filter-calculated covariances**, which are exact if all the modeling assumptions used in the filter derivation hold. In practice this is not always the case and the validity of these filter-calculated estimation accuracies can be tested, as discussed in Section 5.4.

5.2.4 Overview of the Kalman Filter Algorithm

Under the *Gaussian assumption* for the initial state (or initial state error) and all the noises entering into the system, the Kalman filter is the **optimal MMSE state estimator**. If these random variables are *not Gaussian* and one has only their first two moments, then in view of the discussion from Section 3.3 the Kalman filter algorithm is the **best linear state estimator**, that is, the **LMMSE state estimator**.

The flowchart of one cycle of the Kalman filter is presented in Figure 5.2.4-1. Note that at every stage k the entire past is summarized by the *sufficient statistic* $\hat{x}(k|k)$ and the associated covariance $P(k|k)$.

The left-side column represents the true system's evolution from the state at time k to the state at time $k+1$ with the input $u(k)$ and the process noise $v(k)$. The measurement follows from the new state and the noise $w(k+1)$. The known input (e.g., control, platform motion, or sensor pointing) enters (usually) the system with the knowledge of the latest state estimate and is used by the state estimator to obtain the predicted value for the state at the next time.

The state estimation cycle consists of

1. State and measurement prediction (also called **time update**), and
2. State update (also called **measurement update**).

The state update requires the filter gain, obtained in the course of the covariance calculations. The covariance calculations are *independent* of the state (and control — assumed to be known) and can, therefore, be performed *offline*.

The Workhorse of Estimation — The Kalman Filter

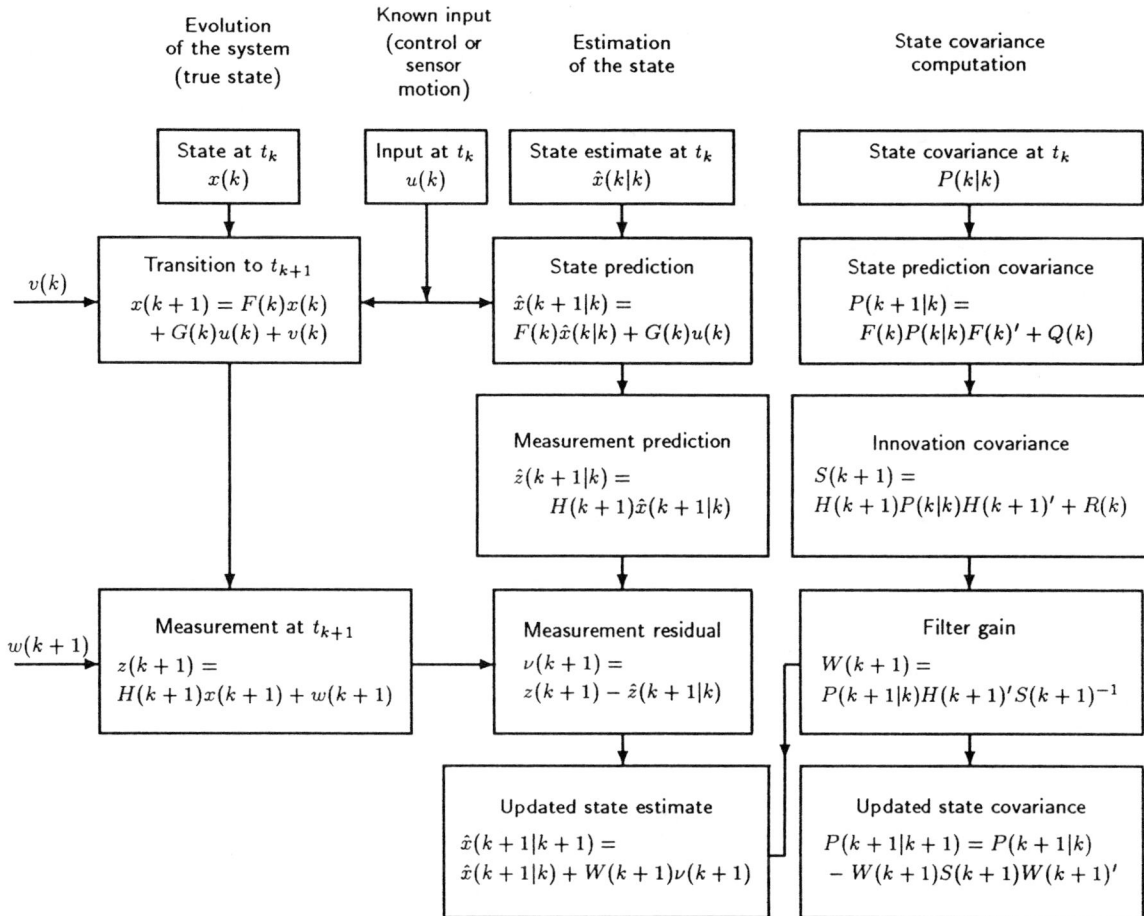

Evolution of the system (true state)	Known input (control or sensor motion)	Estimation of the state	State covariance computation

State at t_k $x(k)$	Input at t_k $u(k)$	State estimate at t_k $\hat{x}(k\|k)$	State covariance at t_k $P(k\|k)$

$v(k)$ →

| Transition to t_{k+1} $x(k+1) = F(k)x(k)$ $+ G(k)u(k) + v(k)$ | | State prediction $\hat{x}(k+1\|k) =$ $F(k)\hat{x}(k\|k) + G(k)u(k)$ | State prediction covariance $P(k+1\|k) =$ $F(k)P(k\|k)F(k)' + Q(k)$ |

| | | Measurement prediction $\hat{z}(k+1\|k) =$ $H(k+1)\hat{x}(k+1\|k)$ | Innovation covariance $S(k+1) =$ $H(k+1)P(k\|k)H(k+1)' + R(k)$ |

$w(k+1)$ →

| Measurement at t_{k+1} $z(k+1) =$ $H(k+1)x(k+1) + w(k+1)$ | | Measurement residual $\nu(k+1) =$ $z(k+1) - \hat{z}(k+1\|k)$ | Filter gain $W(k+1) =$ $P(k+1\|k)H(k+1)'S(k+1)^{-1}$ |

| | | Updated state estimate $\hat{x}(k+1\|k+1) =$ $\hat{x}(k+1\|k) + W(k+1)\nu(k+1)$ | Updated state covariance $P(k+1\|k+1) = P(k+1\|k)$ $- W(k+1)S(k+1)W(k+1)'$ |

Figure 5.2.4-1: One cycle in the state estimation of a linear system.

Summary of the Statistical Assumptions of the Kalman Filter

The initial state has the known mean and covariance

$$E[x(0)|Z^0] = \hat{x}(0|0) \qquad (5.2.4\text{-}1)$$

$$\text{cov}[x(0)|Z^0] = P(0|0) \qquad (5.2.4\text{-}2)$$

where Z^0 denotes the initial (prior) information.

The process and measurement noise sequences are *zero mean and white* with *known covariance matrices*

$$E[v(k)] = 0 \qquad (5.2.4\text{-}3)$$

$$E[v(k)v(j)'] = Q(k)\delta_{kj} \qquad (5.2.4\text{-}4)$$

$$E[w(k)] = 0 \qquad (5.2.4\text{-}5)$$

$$E[w(k)w(j)'] = R(k)\delta_{kj} \qquad (5.2.4\text{-}6)$$

All the above are *mutually uncorrelated*

$$E[x(0)v(k)'] = 0 \qquad (5.2.4\text{-}7)$$

$$E[x(0)w(k)'] = 0 \qquad (5.2.4\text{-}8)$$

$$E[v(k)w(j)'] = 0 \qquad (5.2.4\text{-}9)$$

It can be easily shown that under the Gaussian assumption the whiteness and the uncorrelatedness of the noises imply the following:

$$E[v(k)|Z^k] = E[v(k)] = 0 \qquad (5.2.4\text{-}10)$$

$$E[w(k)|Z^{k-1}] = E[w(k)] = 0 \qquad (5.2.4\text{-}11)$$

Property (5.2.4-10) was used in (5.2.3-2), while (5.2.4-11) was used in (5.2.3-7).

Remark

The dynamic (plant) equation parameters — the matrices F, G — and the measurement equation parameters — the matrix H — are assumed known.

Some Extensions

The assumptions of

- White process noise
- White measurement noise
- Uncorrelatedness between the process and the measurement noise sequences

can be relaxed.

An autocorrelated ("colored") noise has to be modeled as the output of a subsystem driven by white noise, that is, it is to be *prewhitened*, as discussed in Chapter 4. For an *autocorrelated process noise*, the state vector has to be augmented to incorporate this subsystem. An example of prewhitening of an autocorrelated process noise is presented in Section 8.2.

The situation where there is *correlation between the two noise sequences* is discussed in Section 8.3. The filter derivation for an *autocorrelated measurement noise*, which can be done without augmenting the state, is presented in Section 8.4.

Discrete time *smoothing* is presented in Section 8.6.

Continuous time state estimation as an extension of the discrete time results is discussed in Chapter 9.

5.2.5 The Matrix Riccati Equation

As pointed out in Section 3.2, the covariance equations in the static MMSE estimation problem are independent of the measurements. Consequently, the covariance equations for the state estimation problem (in a linear dynamic system), derived in Subsection 5.2.3, can be iterated forward offline.

It can be easily shown that the following recursion can be written for the one-step prediction covariance

$$P(k+1|k) = F(k)\Big\{P(k|k-1) - P(k|k-1)H(k)'[H(k)P(k|k-1)H(k)' + R(k)]^{-1}$$
$$\cdot H(k)P(k|k-1)\Big\}F(k)' + Q(k) \tag{5.2.5-1}$$

This is the **discrete time (difference) matrix Riccati equation**, or just the **Riccati equation**. The above follows by substituting (5.2.3-9) and (5.2.3-11) into (5.2.3-15) and the result into (5.2.3-5).

The solution of the above Riccati equation for a time-invariant system converges to a (finite) **steady-state covariance** if

1. The pair $\{F, H\}$ is *completely observable*.

If, in addition,

2. The pair $\{F, D\}$, where $Q \triangleq DD'$ (D is a square root of Q), is *completely controllable*, then the steady-state covariance is a *unique positive definite matrix*.

The steady-state covariance matrix is the solution of the **algebraic matrix Riccati equation** (or just the **algebraic Riccati equation**

$$P = F[P - PH'(HPH' + R)^{-1}HP]F' + Q \tag{5.2.5-2}$$

and this yields the **steady-state gain** for the Kalman Filter.

The interpretation of the above conditions is:

1. The observability condition on the state guarantees a "steady flow" of information about each state component — this prevents the uncertainty from becoming unbounded. This condition yields the existence of a (not necessarily unique) steady-state solution for the covariance matrix that is positive definite or positive semidefinite (i.e., with finite positive or nonnegative eigenvalues, respectively).

2. The controllability condition states that the process noise enters into each state component and prevents the covariance of the state from converging to zero. This condition causes the covariance to be positive definite (i.e., all the eigenvalues are positive).

Filter Stability

The convergence of the covariance to a *finite* steady state, that is, *the error becoming a stationary process*, is equivalent to **filter stability** (in the bounded input bounded output sense).

Stability of the filter does not require the dynamic system to be stable — only the observability condition (1) is required. As indicated above, observability alone does not guarantee uniqueness — the steady-state solution might depend on the initial covariance — but the existence (finiteness) of the solution is the key.

This is particularly important, since the state models used in tracking are *unstable* — they have an integration (from velocity to position). Stability means "bounded input bounded output" and this condition is not satisfied by an integrator — its continuous time transfer function has a pole at the origin and in discrete time it has a pole at 1.

Remarks

If the state covariance matrix is *positive semidefinite* rather than positive definite, that is, it has some zero eigenvalues that reflect the filter's "belief" that it has "perfectly accurate" estimates of some state components, the gain will be zero for those state components — an *undesirable feature*.

In view of this, in many applications where there is no physical process noise, an **artificial process noise** is assumed (i.e., a matrix Q that will lead to condition (2) to be satisfied).

5.2.6 Properties of the Innovations and the Likelihood Function of the System Model

The Innovations — a Zero-Mean White Sequence

An important property of the **innovation sequence** is that it is an *orthogonal sequence*, that is,

$$\boxed{E[\nu(k)\nu(j)'] = S(k)\delta_{kj}} \qquad (5.2.6\text{-}1)$$

where δ_{kj} is the Kronecker delta function.

This can be seen as follows. Without loss of generality, let $j \leq k-1$. Use will be made of the smoothing property of the conditional expectations (see Subsection 1.4.12)

$$E[\nu(k)\nu(j)'] = E\Big[E[\nu(k)\nu(j)'|Z^{k-1}]\Big] \qquad (5.2.6\text{-}2)$$

Note that $\nu(j)$ is a linear combination of the measurements up to j, that is, given Z^{k-1} it is *not a random variable anymore* and it can thus be taken outside the inner expectation. This yields

$$E[\nu(k)\nu(j)'] = E\Big[E[\nu(k)|Z^{k-1}]\nu(j)'\Big] \qquad (5.2.6\text{-}3)$$

The inside expectation in (5.2.6-3) is, in view of (5.2.2-6),

$$E[z(k) - \hat{z}(k|k-1)|Z^{k-1}] = 0 \qquad (5.2.6\text{-}4)$$

and therefore (5.2.6-1) follows for $k \neq j$.

The uncorrelatedness property (5.2.6-1) of the innovations implies that since they are Gaussian, the innovations are independent of each other and thus the innovation sequence is *strictly white*. Without the Gaussian assumption, the innovation sequence is wide sense white.

Thus the innovation sequence is *zero mean and white*.

Remark

Unlike the innovations, the state estimation errors are not white — they are *correlated in time*. (See also problem 5-11.)

The Likelihood Function of the System Model

The joint pdf of the measurements up to k, denoted as

$$Z^k = \{z(j)\}_{j=1}^k \tag{5.2.6-5}$$

can be written as follows

$$
\begin{aligned}
p[Z^k] &= p[z(k), Z^{k-1}] \\
&= p[z(k)|Z^{k-1}]p[Z^{k-1}] \\
&= \prod_{i=1}^{k} p[z(i)|Z^{i-1}]
\end{aligned}
\tag{5.2.6-6}
$$

where Z^0 is the prior information, shown explicitly only in the expression of (5.2.6-6).

If the above pdfs are Gaussian, then

$$
\begin{aligned}
p[z(i)|Z^{i-1}] &= \mathcal{N}[z(i); \hat{z}(i|i-1), S(i)] \\
&= \mathcal{N}[z(i) - \hat{z}(i|i-1); 0, S(i)] \\
&= \mathcal{N}[\nu(i); 0, S(i)] \\
&= p[\nu(i)]
\end{aligned}
\tag{5.2.6-7}
$$

Using (5.2.6-7) in (5.2.6-6) yields

$$\boxed{p[Z^k] = \prod_{j=1}^{k} p[\nu(i)]} \tag{5.2.6-8}$$

that is, the joint pdf of the sequence of measurements Z^k is equal to the product of the marginal pdfs of the corresponding innovations. This shows the *informational equivalence of the measurements and the innovations.*

Since (5.2.6-8) is the joint pdf of Z^k conditioned on the system model (not indicated explicitly) it is the **likelihood function of the system model**. This will be used in Chapter 11 to evaluate the "goodness" of models in adaptive filtering.

5.2.7   The Innovations Representation

The counterpart of the Riccati equation that yields the recursion of the one-step prediction covariance $P(k+1|k)$ is the recursion of the one-step prediction of the state $\hat{x}(k+1|k)$, called the **innovations representation**.

This is obtained from (5.2.3-2) and (5.2.3-12), without the deterministic input, for simplicity, as

$$\begin{aligned}
\hat{x}(k+1|k) &= F(k)\hat{x}(k|k-1) + F(k)W(k)\nu(k) \\
&= F(k)\hat{x}(k|k-1) + W_i(k)[z(k) - H(k)\hat{x}(k|k-1)] \quad (5.2.7\text{-}1)
\end{aligned}$$

where

$$W_i(k) \triangleq F(k)W(k) \qquad (5.2.7\text{-}2)$$

is the gain in the innovations representation (sometimes called ambiguously the filter gain).

Equation (5.2.7-1) can also be rewritten as the state equation

$$\hat{x}(k+1|k) = [F(k) - W_i(k)H(k)]\hat{x}(k|k-1) + W_i(k)z(k) \qquad (5.2.7\text{-}3)$$

with the input being the (nonwhite) sequence $z(k)$ and the output being the innovation

$$\nu(k) = -H(k)\hat{x}(k|k-1) + z(k) \qquad (5.2.7\text{-}4)$$

This motivates the name innovations representation for the system (5.2.7-3) and (5.2.7-4).

The Kalman Filter as a Whitening System

Note that, while the input to this system is not white, its output is a white sequence. Thus the state estimation filter, written as (5.2.7-3) and (5.2.7-4), can be seen as a *whitening system for the measurement sequence*.

5.2.8 Some Orthogonality Properties

In the static estimation problem the LMMSE estimator was derived based on the principle of orthogonality, which states that the estimation error $\tilde{x}$ has to be orthogonal to the observation(s) z, that is,

$$\tilde{x} \perp z \qquad \Longleftrightarrow \qquad \langle \tilde{x}, z \rangle \overset{\Delta}{=} E[\tilde{x}z] = 0 \qquad (5.2.8\text{-}1)$$

The state estimator for linear dynamic systems — the Kalman filter — while derived under the Gaussian assumption, is (as pointed out earlier) the LMMSE estimator. Therefore, the orthogonality properties carry over.

Note that the estimate is a linear function of the measurements

$$\hat{x}(k|k) = \mathsf{L}_k(Z^k) \qquad \forall k \qquad (5.2.8\text{-}2)$$

or, in a more general manner,

$$\hat{x}(i|k) = \mathsf{L}_i(Z^k) \qquad \forall i, k \qquad (5.2.8\text{-}3)$$

where L denotes a *linear transformation* (because the estimates are linear functions of the measurements) and the measurement set Z^k includes the initial information Z^0.

Thus, the estimation error

$$\tilde{x}(i|k) \overset{\Delta}{=} x(i) - \hat{x}(i|k) \qquad (5.2.8\text{-}4)$$

has the following **orthogonality properties**

$$\tilde{x}(i|k) \perp z(j) \qquad \forall j \leq k \qquad (5.2.8\text{-}5)$$

$$\tilde{x}(i|k) \perp \hat{x}(l|j) \qquad \forall j \leq k, \quad \forall i, l \qquad (5.2.8\text{-}6)$$

With the Gaussian assumption, all the orthogonality properties — which are equivalent to uncorrelatedness — also imply independence.

5.2.9   The Kalman Filter — Summary

The *MMSE state estimation model* for a dynamic system consists of the following:

- *Initial state* — unknown, assumed to be a *random variable* with a certain mean (initial estimate) and covariance (measure of the accuracy of the initial estimate).
- Evolution of the *system's state* — according to a possibly time-varying *linear* difference equation (*plant equation* or *dynamics*) driven by:

  - A known input (the control);
  - An additive random disturbance — the *process noise* — a zero-mean white (uncorrelated) stochastic process with a known, possibly time-varying, covariance.

- *Measurements* — a *linear* function of the state with an additive random disturbance (*measurement noise*), which is a zero-mean white stochastic process with a known, possibly time-varying, covariance.

If, in addition, all the random variables of the problem, that is,

- The initial state,
- The process noises, and
- The measurement noises

are *Gaussian and mutually independent* (i.e., under the *LG assumption*), the MMSE estimate of the state of the system under consideration — the conditional mean of the state given the measurements — is given by the (discrete time) Kalman filter.

The discrete time Kalman filter computes recursively the MMSE estimate of the state of a dynamic system through the following stages:

- Starting from the *current updated state estimate* (estimate of the current state given the observations up to and including the current time) the *predicted value* of the *state* for the next sampling time is computed.
- Using the predicted state, the *predicted value* of the next *measurement* is calculated.
- When the new measurement is obtained, the difference between it and its predicted value — the *innovation* (residual) — is evaluated.
- The *updated state at the next time* is obtained as the sum of the predicted state and the correction term, which is the *filter gain* (obtained separately) multiplying the innovation.

Covariance and filter gain calculation:

- Starting from the current *updated state covariance* the *state prediction covariance* is computed.

- Using the state prediction covariance, the *measurement prediction covariance* (which is the same as the *innovation covariance*) is obtained.

- The *filter gain* is calculated from the state and measurement prediction covariances.

- The *updated covariance* associated with the next state is then computed.

The covariance prediction and update equations combined together result in the *(discrete time) matrix Riccati equation.*

For a *time-invariant system*, if it is *observable*, the state estimation covariance will be finite and the Riccati equation will converge to a steady-state solution. If, in addition, the process noise excites each state component, then the steady-state covariance is also positive definite.

The gain of the filter reflects the relative accuracy of the predicted state versus the new measurement:

- If the new measurement is deemed "more accurate" than the predicted state, then the filter gain will be relatively high;

- If the predicted state is deemed "more accurate" than the new measurement, then the gain will be low.

If the state covariance matrix is positive semidefinite, this reflects the filter's "belief" that it has "perfectly accurate" estimates of some state components — an undesirable feature.

The innovation sequence is *zero mean, white (uncorrelated)*, with covariance equal to the measurement prediction covariance.

The joint pdf of the sequence of measurements, which is the *likelihood function of the model*, is equal to the product of the marginal pdfs of the corresponding innovations, which are Gaussian under the LG assumption.

If the random variables in the state estimation problem for a linear system are *not Gaussian* and one only has their first two moments, then the Kalman filter is the *LMMSE estimator*. In this case the covariances are really the corresponding MSE matrices.

5.3 EXAMPLE OF A FILTER

5.3.1 The Model

Given the system with state

$$x = \begin{bmatrix} \xi \\ \dot{\xi} \end{bmatrix} \tag{5.3.1-1}$$

which evolves according to

$$x(k+1) = \begin{bmatrix} 1 & T \\ 0 & 1 \end{bmatrix} x(k) + \begin{bmatrix} T^2/2 \\ T \end{bmatrix} v(k) \qquad k = 0, 1, \ldots, 99 \tag{5.3.1-2}$$

with initial condition

$$x(0) = \begin{bmatrix} 0 \\ 10 \end{bmatrix} \tag{5.3.1-3}$$

This represents a one-dimensional motion with position ξ and velocity $\dot{\xi}$ sampled at intervals T, which will be assumed as unity in the sequel.

Note that (5.3.1-2) is of the form

$$x(k+1) = Fx(k) + \Gamma v(k) \tag{5.3.1-4}$$

The process noise, a scalar, which models the acceleration, is a zero-mean white sequence with variance

$$E[v(k)^2] = q \tag{5.3.1-5}$$

The measurements consist of position corrupted by additive noise

$$z(k) = [1 \quad 0] x(k) + w(k) \qquad k = 1, \ldots, 100 \tag{5.3.1-6}$$

where the measurement noise is a zero-mean white sequence with variance

$$E[w(k)^2] = r = 1 \tag{5.3.1-7}$$

The two noise sequences are mutually independent.

The filter was initialized according to the "two-point differencing" procedure discussed later in Section 5.5 in (5.5.3-3) to (5.5.3-5).

5.3.2   Results for a Kalman Filter

Figures 5.3.2-1 through 5.3.2-3 present

1. The true and estimated trajectories of the system in the state space (position-velocity);
2. The variances of the predicted position, $P_{11}(k|k-1)$, and updated position, $P_{11}(k|k)$;
3. The variances of the predicted velocity, $P_{22}(k|k-1)$, and updated velocity, $P_{22}(k|k)$;

for three cases with different values of the process noise variance q.

In the first case, with $q = 0$, the controllability condition (2) of Subsection 5.2.5 does not hold and the state estimation covariance matrix is seen from Figures 5.3.2-1b and 5.3.2-1c to converge to zero. In this case the filter gain also converges to zero. The reason this is an undesirable feature is that the filter's "belief" that it has a *perfect state estimate* (which "shuts it off") hinges on the *assumed perfect noiseless constant velocity motion*. In practice such an assumption usually does not hold, except for short periods of time.

The filter for $q = 0$ is equivalent to the LS estimation of the initial position and velocity (straight line fitting), as discussed in Subsection 3.5.1.

For the nonzero values of q, the filter is seen from Figures 5.3.2-2 and 5.3.2-3 to reach steady state quite rapidly, in a few time steps.

In Figure 5.3.2-2, which shows the results on a motion with $q = 1$, the steady-state filter gain is $[0.75 \ \ 0.50]'$. In Figure 5.3.2-3, which corresponds to a strong process noise ($q = 9$) that models a "highly unpredictable" motion, the (optimal) steady-state gain is $[0.90 \ \ 0.94]'$ (i.e., higher). Note how the case with higher gain (due to the higher process noise level) leads to a larger state variance — less accurate state estimates.

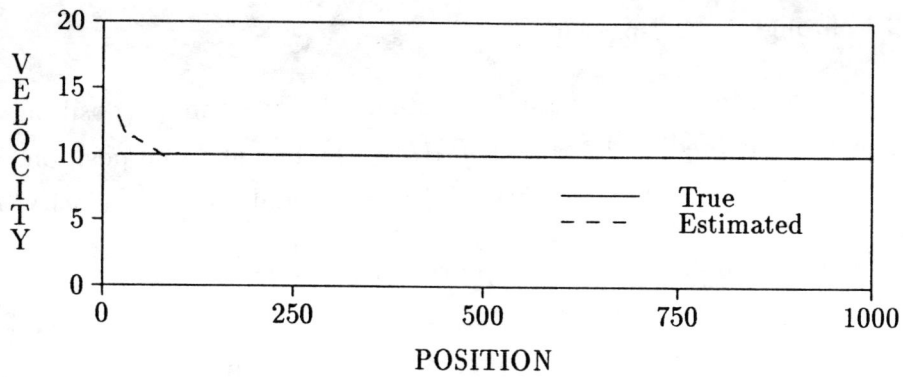

(a) State trajectory

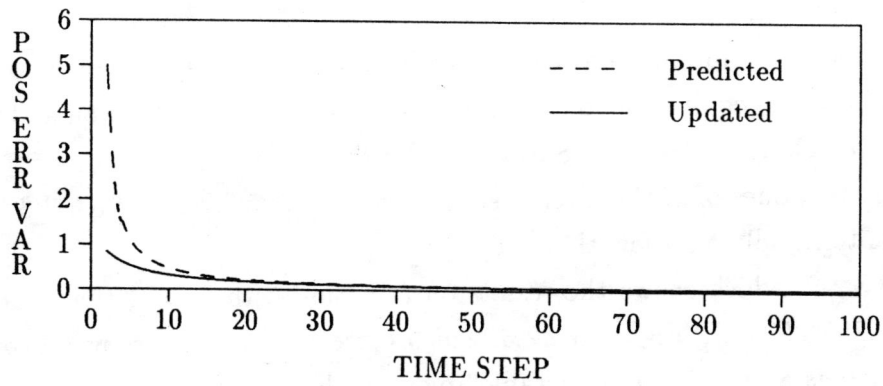

(b) position error variance

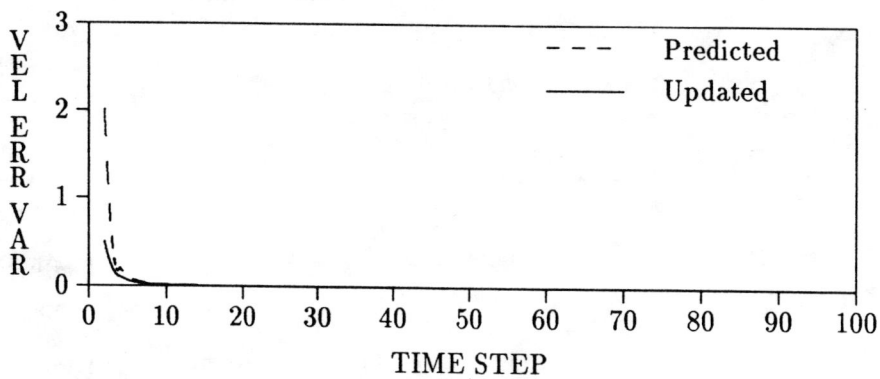

(c) velocity error variance

Figure 5.3.2-1: State trajectory and error variances for $q = 0$.

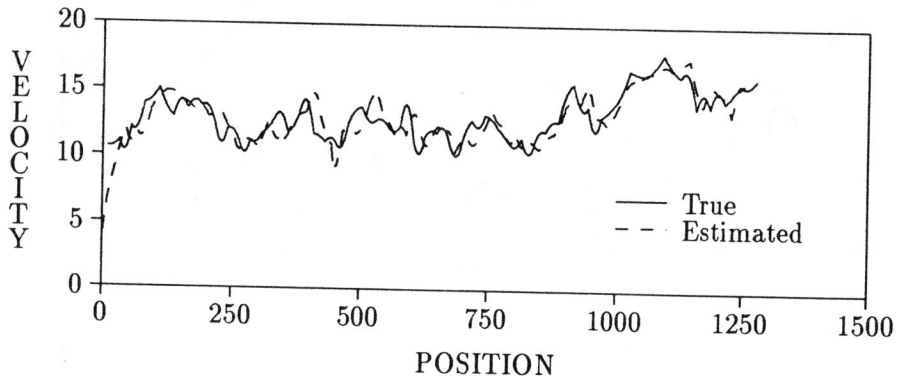

(a) State trajectory

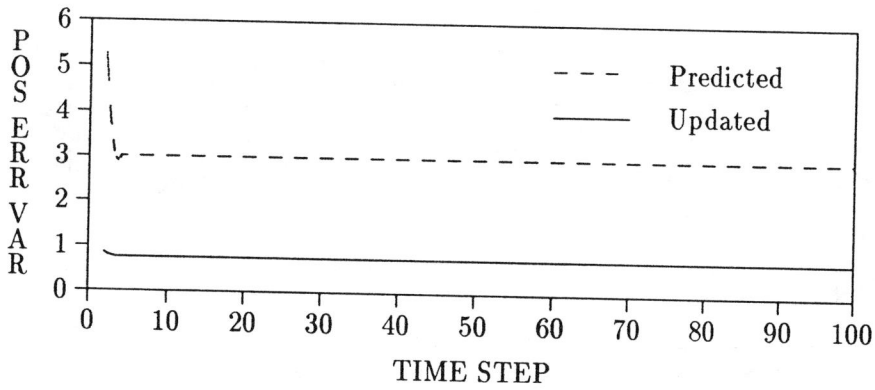

(b) position error variance

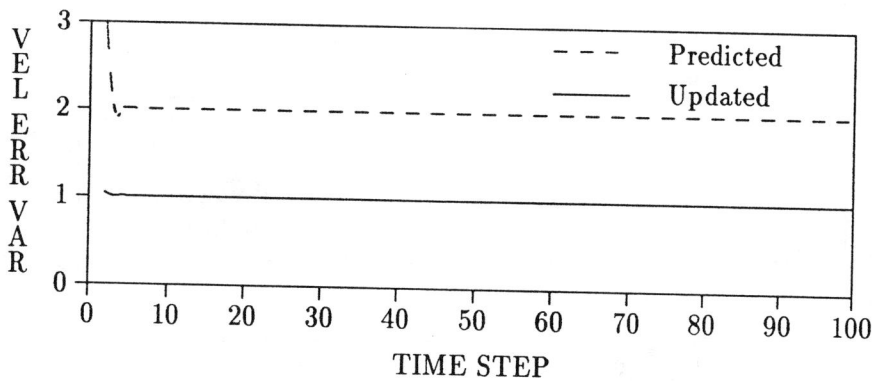

(c) velocity error variance

Figure 5.3.2-2: State trajectory and error variances for $q = 1$.

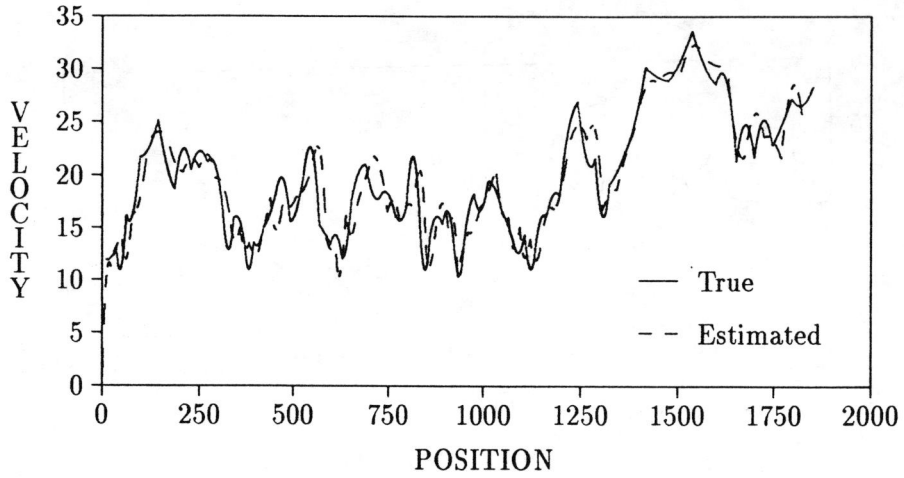

(a) State trajectory

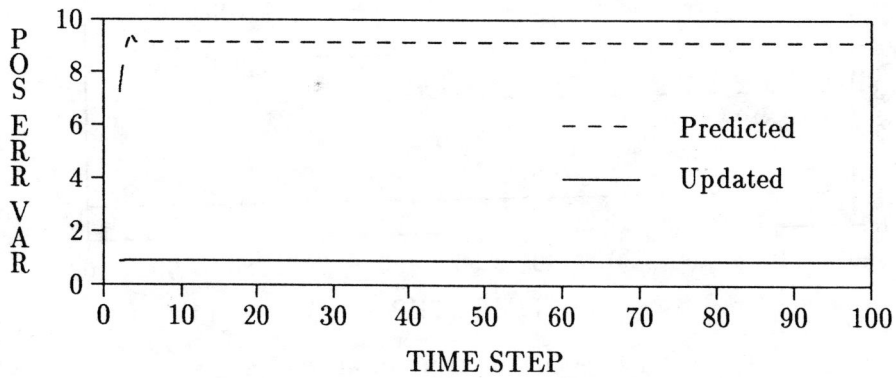

(b) position error variance

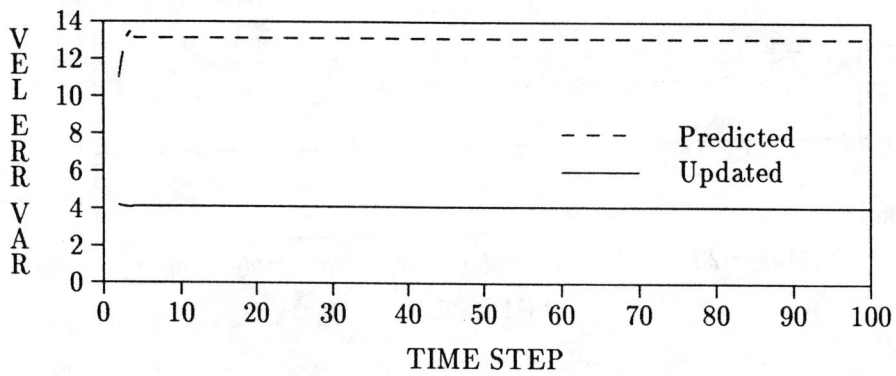

(c) velocity error variance

Figure 5.3.2-3: State trajectory and error variances for $q = 9$.

5.4 CONSISTENCY OF STATE ESTIMATORS

5.4.1 The Problem of Filter Consistency

In the problem of estimating a parameter, which is constant, consistency of an estimator was defined as *convergence of the estimate to the true value.* This implies that there is a steadily increasing amount of information (in the sense of Fisher) about the parameter that asymptotically reduces to zero the uncertainty about its true value.

When estimating the state of a system, in general no convergence of its estimate occurs. What one has, in addition to the "current" estimate of the state, $\hat{x}(k|k)$, is the associated covariance matrix, $P(k|k)$.

The phenomenon of **divergence** has been observed: sometimes the filter yields unacceptably large state estimation errors. [2] These can be due to one (or more) of the following:

- Modeling errors
- Numerical errors
- Programming errors.

The question of what is an **acceptable estimation error** will be discussed in the sequel.

Under the *Linear-Gaussian (LG)* assumption, the conditional pdf of the state $x(k)$ at time k is

$$p[x(k)|Z^k] = \mathcal{N}[x(k); \hat{x}(k|k), P(k|k)] \qquad (5.4.1-1)$$

The modeling of the system consists of the dynamic equation, the measurement equation, and the statistical properties of the random variables entering into these equations. If all these are completely accurate, then (5.4.1-1) holds exactly. Since in practice all models contain some approximations, it is of interest to what extent one can verify (5.4.1-1) in practice.

---

[2]In the sixties, when one of the first Kalman filters performed poorly in an avionics application, it was called "the worst invention of the decade." At that time the design of state estimators was still in the domain of "black magic."

Practical Evaluation of Consistency

The **statistical characterization of the disturbances** is usually done with moments up to second order and the resulting filter will then (hopefully) give approximate first and second order moments of the state.

In view of this, (5.4.1-1) is replaced by the two moment conditions

$$E[x(k) - \hat{x}(k|k)] \triangleq E[\tilde{x}(k|k)] = 0 \qquad (5.4.1\text{-}2)$$

$$
\begin{aligned}
E\Big[[x(k) - \hat{x}(k|k)][x(k) - \hat{x}(k|k)]'\Big] &\triangleq E[\tilde{x}(k|k)\tilde{x}(k|k)'] \\
&= P(k|k) \qquad (5.4.1\text{-}3)
\end{aligned}
$$

that the filter should satisfy in spite of its inherent approximations.

Condition (5.4.1-2) is the **unbiasedness** requirement for the estimates (i.e., zero-mean estimation error), while (5.4.1-3) is the **covariance matching** requirement, i.e., that the **actual MSE** (left hand side) matches the **filter-calculated covariance** (right hand side).

Note that if there is a bias, this will increase the MSE, which is the bias squared plus the variance in the scalar case. Thus the test to be discussed in the next subsection will deal with the MSE. This test will be based on the fact that, under the LG assumption, one has

$$E[\tilde{x}(k|k)' P(k|k)^{-1} \tilde{x}(k|k)] = n_x \qquad (5.4.1\text{-}4)$$

that is, the average of the **squared norm of the estimation error** indicated above, has to be equal to the dimension of the corresponding vector since it is chi-square distributed (see Subsection 1.4.17).

Consistency and Optimality

Since the filter gain is based on the filter-calculated error covariances, it follows that *consistency is necessary for filter optimality*: wrong covariances yield wrong gain.

This is why **consistency evaluation** is vital for verifying a filter design — it amounts to **evaluation of estimator optimality**.

5.4.2 Definition and the Statistical Tests for Filter Consistency

A state estimator (filter) is called **consistent** if its state estimation errors satisfy (5.4.1-2) and (5.4.1-3). This is a **finite-sample consistency** property, that is, the estimation errors based on a finite number of samples (measurements) should be consistent with their theoretical statistical properties:

1. Have mean zero (i.e., the estimates are unbiased).
2. Have covariance matrix as calculated by the filter.

In contra-distinction, the parameter estimator consistency is an asymptotic (infinite size sample) property.

The **consistency criteria of a filter** are:

(a) The state errors should be acceptable as zero mean and have magnitude commensurate with the state covariance as yielded by the filter.
(b) The innovations should also have the same property.
(c) The innovations should be acceptable as white.

The last two criteria are the only ones that can be tested in *real data* applications. The first criterion, which is really the most important, can be tested only in simulations.

Using the notation

$$\tilde{x}(k|k) = x(k) - \hat{x}(k|k) \tag{5.4.2-1}$$

define the **normalized (state) estimation error squared (NEES)**

$$\boxed{\epsilon(k) = \tilde{x}(k|k)'P(k|k)^{-1}\tilde{x}(k|k)} \tag{5.4.2-2}$$

The test to be presented next is based on the above quadratic form and it can verify simultaneously both properties (1) and (2).

Under hypothesis H_0 that the filter is consistent and the LG assumption, $\epsilon(k)$ is chi-square distributed with n_x degrees of freedom, where n_x is the dimension of x. Then

$$E[\epsilon(k)] = n_x \tag{5.4.2-3}$$

and the test is whether (5.4.2-3) can be accepted.

237

Monte Carlo Simulation Based Tests

The test will be based on the results of **Monte Carlo simulations (runs)** that provide N independent samples $\epsilon^i(k)$, $i = 1, \ldots, N$, of the random variable $\epsilon(k)$. Let the sample average of $\epsilon(k)$ — the (N-run) **average NEES** — be

$$\bar{\epsilon}(k) = \frac{1}{N} \sum_{i=1}^{N} \epsilon^i(k) \tag{5.4.2-4}$$

Then $N\bar{\epsilon}(k)$ will have, under H_0, a chi-square density with Nn_x degrees of freedom.

Hypothesis (5.4.2-3), that the *state estimation errors are consistent with the filter-calculated covariances* — criterion (a), also called the **chi-square test** — is accepted if

$$\bar{\epsilon}(k) \in [r_1, r_2] \tag{5.4.2-5}$$

where the **acceptance interval** is determined such that

$$P\left\{\bar{\epsilon}(k) \in [r_1, r_2] | H_0\right\} = 1 - \alpha \tag{5.4.2-6}$$

For example, with $\alpha = 0.05$, $n_x = 2$, and $N = 50$, one has from (1.5.4-6), for a two-sided interval, $r_1 = 1.5$ and $r_2 = 2.6$. The interval given in (5.4.2-5) is then the (two-sided) 95% **probability concentration region** for $\bar{\epsilon}(k)$.

If $N = 1$ (i.e., a single run), one can also use this test. The two-sided 95% interval is in this case $[0.05, 7.38]$. Note the much narrower range of the interval corresponding to $N = 50$ Monte Carlo runs — this illustrates the **variability reduction** in such repeated simulations.

Note that a bias in the state estimation error will increase (5.4.2-2) and, if significant, it will yield unacceptably large values for the statistic (5.4.2-4).

If (5.4.2-5) is not satisfied, then a separate bias test using the sample mean of (5.4.2-1) should be carried out to identify the source of the problem. This can be done by taking each component of the state error, divided by its standard deviation, which makes it (under ideal conditions) $\mathcal{N}(0, 1)$, and testing to see if its mean can be accepted as zero.

The commensurateness of the innovations with their filter-calculated covariances — criterion (b) — is tested in a similar manner.

Under the hypothesis that the filter is consistent, the **normalized innovation squared (NIS)**

$$\epsilon_\nu(k) = \nu(k)'S(k)^{-1}\nu(k)$$

(5.4.2-7)

has a chi-square distribution with n_z degrees of freedom, where n_z is the dimension of the measurement.

From N independent samples $\epsilon_\nu^i(k)$ one calculates the (N-run) **average NIS**

$$\bar{\epsilon}_\nu(k) = \frac{1}{N}\sum_{i=1}^{N}\epsilon_\nu^i(k)$$

(5.4.2-8)

which is then tested as in (5.4.2-5) but with acceptance region determined based on the fact that $N\bar{\epsilon}_\nu(k)$ is chi-square distributed with Nn_z degrees of freedom.

Similarly to the procedure for state errors, if (5.4.2-8) is too large, then a bias test (i.e., whether the mean of the innovations is nonzero) has to be carried out.

The **whiteness test** for the innovations can be done as follows.

The following (N-run) **sample autocorrelation** statistic is used

$$\bar{\rho}(k,j) = \sum_{i=1}^{N}\nu^i(k)'\nu^i(j)\left[\sum_{i=1}^{N}\nu^i(k)'\nu^i(k)\sum_{i=1}^{N}\nu^i(j)'\nu^i(j)\right]^{-1/2}$$

(5.4.2-9)

For N large enough, a normal approximation of the density of (5.4.2-9) for $k \neq j$ is convenient (and reasonable in view of the central limit theorem). If the innovations are zero mean and white, then the mean of (5.4.2-9) is zero and its variance is $1/N$. (See problem 4-4.)

Denoting by ξ a zero-mean unity-variance normal random variable, let r_1 be such that

$$P\{\xi \in [-r_1, r_1]\} = 1 - \alpha$$

(5.4.2-10)

where, say, $\alpha = 0.05$. Then, since the standard deviation of $\bar{\rho}$ is $1/\sqrt{N}$, the corresponding $1 - \alpha$ probability region for $\bar{\rho}$ will be $[-r, r]$ where $r = r_1/\sqrt{N}$. For the above value of α one has the two-sided 95% region given by $r_1 = 1.96$.

Thus, using this acceptance region based on the normal density, the hypothesis that the true correlation of the innovation sequence is zero — criterion (c) — is accepted if

$$\bar{\rho}(k,j) \in [-r, r]$$

(5.4.2-11)

Real Time (Single-Run) Tests

All the above tests assume that N **independent runs** have been made. While they can be used on a single run ($N = 1$), they have a very high variability in this case, as illustrated above. The question is whether one can achieve a low variability of the test statistic based on a single run, as in a real time implementation, that is, having **real time consistency tests**.

Test (a) requires that such independent simulations be made, however, criteria (b) and (c) can be tested on a **single run** in time as follows.

These tests are based on replacing the **ensemble averages** by **time averages** based on the *ergodicity* of the innovation sequence.

The whiteness test statistic for innovations l steps apart from a single run can be written as the **time-average autocorrelation**

$$\bar{\rho}(l) = \sum_{k=1}^{K} \nu(k)'\nu(k+l) \left[\sum_{k=1}^{K} \nu(k)'\nu(k) \sum_{k=1}^{K} \nu(k+l)'\nu(k+l) \right]^{-1/2} \qquad (5.4.2\text{-}12)$$

This statistic is, for large enough K, in view of the central limit theorem, normally distributed. Furthermore, its variance can be shown to be $1/K$ (see problem 4-4).

Criterion (b) can be tested with the **time-average normalized innovation squared statistic**

$$\bar{\epsilon}_\nu = \frac{1}{K} \sum_{k=1}^{K} \nu(k)'S(k)^{-1}\nu(k) \qquad (5.4.2\text{-}13)$$

If the innovations are white, zero mean, and with covariance $S(k)$, then $K\bar{\epsilon}_\nu$ has a chi-square distribution with Kn_z degrees of freedom.

The probability regions for (5.4.2-12) and (5.4.2-13) for acceptance of the "consistent filter" hypothesis are then set up as before.

5.4.3 Examples of Filter Consistency Testing

The example of Section 5.3 is continued to illustrate the use of the **consistency tests**. The following tests are carried out:

- Offline single-run (simulation) tests;
- Offline multiple run (Monte Carlo simulation) tests;
- Online single-run (real time) tests.

Two cases will be considered:

1. A filter which is based on exactly the same model as the process, that is, **matched** to the system;
2. A filter which is based on a different model than the system, that is, **mismatched** to the system.

Single-Run Simulation Tests

Figure 5.4.3-1 shows, for a *single run*, the behavior of the state's **NEES** (5.4.2-2) for various values of the process noise variance q for filters that are perfectly matched. Out of 100 points, 3 to 6 are found outside the 95% probability region.

In this case a one-sided region was considered. The upper limit of this probability region is approximately 6 since, for a $n_x = 2$ degrees of freedom chi-square random variable, the 5% tail point is

$$\chi_2^2(0.95) = 5.99 \tag{5.4.3-1}$$

Note that in this case the two-sided 95% region is $[0.05, 7.38]$. Since the lower limit is practically zero, only the upper limit is of interest and it was taken for the 5% tail rather than for the 2.5% tail, which is 7.38. It should be noted that taking a 5% or a 2.5% (or a 1%) tail is rather arbitrary.

241

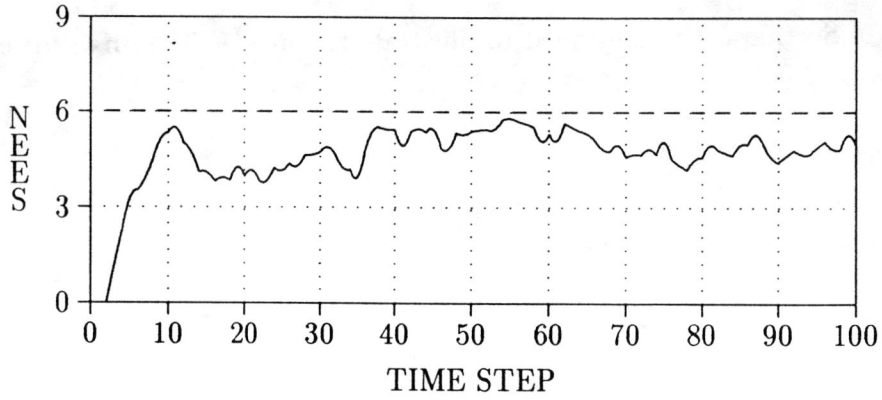

(a) $q = 0$

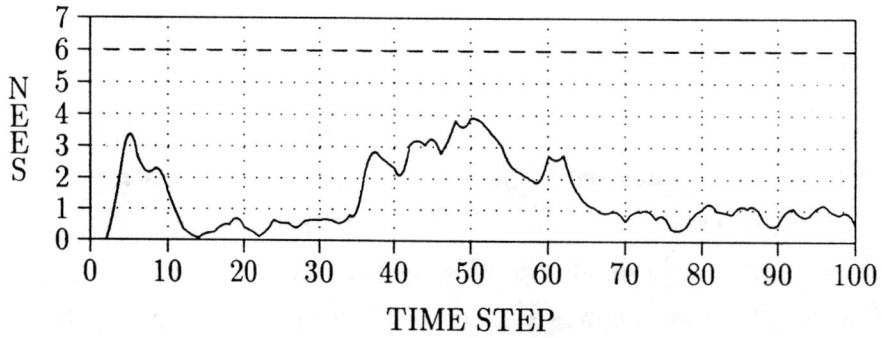

(b) $q = 1$

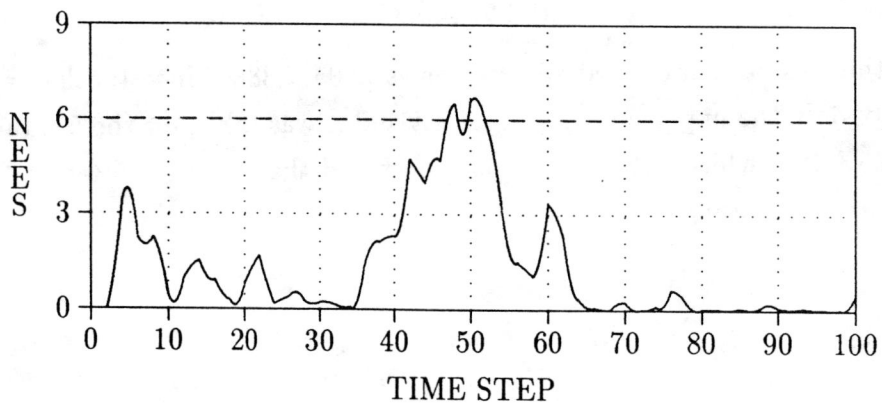

(c) $q = 9$

Figure 5.4.3-1: Normalized state estimation error squared from a single run with its 95% probability region.

Monte Carlo Simulation Tests

Figures 5.4.3-2(a) to (c) illustrate the test statistics obtained from $N = 50$ Monte Carlo runs. Two-sided probability regions are used in the sequel.

Figure 5.4.3-2a shows the state's N-run **average NEES** (5.4.2-4). The two-sided 95% region for a 100 degrees of freedom chi-square random variable is

$$[\chi^2_{100}(0.025), \chi^2_{100}(0.975)] = [74.2, 129.6] \tag{5.4.3-2}$$

Dividing the above by $N = 50$, the 95% probability region (5.4.2-5) for the average normalized state estimation error squared becomes $[1.5, 2.6]$. Note that 6 out of the 100 points fall outside this 95% region, which is acceptable.

Figure 5.4.3-2b shows the (N-run) **average NIS** (5.4.2-8). Noting that the innovations are scalar, the 95% probability region will be based on the 50 degrees of freedom chi-square distribution, and is

$$[\chi^2_{50}(0.025), \chi^2_{50}(0.975)] = [32.3, 71.4] \tag{5.4.3-3}$$

Dividing by $N = 50$, the region becomes $[0.65, 1.43]$. As the plot shows, 4 out of the 100 points are outside the 95% region, again an acceptable situation.

Figure 5.4.3-2c shows the (N-run) **sample autocorrelation** of the innovations (5.4.2-9) one step apart ($k - j = 1$). The 95% region $[-1.96\sigma, 1.96\sigma]$ is, for $\sigma = \frac{1}{\sqrt{N}} = 0.141$, the interval $[-0.277, 0.277]$. In this case 8 out of the 100 points fall outside the 95% region, which is also acceptable.

Remark

These probability regions are also called **acceptance regions** because, if the test statistics fall in these regions, then one can *accept* the hypothesis that the *filter is consistent*.

A Mismatched Filter

Next, a **mismatched filter** is examined. For this purpose it is assumed that the true (i.e., system) process noise has variance $q = 9$ while the model (i.e., filter) process noise has variance $q_F = 1$.

The normalized state estimation error squared is shown in Figure 5.4.3-3. The mismatch caused 41 points out of 100 to be outside the 95% probability region in a single run, clearly an unacceptable situation. The results of the Monte Carlo runs ($N = 50$) shown in Figures 5.4.3-3b and 5.4.3-4 all show the serious mismatch — all points are outside the 95% region — actually they are all in the upper 2.5% tail region.

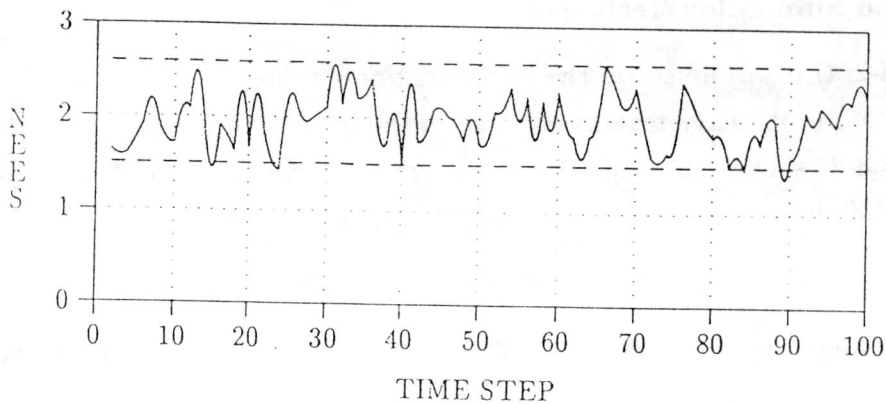

(a)

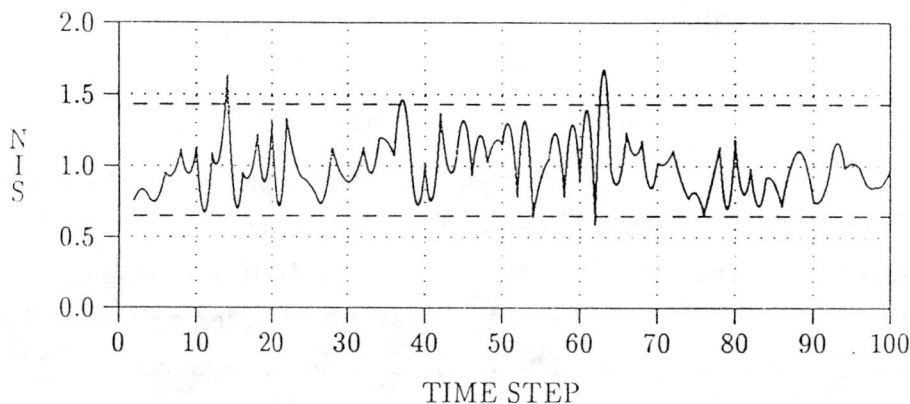

(b)

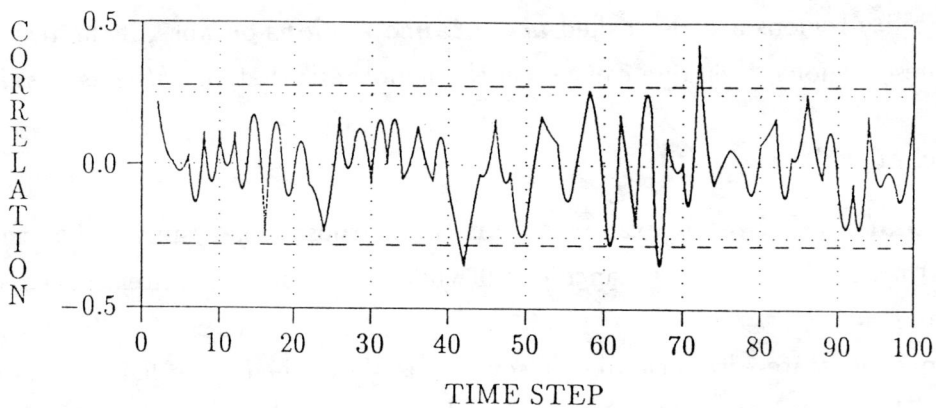

(c)

Figure 5.4.3-2: Normalized state estimation error squared (a), normalized innovation squared (b), and innovation autocorrelation (c), from 50 Monte Carlo runs with their 95% probability regions for $q = 1$.   244

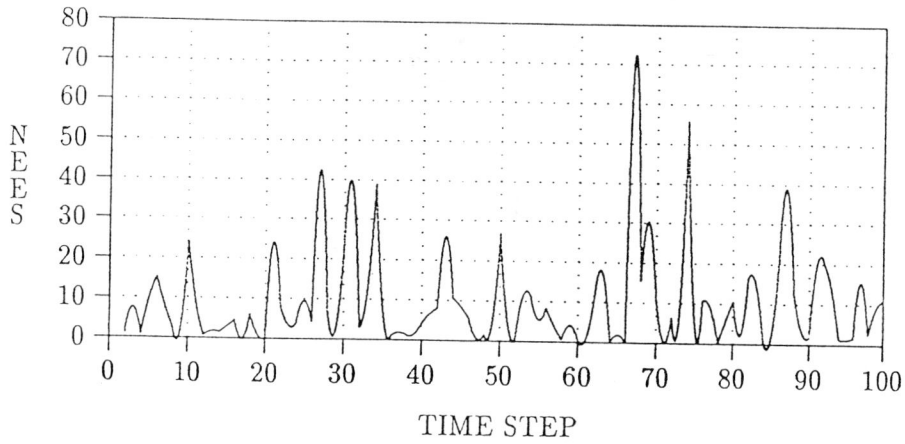

(a)

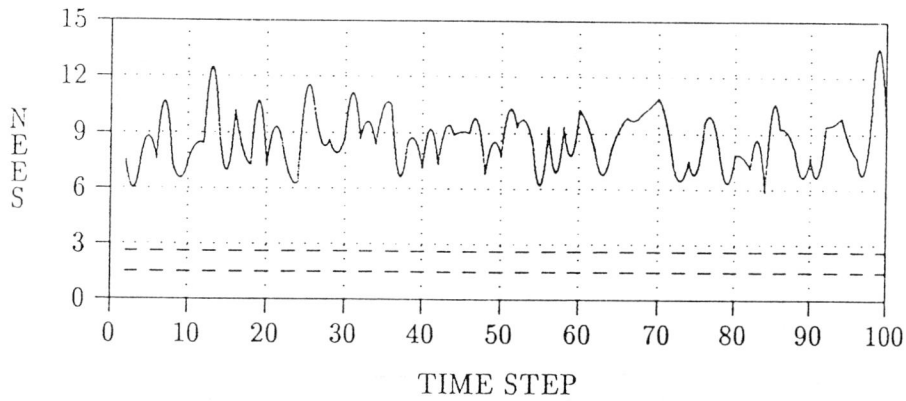

(b)

Figure 5.4.3-3: Mismatched filter: normalized state estimation error squared from a single run (a), and 50-run Monte Carlo average with its 95% probability region (b).

(a)

(b)

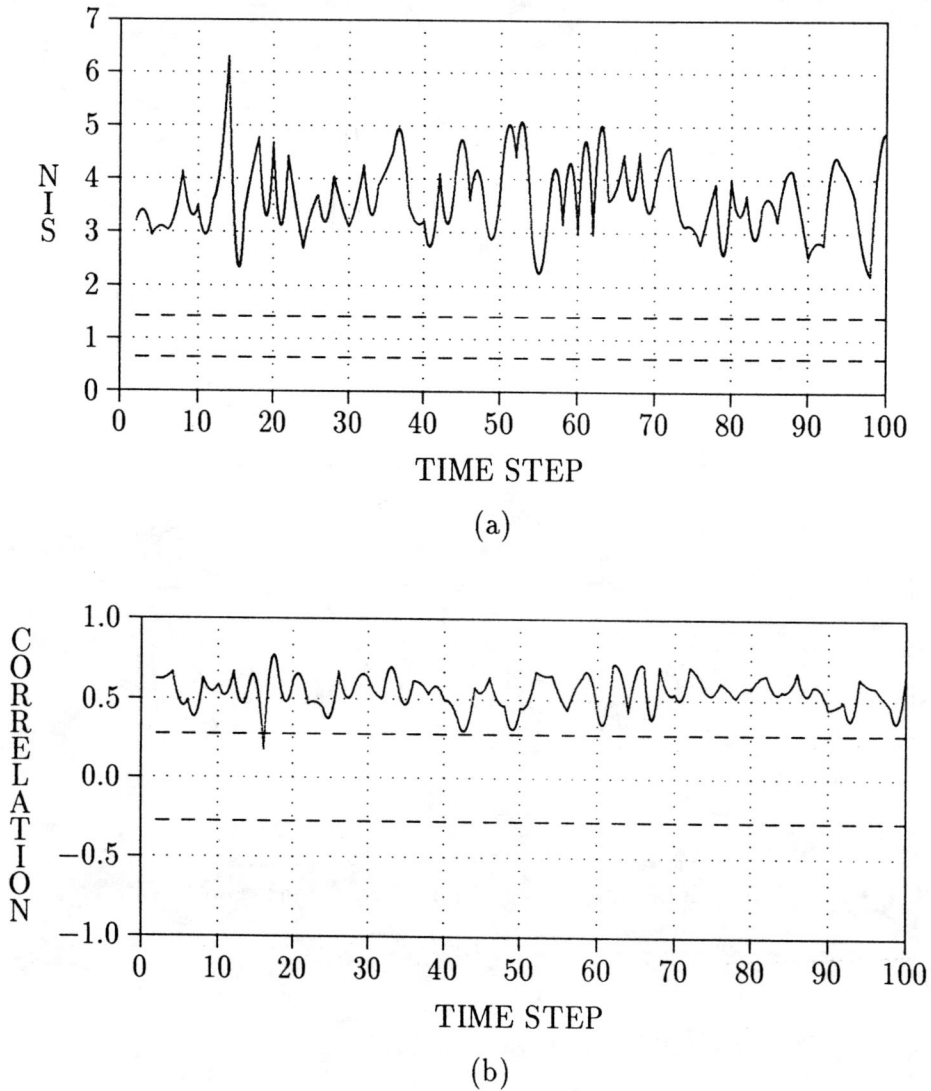

Figure 5.4.3-4: Mismatched filter: normalized innovation squared (a), and innovation auto-correlation (b), from 50 Monte Carlo runs with their 95% confidence regions.

Real Time Tests

Finally, the **real time consistency tests**, that is, the *single-run tests that can be performed in real time* are presented. For the correct filter the **time-average autocorrelation** of the innovations (5.4.2-12) obtained from 100 samples in time was

$$\bar{\rho}(1) = 0.152 \tag{5.4.3-4}$$

Under the assumption that the filter is correct, the error in the above estimate is normally distributed with mean zero and variance $1/100$

$$\bar{\rho} - \rho \sim \mathcal{N}(0, 0.1^2) \tag{5.4.3-5}$$

and its 95% probability region is $[-0.196, 0.196]$. The estimate (5.4.3-4) falls in this region, as expected, since the filter is matched to the system.

In the mismatched case the estimate was obtained as

$$\bar{\rho}(1) = 0.509 \tag{5.4.3-6}$$

which is clearly much too large and outside the region.

The other single-run test is for the **time-average normalized innovation squared** (5.4.2-13), also over 100 time steps.

For the matched filter, the result was

$$\bar{\epsilon}_\nu = 0.936 \tag{5.4.3-7}$$

(the ideal value is 1). The 95% confidence region is, based on the 100 degrees of freedom chi-square distribution, the interval $[0.74, 1.3]$.

For the mismatched filter, the result was

$$\bar{\epsilon}_\nu = 2.66 \tag{5.4.3-8}$$

which is clearly unacceptable.

The above illustrates how one can detect filter mismatch in a single run using suitable statistics and their confidence regions.

Filter Tuning

The process noise is used in practice to model disturbances, e.g., unknown inputs like target maneuvers in tracking. The procedure to match the process noise variance to suitably model such disturbances is called **filter tuning**. Since such inputs are not real noise in the probabilistic sense, it is said that the filter uses **pseudo-noise** or **artificial noise**.

The procedure for tuning is to make the filter consistent, that is, the three criteria (a)–(c) from Subsection 5.4.2 should be satisfied. While this is easier said than done, in practice one has to strive to make the filter as close to being consistent as feasible while at the same time achieving small RMS estimation errors. Note that the RMS errors are **unnormalized errors**, while the test statistics are normalized, that is, divided (in a matrix sense) by the filter-calculated variances.

This is discussed in more detail in Subsection 11.6.7, where the design of several filters for a realistic situation is illustrated.

5.4.4 Filter Consistency — Summary

Consistency testing is crucial for *estimator optimality evaluation.*

A state estimator is *consistent* if the first and second order moments of its estimation errors are as the theory predicts:

- Their means are zero — the estimates are unbiased.
- Their covariance matrices are as calculated by the filter.

The statistic that tests the mean and the covariance is the *normalized estimation error squared (NEES).* This is done for

- the state, and
- the innovations (measurement prediction errors).

Under the Gaussian assumption, these statistics are chi-square distributed and should be with a high probability in certain intervals — the corresponding probability regions. These tests are also called "chi-square" tests.

Bias in the estimates, if significant, or error magnitudes too large compared to the filter-calculated standard deviations will be detected by this statistic.

This test can be used

- Offline (in simulations) for state estimation errors — the truth is available for comparison.
- Offline or online (in real time) for the innovation.

The test for the *innovation's whiteness* is the *sample autocorrelation* whose magnitude has to be below a certain threshold. This can be carried out offline or online.

These tests become very powerful, that is, they can detect *inconsistent (mismatched)* filters when used in Monte-Carlo runs.

Monte Carlo runs evaluate *ensemble averages* of the test statistics. The averages obtained from *independent runs* — with *all the random variables independent from run to run* — decrease the variability and therefore increase the power of the tests.

Acceptance regions — *upper and lower bounds* — within which the test statistics should be for consistent filters, have been established.

If the normalized error statistic exceeds the upper bound then

- there is *significant bias* in the estimates, or
- the *errors are too large* compared to the filter-calculated covariance, or
- the *covariance is too small.*

In this case the filter is **optimistic**.

If the normalized error statistic is below the lower bound, then the *covariance is too large.* In this case the filter can be said to be **pessimistic**.

For *real time filter performance monitoring*, one can implement tests based on *time averaging* for innovation magnitude and whiteness.

5.5 INITIALIZATION OF STATE ESTIMATORS

5.5.1 Initialization and Consistency

A state estimation filter is called consistent if its estimation errors are "commensurate" or "compatible" with the filter-calculated covariances. At **initialization**, it is just as important that the covariance associated with the initial estimate reflects realistically its accuracy.

According to the Bayesian model, the *true initial state* is *a random variable*, assumed to be normally distributed with a known mean — the initial estimate — and a given covariance matrix

$$x(0) \sim \mathcal{N}[\hat{x}(0|0), P(0|0)] \qquad (5.5.1\text{-}1)$$

The norm ("chi-square") test for the initial estimation error is

$$\tilde{x}(0|0)' P(0|0)^{-1} \tilde{x}(0|0) \leq c_1 \qquad (5.5.1\text{-}2)$$

where c_1 is the upper limit of the, say, 95% confidence region from the chi-square distribution with the corresponding number of degrees of freedom.

Sometimes one has to choose the initial covariance — the "choice" has to be such that (5.5.1-2) is satisfied. A large error in the initial estimate, if the latter is deemed highly accurate, will persist a long time because it leads to a low filter gain and thus the new information from the measurements receives weighting that is too low. (See problem 5-6.)

For the one-dimensional case, (5.5.1-2) can be stated as follows: The initial error should be not more than, say, two times the associated standard deviation — this is called a "2σ" error. If the initial variance is such that the initial error is "1σ" and the subsequent measurements are accurate, then the initial error will decrease rapidly.

5.5.2 Initialization in Simulations

As discussed before, the Bayesian model for initialization is

- the initial state is a random variable; and
- the (prior) pdf of the initial state is known and assumed Gaussian.

Thus, in simulations one should generate the initial state with a random number generator according to (5.5.1-1). The initial estimate is, again according to the Bayesian model, a *known quantity* together with the initial covariance matrix.

This approach, while rigorous according to the assumptions of the filter, is not so appealing because the true initial state, which defines the **scenario**, will be different in each simulation.

A more appealing approach is the following:

- choose the initial true state; and
- generate the initial estimate according to

$$\hat{x}(0|0) \sim \mathcal{N}[x(0), P(0|0)] \tag{5.5.2-1}$$

The above relies on the fact that, switching the roles of the initial state $x(0)$ and the initial estimate $\hat{x}(0|0)$, one has the algebraic identity

$$
\begin{aligned}
p[x(0)|\hat{x}(0|0)] &= \mathcal{N}[x(0); \hat{x}(0|0), P(0|0)] \\
&= \mathcal{N}[\hat{x}(0|0); x(0), P(0|0)] \\
&= p[\hat{x}(0|0)|x(0)]
\end{aligned}
\tag{5.5.2-2}
$$

in view of the special form of the normal distribution. In other words, the *scenario* $x(0)$ *is fixed* and the *initial condition* $\hat{x}(0|0)$ *of the filter is random.*

5.5.3 A Practical Implementation in Tracking

The practical implementation of the initialization (5.5.2-1) can be done as follows. Consider two state components, say, position ξ and velocity $\dot{\xi}$ in a given coordinate. If only position measurements

$$z(k) = \xi(k) + w(k) \tag{5.5.3-1}$$

are available, then for the true values $\xi(k)$, $k = -1, 0$, one generates the corresponding measurement noises, say

$$w(k) \sim \mathcal{N}[0, R] \tag{5.5.3-2}$$

Then, denoting by T the sampling interval, one has

$$\hat{\xi}(0|0) = z(0) \tag{5.5.3-3}$$

$$\hat{\dot{\xi}}(0|0) = \frac{z(0) - z(-1)}{T} \tag{5.5.3-4}$$

and the corresponding 2×2 block of the initial covariance matrix is then

$$P(0|0) = \begin{bmatrix} R & R/T \\ R/T & 2R/T^2 \end{bmatrix} \tag{5.5.3-5}$$

This method, called **two-point differencing**, guarantees consistency of the initialization of the filter, which starts updating the state at $k = 1$.

If several (Monte Carlo) runs are made, then the same initialization procedure has to be followed with new (*independent*) noises generated in every run according to (5.5.3-2). "Reuse" of the same initial conditions in Monte Carlo runs will lead to biased estimates (see problem 5-6).

Remark

The above amounts to a **first order polynomial fitting** of the first two measurements. If (and only if) there are significant higher derivatives, then one should use more than two points and a **higher order polynomial fitting** via LS is to be carried out.

5.5.4 Filter Initialization — Summary

The error in the initial state estimate has to be *consistent* with the initial state covariance. When an initial covariance is "chosen," it should be such that the error is at most 2 times the corresponding standard deviation.

According to the Bayesian assumptions in the Kalman filter

- *the true initial state is a random variable,* and
- *the initial estimate is a fixed known quantity.*

In simulations one can reverse this point of view by fixing the initial state (the scenario) and generating the initial estimate with a random number generator as follows:

The initial estimate is generated with mean equal to the true initial state and with covariance equal to the initial state covariance.

In practice the initialization can be done from two consecutive position measurements by

1. using the latest measurement as initial position estimate,
2. differencing them to obtain the velocity estimate,
3. calculating the corresponding covariance matrix.

This amounts to a first order polynomial fit. Higher order polynomial fits can also be used for initializing the estimation of states that have higher derivatives.

This initialization method should also be followed in Monte Carlo simulations where, with *noises independent from run to run*, one will then obtain initial errors also independent from run to run.

5.6 NOTES AND PROBLEMS

5.6.1 Bibliographical Notes

The Kalman filter originated with the work of Kalman and Bucy [Kal60, KB61]. The idea of recursive estimation appeared earlier in the work of Swerling [Swe59]. The topic of the Kalman filtering is covered in many texts, for instance, [Bal84, Med69, SM71, Gel74, AM79, Kai81, May79, May82, Lew86]. The proof of the sequential estimation for dynamic systems as a direct extension of the static case presented in Section 5.2 is simpler than the many different proofs in the literature. Issues of observability and controllability and their implications on the Kalman filter are discussed in detail in [AM79].

The concept of consistency of a state estimator (Section 5.4) has been mentioned in the literature under the name of "covariance matching." The discussion of Section 5.4 is based on [BB83]. The lack of consistency of a Kalman filter has been called "divergence" (that is how severe it was in some cases) and has been the subject of extensive investigations [Sor85]. Sensitivity analysis of mismatched filters and the use of reduced order models are discussed in detail in [Gel74, Lew86].

The technique of initialization of filters presented in Section 5.5 has been known for many years, but overlooked in many instances.

The models considered here had all known "parameters" — the matrices F, G, H, Q, and R. Techniques for the "identification" of these system parameters can be found in [Lju87].

Treatment of a constant bias in recursive filtering using state augmentation is discussed in [Fri69].

5.6.2 Problems

5-1 **Filter implementation and consistency checking.** Consider the scalar system

$$x(k+1) = fx(k) + u(k) + v(k)$$

where $u(k) = 1$ and $v(k) \sim \mathcal{N}(0, q)$ and white, with measurement

$$z(k) = hx(k) + w(k)$$

where $w(k) \sim \mathcal{N}(0, r)$ and white. The initial condition for the system is $x(0)$.

1. Find the expression of the steady-state variance

$$P_\infty = \lim_{k \to \infty} P(k|k)$$

of the estimated state $\hat{x}(k|k)$.

2. With the parameters of the problem $f = h = 1$, $q = 0.01$, $r = 1$, simulate a trajectory using a random number generator starting from $x(0) = 0$ for $k = 1, \ldots, 50$.

3. Let the initial estimate of the state be $\hat{x}(0|0) = z(0)/h$. Determine the corresponding $P(0|0)$ as a function of the parameters of the problem (h, r).

4. For the values given in (2), estimate the state up to $k = 50$, starting as in (3). List the following:

$$k, x(k), v(k), w(k), z(k), \hat{x}(k|k-1), P(k|k-1), \hat{z}(k|k-1), S(k),$$
$$\nu(k), \nu(k)/\sqrt{S(k)}, W(k), \hat{x}(k|k), P(k|k), \tilde{x}(k|k)/\sqrt{P(k|k)}$$

5. Compare the values of $P(k|k)$ from (4) to the result of (1).

6. List $P(k|k)$ for $k = 0, 1, \ldots, 50$ for the following values of the initial variance: $P(0|0) = 0, 1, 10$.

7. One desires to run the filter as in (4) with the various values of $P(0|0)$ as in (6). What should be changed in the filter simulation?

5-2 **Random number generator testing.** Describe the tests for a "correct" $\mathcal{N}(0,1)$ random number generator from which we have n numbers. Indicate the distributions and confidence regions for the

1. Sample mean.

2. Sample variance.

3. Sample correlation.

5-3 **Covariance of the state versus covariance of the estimation error.** Prove that, with $\tilde{x} \overset{\Delta}{=} x - E[x|z]$, one has $\text{cov}[x|z] = \text{cov}[\tilde{x}|z]$.

5-4 **Asymptotic distribution of the sample correlation.** Show that the sample correlation (5.4.2-9) tends to $\mathcal{N}(0, 1/N)$. Hint: use the law of large numbers and the central limit theorem.

5-5 **MSE of the state for an update with an arbitrary gain.**

1. Prove that the Joseph form covariance update (5.2.3-18) holds for *arbitrary* gain at time $k+1$. Hint: write the propagation equation of the error from $\tilde{x}(k|k)$ to $\tilde{x}(k+1|k+1)$.

2. State the condition for its stability.

3. Check this stability condition on the results of problem 5-1.

5-6 **Initialization of a filter (how NOT to do it) — a real story.** A tracking filter was initialized in a set of Monte Carlo runs with the target's initial range of 80kft, initial range estimate of 100kft and initial range variance of 10^6ft^2.

1. Characterize the assumed quality of the initial estimate.

2. How will the average estimation error over the Monte Carlo runs behave?

3. You are a "young Turk" engineer who wants to prove mastery of estimation: suggest a simple fix to the above initialization procedure that involves changing only the initial variance.

5-7 Orthogonality of the innovations to the state prediction. Show that

$$\nu(k) \perp \hat{x}(j|k-1) \qquad \forall j > k - 1$$

5-8 Orthogonality of estimation error to previous estimates. Show that

$$\tilde{x}(i|k) \perp \hat{x}(i|j) \qquad \forall j \leq k$$

5-9 Alternative derivation of the Kalman filter gain. Show that the minimization of the trace of (5.2.3-18) with respect to the filter gain yields (5.2.3-11). Hint: use the formulas from problem 1-10.

5-10 State estimation errors' autocorrelation. Prove that the state estimation errors are not white:

$$E[\tilde{x}(k+1|k+1)\tilde{x}(k|k)'] = [I - W(k+1)H(k+1)]F(k)P(k|k)$$

5-11 Kalman filter with nonzero noise means. Derive the Kalman filter equations for the formulation from Subsection 5.2.1 with the following modifications:

$$E[v(k)] = \bar{v}(k) \qquad\qquad E[w(k)] = \bar{w}(k)$$

where the nonzero noise means are known. Their known covariances are Q and R. All the remaining assumptions are the same. Provide the derivations only for the equations that will be different than those in Subsection 5.2.4. Indicate which equations are not modified and why.

5-12 Bias in the measurements. Consider the problem from Section 5.3 with the modification that the measurement noise has an *unknown mean* $\bar{w}$ (the sensor bias). Append this to the state as an extra component assuming it to be constant in time.

1. Indicate the new state space representation with the augmented state and *zero-mean noises* (specify the matrices F, Γ, and H).

2. What happens if we run a Kalman filter on this representation? Can one estimate this sensor bias? Justify mathematically your answer.

Chapter 6

ESTIMATION FOR KINEMATIC MODELS

6.1 INTRODUCTION

6.1.1 Outline

This chapter discusses a class of widely used models derived from simple equations of motion — constant velocity and constant acceleration. These models are more general in the sense that the corresponding (second and third order) derivatives of the position are not zero, but a zero-mean random process.

Section 6.2 presents the discrete time kinematic model obtained by discretizing the continuous time state space representation driven by white noise. The state model defined directly in discrete time using a piecewise constant white random sequence as process noise is presented in Section 6.3.

Section 6.4 presents explicit filters for noiseless kinematic models.

For noisy kinematic models explicit steady-state filters are derived in Section 6.5. These filters corresponding to second and third order models are known as the α-β and α-β-γ filters and their gains are expressed in terms of the **target maneuvering index** — the ratio of the motion and the observation uncertainties. Since the statistical characterization of the process noise is a key **filter design parameter**, this is discussed in detail.

The models are presented for a single coordinate. For motion in several coordinates, it is customary to use such models assumed independent across coordinates — this leads to "decoupled" filtering. Finally, some intuitive insight into filter design is presented.

6.1.2 Kinematic Models — Summary of Objectives

Define the following kinematic models

- White noise acceleration (second order model)
- Wiener process acceleration (third order model).

Derive discrete time kinematic models by

- Discretizing the continuous time state space representation driven by white noise
- Directly defining the state model in discrete time using a piecewise-constant white random sequence as process noise.

Present

- Explicit filters for noiseless kinematic models
- Explicit steady-state filters for noisy kinematic models
  - α-β
  - α-β-γ

with their gains expressed in terms of the target maneuvering index.

Discuss filter design and "noise reduction."

6.2 DISCRETIZED CONTINUOUS TIME KINEMATIC MODELS

6.2.1 Introduction

Kinematic state models are defined by setting a certain derivative of the position to zero. In the absence of any random inputs they yield motion characterized by a polynomial in time. Such models are also called **polynomial models** and the corresponding state estimation filters are sometimes referred to as **polynomial filters**.

Since it is not realistic to assume that there are no disturbances, one can model them as random inputs. One way of modeling this is via a *continuous time white process noise*.

Since the tracking is done in discrete time, the corresponding discrete time state equations are needed. Subsection 6.2.2 presents the **white noise acceleration** state model, which is two-dimensional per coordinate. The **Wiener process acceleration** state model, which is three-dimensional per coordinate, is presented in Subsection 6.2.3.

In many applications the same model is used for each coordinate. In some applications, for instance, in air traffic control, one can use two third order models for the horizontal motion and a second order model for the (more benign) vertical motion.

In general the motion along each coordinate is assumed "decoupled" from the other coordinates. The noises entering into the various coordinates are also assumed to be mutually independent with possibly different variances. The discussion in this section will deal with kinematic models in *one generic coordinate*.

6.2.2 White Noise Acceleration Model

A **constant velocity object** moving in a generic coordinate ξ is described by the equation

$$\ddot{\xi}(t) = 0 \qquad (6.2.2\text{-}1)$$

Since the position $\xi(t)$ evolves in the absence of noise according to a polynomial in time (in this case, of second order), this model is also called a **polynomial model**.

In practice the velocity undergoes at least slight changes. This can be modeled by a continuous time zero-mean white noise $\tilde{v}$ as follows

$$\ddot{\xi}(t) = \tilde{v}(t) \qquad (6.2.2\text{-}2)$$

where

$$E[\tilde{v}(t)] = 0 \qquad (6.2.2\text{-}3)$$

$$E[\tilde{v}(t)\tilde{v}(\tau)] = \tilde{q}(t)\delta(t - \tau) \qquad (6.2.2\text{-}4)$$

The continuous time process noise intensity $\tilde{q}$, which is its power spectral density, is, in general, a design parameter for the estimation filter based on this model. This will be discussed later in more detail in Section 6.5.

The state vector corresponding to (6.2.2-2), which is two-dimensional per coordinate, is

$$x = \begin{bmatrix} \xi & \dot{\xi} \end{bmatrix}' \qquad (6.2.2\text{-}5)$$

Thus, this model will be called the **white noise acceleration model** or **second order kinematic model**. Note that the velocity in this model is a Wiener process — the integral of white noise.

The continuous time state equation is

$$\dot{x}(t) = Ax(t) + D\tilde{v}(t) \qquad (6.2.2\text{-}6)$$

where

$$A = \begin{bmatrix} 0 & 1 \\ 0 & 0 \end{bmatrix} \qquad (6.2.2\text{-}7)$$

$$D = \begin{bmatrix} 0 \\ 1 \end{bmatrix} \qquad (6.2.2\text{-}8)$$

The Discretized State Equation

The discrete time state equation with sampling period T is

$$x(k+1) = Fx(k) + v(k) \qquad (6.2.2\text{-}9)$$

where (see Subsection 4.3.1)

$$F = e^{AT} = \begin{bmatrix} 1 & T \\ 0 & 1 \end{bmatrix} \qquad (6.2.2\text{-}10)$$

and the discrete time process noise relates to the continuous time one as follows

$$v(k) = \int_0^T e^{A(T-\tau)} D\tilde{v}(kT + \tau)d\tau \qquad (6.2.2\text{-}11)$$

From the above, the covariance of the discrete time process noise $v(k)$, assuming $\tilde{q}$ to be constant and using (6.2.2-4), is

$$
\begin{aligned}
Q &= E[v(k)v(k)'] \\
&= \int_0^T \begin{bmatrix} T - \tau \\ 1 \end{bmatrix} [T - \tau \quad 1]\tilde{q}d\tau \\
&= \begin{bmatrix} \frac{1}{3}T^3 & \frac{1}{2}T^2 \\ \frac{1}{2}T^2 & T \end{bmatrix} \tilde{q} \qquad (6.2.2\text{-}12)
\end{aligned}
$$

Guideline for Choice of Process Noise Intensity

The changes in the velocity over a sampling period T are of the order of

$$\sqrt{Q_{22}} = \sqrt{\tilde{q}T} \qquad (6.2.2\text{-}13)$$

This can serve as a guideline for **process noise intensity choice** — the choice of the power spectral density $\tilde{q}$ of the process noise in this model.

A **nearly constant velocity model** is obtained by the choice of a "small" intensity $\tilde{q}$ in the following sense: the changes in the velocity have to be small compared to the actual velocity.

6.2.3 Wiener Process Acceleration Model

The motion of a constant acceleration object for a generic coordinate ξ is described by the equation

$$\dddot{\xi}(t) = 0 \qquad (6.2.3\text{-}1)$$

Similarly to (6.2.2-2), the acceleration is not exactly constant and its changes can be modeled by a continuous time zero-mean white noise as follows

$$\dddot{\xi}(t) = \tilde{v}(t) \qquad (6.2.3\text{-}2)$$

Note that in this case the acceleration is a Wiener process — hence the name **Wiener process acceleration model**. Since the derivative of the acceleration is the jerk, this model can also be called the **white noise jerk model**.

The state vector corresponding to the above is

$$x = \begin{bmatrix} \xi & \dot{\xi} & \ddot{\xi} \end{bmatrix}' \qquad (6.2.3\text{-}3)$$

and its continuous time state equation is

$$\dot{x}(t) = Ax(t) + D\tilde{v}(t) \qquad (6.2.3\text{-}4)$$

where

$$A = \begin{bmatrix} 0 & 1 & 0 \\ 0 & 0 & 1 \\ 0 & 0 & 0 \end{bmatrix} \qquad D = \begin{bmatrix} 0 \\ 0 \\ 1 \end{bmatrix} \qquad (6.2.3\text{-}5)$$

This is a **third order model** with three integrations: all three eigenvalues of A — the poles of the continuous time transfer function — are zero.

Guideline for Choice of Process Noise Intensity

The changes in the acceleration over a sampling period T are of the order of

$$\sqrt{Q_{33}} = \sqrt{\tilde{q}T} \qquad (6.2.3\text{-}6)$$

This can serve as a guideline in the **process noise intensity choice** — the choice of the power spectral density $\tilde{q}$ of the continuous time process noise $\tilde{v}$ for "tuning" this model to the actual motion of the object of interest.

A **nearly constant acceleration model** is obtained by choosing a "small" intensity $\tilde{q}$ in the following sense: the changes in the acceleration should be small relative to the actual acceleration levels.

Remark

One can have other third order models, for instance, with the acceleration having an exponentially decaying autocorrelation, rather than being a Wiener process. The **exponentially autocorrelated acceleration model** is presented later in Subsection 8.2.2.

The Discretized State Equation

The discrete time state equation with sampling period T is

$$x(k+1) = Fx(k) + v(k) \qquad (6.2.3\text{-}7)$$

with the transition matrix

$$F = \begin{bmatrix} 1 & T & \frac{1}{2}T^2 \\ 0 & 1 & T \\ 0 & 0 & 1 \end{bmatrix} \qquad (6.2.3\text{-}8)$$

and the covariance matrix of $v(k)$ given by

$$\begin{aligned} Q &= E[v(k)v(k)'] \\ &= \begin{bmatrix} \frac{1}{20}T^5 & \frac{1}{8}T^4 & \frac{1}{6}T^3 \\ \frac{1}{8}T^4 & \frac{1}{3}T^3 & \frac{1}{2}T^2 \\ \frac{1}{6}T^3 & \frac{1}{2}T^2 & T \end{bmatrix} \tilde{q} \end{aligned} \qquad (6.2.3\text{-}9)$$

Note the three unity eigenvalues of the transition matrix F in (6.2.3-8) — the poles of the discrete time transfer function — corresponding to the three integrations.

6.3 DIRECT DISCRETE TIME KINEMATIC MODELS

6.3.1 Introduction

The discrete time plant equation for the continuous time white noise acceleration and white noise jerk (Wiener process acceleration) were given in the previous section.

Another common kinematic model is directly defined in discrete time as follows. The discrete time process noise $v(k)$ is a scalar-valued *zero-mean white sequence*

$$E[v(k)v(j)] = \sigma_v^2 \delta_{kj} \qquad (6.3.1\text{-}1)$$

and enters into the dynamic equation as follows

$$x(k+1) = Fx(k) + \Gamma v(k) \qquad (6.3.1\text{-}2)$$

where the **noise gain** Γ is an n_x-dimensional vector.

The assumption in the second order model is that the target undergoes a *constant acceleration* during each sampling period (of length T)

$$\tilde{v}(t) = v(k) \qquad t \in [kT, (k+1)T) \qquad (6.3.1\text{-}3)$$

and that these accelerations are *uncorrelated from period to period*. The above indicates a **piecewise constant acceleration**.

Remark

It is clear that if the above assumption is correct for a given sampling period T_1 then it cannot be correct for any other T_2 (except integer multiples of T_1). Neither this *piecewise constant white noise* assumption nor the *continuous time white noise* (6.2.2-4) are completely realistic — both are approximations.

6.3.2 Piecewise Constant White Acceleration Model

If $v(k)$ is the *constant acceleration* during the k-th sampling period (of length T), the increment in the velocity during this period is $v(k)T$, while the effect of this acceleration on the position is $v(k)T^2/2$.

The state equation for the **piecewise constant white acceleration model**, which is of second order, is

$$x(k+1) = Fx(k) + \Gamma v(k) \qquad (6.3.2\text{-}1)$$

with the process noise $v(k)$ a *zero-mean white* acceleration sequence.

The transition matrix is

$$F = \begin{bmatrix} 1 & T \\ 0 & 1 \end{bmatrix} \qquad (6.3.2\text{-}2)$$

and the vector gain multiplying the scalar process noise is given, in view of the above discussion, by

$$\Gamma = \begin{bmatrix} \frac{1}{2}T^2 \\ T \end{bmatrix} \qquad (6.3.2\text{-}3)$$

The covariance of the process noise multiplied by the gain, $\Gamma v(k)$, is

$$\begin{aligned} Q &= E[\Gamma v(k)v(k)\Gamma'] \\ &= \Gamma \sigma_v^2 \Gamma' \\ &= \begin{bmatrix} \frac{1}{4}T^4 & \frac{1}{2}T^3 \\ \frac{1}{2}T^3 & T^2 \end{bmatrix} \sigma_v^2 \end{aligned} \qquad (6.3.2\text{-}4)$$

Note the difference between this and (6.2.2-12).

Guideline for Choice of Process Noise Variance

For this model, σ_v should be of the order of the maximum acceleration magnitude a_M. A practical range is $0.5a_M \leq \sigma_v \leq a_M$.

Note on the Multidimensional Case

When motion is in several coordinates, then with decoupled filtering (6.3.2-4) is a block of the overall Q, which is then block diagonal.

6.3.3 Piecewise Constant Wiener Process Acceleration Model

For the **piecewise constant Wiener process acceleration model** the (third order) state equation is

$$x(k+1) = Fx(k) + \Gamma v(k) \qquad (6.3.3\text{-}1)$$

where

$$F = \begin{bmatrix} 1 & T & \frac{1}{2}T^2 \\ 0 & 1 & T \\ 0 & 0 & 1 \end{bmatrix} \qquad (6.3.3\text{-}2)$$

$$\Gamma = \begin{bmatrix} \frac{1}{2}T^2 \\ T \\ 1 \end{bmatrix} \qquad (6.3.3\text{-}3)$$

In this model the white process noise $v(k)$ is the **acceleration increment** during the k-th sampling period and it is assumed to be a *zero-mean white sequence* — the *acceleration is a discrete time Wiener process*. The formulation in terms of acceleration increment is more convenient than the one in terms of the third order derivative (jerk).

The covariance of the process noise multiplied by the gain Γ is

$$\begin{aligned} Q &= \Gamma \sigma_v^2 \Gamma' \\ &= \begin{bmatrix} \frac{1}{4}T^4 & \frac{1}{2}T^3 & \frac{1}{2}T^2 \\ \frac{1}{2}T^3 & T^2 & T \\ \frac{1}{2}T^2 & T & 1 \end{bmatrix} \sigma_v^2 \end{aligned} \qquad (6.3.3\text{-}4)$$

Guideline for Choice of Process Noise Variance

For this model, σ_v should be of the order of the magnitude of the maximum acceleration increment over a sampling period, Δa_M. A practical range is $0.5\Delta a_M \leq \sigma_v \leq \Delta a_M$.

The Process Noise Variance for Different Sampling Periods

Note that if the sampling period is changed, one has to carry out a **rescaling of the variance of the process noise.** (See problems 6-1 through 6-4.)

6.3.4 Kinematic Models — Summary

Kinematic (polynomial) model of order n: the n-th derivative of the position is equal to

- zero — noiseless model,
- white noise — noisy model.

There are two major classes of noisy discrete time kinematic models:

1. Obtained from discretization of the continuous time model, driven by *continuous time white noise*, for a given sampling period.

2. Obtained by direct definition of the process noise in discrete time as a *piecewise constant white sequence* — the process noise is assumed to be constant over each sampling period and independent between periods.

Within each class the following models were discussed in detail:

- White noise acceleration — second order model.
- Wiener process acceleration — third order model.

The resulting process noise covariance matrices for the two classes of models are different in their dependence on the sampling period.

The process noise covariance matrices in the direct discrete time models are positive semidefinite of rank 1 while their counterparts from the discretized continuous time models are of full rank.

Both models are, obviously, approximations.

The more commonly used model is (2).

6.4 EXPLICIT FILTERS FOR NOISELESS KINEMATIC MODELS

6.4.1 LS Estimation for Noiseless Kinematic Models

Consider an object moving with constant velocity (without process noise). Such a case was treated in Subsection 1.5.1 where, using least squares, the estimate of the *initial position and the (constant) velocity* were obtained, both in batch as well as recursive form.

Using these results, the recursion for the estimate of the *current state* $x(k) \triangleq x(t_k)$ of a second order **noiseless kinematic model** — the **constant velocity model** — will be obtained.

Similarly to (6.2.2-9), but without the process noise, one has

$$x(t) = \begin{bmatrix} 1 & t \\ 0 & 1 \end{bmatrix} x(0) \qquad (6.4.1\text{-}1)$$

Assuming a uniform sampling rate with sampling period T,

$$\hat{x}(t_{k+1}|t_{k+1}) \triangleq \hat{x}(k+1|k+1)$$
$$= \begin{bmatrix} 1 & (k+1)T \\ 0 & 1 \end{bmatrix} \hat{x}(0|k+1) \qquad (6.4.1\text{-}2)$$

In the above equation $\hat{x}(0|k+1)$ is the estimate of the *initial state*, with the initial time 0, to which it pertains, now explicitly indicated. This estimate of the initial state was obtained via recursive LS in (1.5.1-8), where it was denoted, without the first time index, as $\hat{x}(k+1)$.

6.4.2 The KF for Noiseless Kinematic Models

The recursion for the estimate of the current state is

$$\hat{x}(k+1|k+1) = \hat{x}(k+1|k) + W(k+1)[z(k+1) - \hat{z}(k+1|k)] \qquad (6.4.2\text{-}1)$$

where

$$\hat{x}(k+1|k) = \begin{bmatrix} 1 & T \\ 0 & 1 \end{bmatrix} \hat{x}(k|k) \qquad (6.4.2\text{-}2)$$

$$\hat{z}(k+1|k) = [1 \quad 0]\, \hat{x}(k+1|k) \qquad (6.4.2\text{-}3)$$

Note that (6.4.2-1) is actually a recursive LS estimator "disguised" as a Kalman filter. In view of this, the gain $W(k)$ in (6.4.2-1) can be obtained directly from the gain (1.5.1-18), denoted now as $W_0(k)$ (since it pertains to the initial state), by multiplying it with the same matrix as in (6.4.1-2)

$$
\begin{aligned}
W(k+1) &= \begin{bmatrix} 1 & (k+1)T \\ 0 & 1 \end{bmatrix} W_0(k+1) \\
&= \begin{bmatrix} 1 & (k+1)T \\ 0 & 1 \end{bmatrix} \begin{bmatrix} -\frac{2}{(k+1)} \\ \frac{6}{(k+1)(k+2)T} \end{bmatrix} = \begin{bmatrix} \frac{4k+2}{(k+1)(k+2)} \\ \frac{6}{(k+1)(k+2)T} \end{bmatrix} \qquad (6.4.2\text{-}4)
\end{aligned}
$$

It can be easily shown that the covariance associated with the estimate (6.4.1-2) is, using (1.5.1-19), given by

$$
\begin{aligned}
P(k+1|k+1) &= \begin{bmatrix} 1 & (k+1)T \\ 0 & 1 \end{bmatrix} P(k+1) \begin{bmatrix} 1 & (k+1)T \\ 0 & 1 \end{bmatrix}' \\
&= \frac{2\sigma^2}{(k+1)(k+2)} \begin{bmatrix} 2k+1 & \frac{3}{T} \\ \frac{3}{T} & \frac{6}{kT^2} \end{bmatrix} \qquad (6.4.2\text{-}5)
\end{aligned}
$$

Summary

For noiseless kinematic (polynomial) models one can obtain explicit expressions of the Kalman filter gain and the state covariance.

Similar results can be obtained for the **constant acceleration model**.

Due to the absence of process noise, the state covariance converges to zero and so does the filter gain.

6.5 STEADY-STATE FILTERS FOR NOISY KINEMATIC MODELS

6.5.1 The Problem

As indicated in Subsection 5.2.5, the state estimation covariance for a time-invariant system (with constant coefficients in the state and measurement equations) will converge under suitable conditions to a steady-state value.

These conditions are satisfied for the kinematic models described in Sections 6.2 and 6.3. Furthermore, *explicit expressions of the steady-state covariance and filter gain* can be obtained.

It will be assumed that only position measurements are available, that is,

$$z(k) = Hx(k) + w(k) \qquad (6.5.1\text{-}1)$$

where for the white noise acceleration (second order) model

$$H = [1 \quad 0] \qquad (6.5.1\text{-}2)$$

and for the Wiener process acceleration (third order) model

$$H = [1 \quad 0 \quad 0] \qquad (6.5.1\text{-}3)$$

The measurement noise autocorrelation function is

$$\begin{aligned} E[w(k)w(j)] &= R\delta_{kj} \\ &= \sigma_w^2 \delta_{kj} \end{aligned} \qquad (6.5.1\text{-}4)$$

The resulting **steady-state filters for noisy kinematic models** are known as **alpha-beta** and **alpha-beta-gamma** filters for the second and third order models, respectively. The coefficients α, β, and γ yield the filter steady-state gain vector components.

Subsection 6.5.2 presents the methodology of the derivation of the α-β filter. This is used for the direct discrete time and the discretized continuous time second order models in Subsections 6.5.3 and 6.5.4, respectively. Subsection 6.5.5 presents the α-β-γ filter for the direct discrete time third order model.

6.5.2 Derivation Methodology for the Alpha-Beta Filter

The steady-state filter for the 2-dimensional kinematic model is obtained as follows. The plant equation is

$$x(k+1) = Fx(k) + v(k) \qquad (6.5.2\text{-}1)$$

where

$$F = \begin{bmatrix} 1 & T \\ 0 & 1 \end{bmatrix} \qquad (6.5.2\text{-}2)$$

and the (vector-valued) process noise has the autocorrelation function

$$E[v(k)v(j)'] = Q\delta_{kj} \qquad (6.5.2\text{-}3)$$

The measurement is given by (6.5.1-1), (6.5.1-2), and (6.5.1-4). The variance of the (scalar) measurement noise will be denoted as $\sigma_w^2 \triangleq R$.

The steady-state values of the components of the state estimation covariance matrix will be denoted as

$$\lim_{k \to \infty} P(k|k) = [p_{ij}] \qquad (6.5.2\text{-}4)$$

The components of the one-step prediction covariance are denoted as

$$\lim_{k \to \infty} P(k+1|k) = [m_{ij}] \qquad (6.5.2\text{-}5)$$

while for the **alpha-beta filter gain** the notation will be

$$\lim_{k \to \infty} W(k) \triangleq [g_1 \ \ g_2]' \triangleq \left[\alpha \ \ \frac{\beta}{T}\right]' \qquad (6.5.2\text{-}6)$$

Note that, as defined, α and β are dimensionless.

Note

The existence and positive definiteness of (6.5.2-4) are guaranteed since the required observability and controllability conditions are satisfied (see problem 6-6).

The expression of the innovation covariance (5.2.3-9) yields

$$S = H \begin{bmatrix} m_{11} & m_{12} \\ m_{12} & m_{22} \end{bmatrix} H' + R$$
$$= m_{11} + \sigma_w^2 \qquad (6.5.2\text{-}7)$$

where the notation $\sigma_w^2 = R$ is now used.

The filter gain given by (5.2.3-11) becomes

$$W = \begin{bmatrix} m_{11} & m_{12} \\ m_{12} & m_{22} \end{bmatrix} H' S^{-1}$$
$$= \begin{bmatrix} \dfrac{m_{11}}{m_{11} + \sigma_w^2} & \dfrac{m_{12}}{m_{11} + \sigma_w^2} \end{bmatrix}' \qquad (6.5.2\text{-}8)$$

From (6.5.2-6) and (6.5.2-8) it follows that

$$g_1 = \frac{m_{11}}{m_{11} + \sigma_w^2} \qquad (6.5.2\text{-}9)$$

$$g_2 = \frac{m_{12}}{m_{11} + \sigma_w^2} = g_1 \frac{m_{12}}{m_{11}} \qquad (6.5.2\text{-}10)$$

The covariance update equation (5.2.3-15) becomes, using (6.5.2-8) to (6.5.2-10),

$$\begin{bmatrix} p_{11} & p_{12} \\ p_{12} & p_{22} \end{bmatrix} = (I - WH) \begin{bmatrix} m_{11} & m_{12} \\ m_{12} & m_{22} \end{bmatrix}$$
$$= \begin{bmatrix} (1 - g_1)m_{11} & (1 - g_1)m_{12} \\ (1 - g_1)m_{12} & m_{22} - g_2 m_{12} \end{bmatrix} \qquad (6.5.2\text{-}11)$$

The covariance prediction equation (5.2.3-5) is rewritten as follows:

$$P(k|k) = F^{-1}[P(k+1|k) - Q](F^{-1})' \qquad (6.5.2\text{-}12)$$

where, from (6.5.2-2), one has

$$F^{-1} = \begin{bmatrix} 1 & -T \\ 0 & 1 \end{bmatrix} \qquad (6.5.2\text{-}13)$$

The steady-state solution for the covariance and gains is obtained from the set of nonlinear equations (6.5.2-9) to (6.5.2-12) using the suitable expression of the process noise covariance Q in (6.5.2-12). The expression for the direct discrete time model is (6.3.2-4), while for the discretized continuous time model it is (6.2.2-12).

6.5.3 The Alpha-Beta Filter for the Piecewise Constant White Acceleration Model

Using the process noise covariance (6.3.2-4), which corresponds to a *piecewise constant white process noise*, that is, the **piecewise constant acceleration model** — accelerations that are constant over each sampling period and uncorrelated from period to period — yields in (6.5.2-12)

$$
\begin{bmatrix} p_{11} & p_{12} \\ p_{12} & p_{22} \end{bmatrix} = \begin{bmatrix} m_{11} - 2Tm_{12} + T^2 m_{22} - \frac{1}{4}T^4 \sigma_v^2 & m_{12} - Tm_{22} + \frac{1}{2}T^3 \sigma_v^2 \\ m_{12} - Tm_{22} + \frac{1}{2}T^3 \sigma_v^2 & m_{22} - T^2 \sigma_v^2 \end{bmatrix} \tag{6.5.3-1}
$$

Equating the terms of (6.5.2-11) and (6.5.3-1) yields, after some cancellations,

$$
g_1 m_{11} = 2Tm_{12} - T^2 m_{22} + \frac{T^4}{4} \sigma_v^2 \tag{6.5.3-2}
$$

$$
g_1 m_{12} = Tm_{22} - \frac{T^3}{2} \sigma_v^2 \tag{6.5.3-3}
$$

$$
g_2 m_{12} = T^2 \sigma_v^2 \tag{6.5.3-4}
$$

Equations (6.5.2-9), (6.5.2-10), and (6.5.3-2) to (6.5.3-4) with the five unknowns g_1, g_2, m_{11}, m_{12}, and m_{22} are solved next.

From (6.5.2-9) and (6.5.2-10) one has

$$
m_{11} = \frac{g_1}{1 - g_1} \sigma_w^2 \tag{6.5.3-5}
$$

$$
m_{12} = \frac{g_2}{1 - g_1} \sigma_w^2 \tag{6.5.3-6}
$$

From (6.5.3-3) and (6.5.3-4) one obtains

$$
m_{22} = \frac{g_1 m_{12}}{T} + \frac{T^2}{2} \sigma_v^2 = \left(\frac{g_1}{T} + \frac{g_2}{2} \right) m_{12} \tag{6.5.3-7}
$$

Using (6.5.3-4) to (6.5.3-7) in (6.5.3-2) yields

$$
\frac{g_1^2}{1 - g_1} \sigma_w^2 = 2T \frac{g_2}{1 - g_1} \sigma_w^2 - T^2 \left(\frac{g_1}{T} + \frac{g_2}{2} \right) \frac{g_2}{1 - g_1} \sigma_w^2 + \frac{T^2}{4} \frac{g_2^2}{1 - g_1} \sigma_w^2 \tag{6.5.3-8}
$$

which, after cancellations, becomes

$$
g_1^2 - 2Tg_2 + Tg_1 g_2 + \frac{T^2}{4} g_2^2 = 0 \tag{6.5.3-9}
$$

275

With the dimensionless variables α and β one has

$$\alpha^2 - 2\beta + \alpha\beta + \frac{\beta^2}{4} = 0 \qquad (6.5.3\text{-}10)$$

which yields the first equation for α and β

$$\alpha = \sqrt{2\beta} - \frac{\beta}{2} \qquad (6.5.3\text{-}11)$$

The second equation for α and β follows immediately from (6.5.3-4) and (6.5.3-6) as

$$m_{12} = \frac{T^2 \sigma_v^2}{T/\beta} = \frac{T/\beta}{1-\alpha}\sigma_w^2 \qquad (6.5.3\text{-}12)$$

or

$$\frac{\beta^2}{1-\alpha} = \frac{T^4 \sigma_v^2}{\sigma_w^2} \triangleq \lambda^2 \qquad (6.5.3\text{-}13)$$

The quantity

$$\boxed{\lambda \triangleq \frac{\sigma_v T^2}{\sigma_w}} \qquad (6.5.3\text{-}14)$$

is called the **target maneuvering index** since it is proportional to the ratio of

- the **motion uncertainty** — the RMS value of the process noise (acceleration) effect on the position, which is $\sigma_v T^2/2$ — see (6.5.2-4);
- the **observation uncertainty** — the measurement noise RMS value σ_w.

Eliminating α from (6.5.3-14) with (6.5.3-11) yields

$$\frac{\beta^2}{1-\alpha} = \frac{\beta^2}{1 - \sqrt{2\beta} + \beta/2} = \frac{\beta^2}{\left(1 - \sqrt{\beta/2}\right)^2} = \lambda^2 \qquad (6.5.3\text{-}15)$$

or

$$\beta + \frac{\lambda}{\sqrt{2}}\sqrt{\beta} - \lambda = 0 \qquad (6.5.3\text{-}16)$$

The positive solution for $\sqrt{\beta}$ from the above is

$$\sqrt{\beta} = \frac{1}{2\sqrt{2}}\left(-\lambda + \sqrt{\lambda^2 + 8\lambda}\right) \qquad (6.5.3\text{-}17)$$

The expression of the **velocity gain coefficient** β in terms of λ is

$$\boxed{\beta = \frac{1}{4}\left(\lambda^2 + 4\lambda - \lambda\sqrt{\lambda^2 + 8\lambda}\right)} \qquad (6.5.3\text{-}18)$$

Using (6.5.3-18) in (6.5.3-11) gives the **position gain** α in terms of λ as

$$\alpha = -\frac{1}{8}\left(\lambda^2 + 8\lambda - (\lambda + 4)\sqrt{\lambda^2 + 8\lambda}\right) \tag{6.5.3-19}$$

The elements of the state estimation covariance matrix are, using (6.5.2-11),

$$\begin{aligned} p_{11} &= (1 - g_1)m_{11} \\ &= g_1\sigma_w^2 \end{aligned} \tag{6.5.3-20}$$

$$\begin{aligned} p_{12} &= (1 - g_1)m_{12} \\ &= g_2\sigma_w^2 \end{aligned} \tag{6.5.3-21}$$

$$\begin{aligned} p_{22} &= \left(\frac{g_1}{T} + \frac{g_2}{2}\right)m_{12} - g_2 m_{12} \\ &= \left(\frac{g_1}{T} - \frac{g_2}{2}\right)m_{12} \end{aligned} \tag{6.5.3-22}$$

The expressions of these (steady-state) error covariance matrix elements can be rewritten using (6.5.3-5) to (6.5.3-7) as

$$p_{11} = \alpha\sigma_w^2 \tag{6.5.3-23}$$

$$p_{12} = \frac{\beta}{T}\sigma_w^2 \tag{6.5.3-24}$$

$$p_{22} = \frac{\beta}{T^2}\frac{\alpha - \beta/2}{1 - \alpha}\sigma_w^2 \tag{6.5.3-25}$$

The **position estimation improvement**, or the **noise reduction factor**, with respect to a single observation is seen from (6.5.3-23) to be α ($0 \le \alpha \le 1$), that is, the same as the *optimal* position gain of the filter.

The innovation variance is, in terms of the position prediction variance and the measurement noise variance,

$$s = m_{11} + \sigma_w^2 \tag{6.5.3-26}$$

Using (6.5.3-5) with g_1 replaced by α yields

$$s = \frac{\sigma_w^2}{1 - \alpha} \tag{6.5.3-27}$$

Figure 6.5.3-1 presents the **alpha-beta filter** gain coefficients α and β as a function of the maneuvering index λ in semi-log and log-log scales.

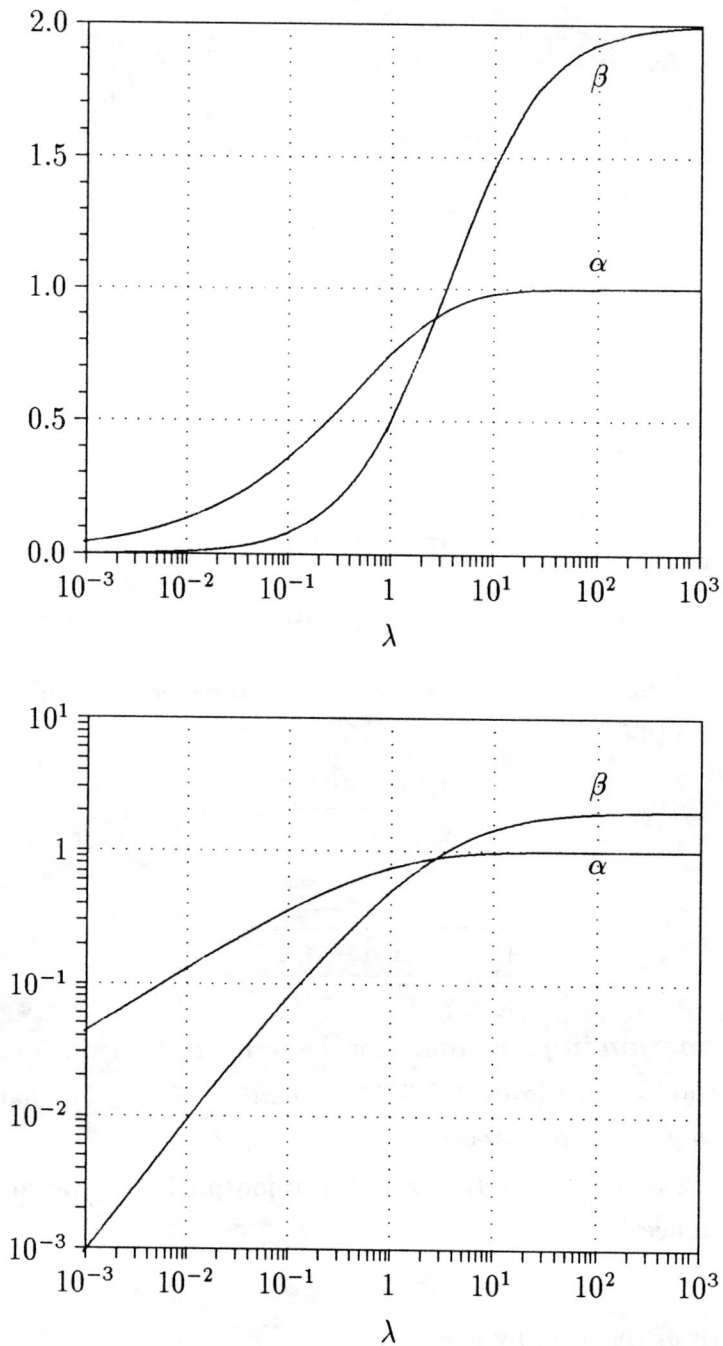

Figure 6.5.3-1: Steady-state filter gain coefficients for the piecewise constant white acceleration model.

The **velocity estimation improvement** — compared to the differencing of two adjacent observations — is

$$\eta \triangleq \frac{p_{22}}{2\sigma_w^2/T^2} = \frac{\beta}{2}\frac{\alpha - \beta/2}{1 - \alpha}$$

(6.5.3-28)

Note that this ignores the process noise — it is not meaningful for significant levels of the maneuvering index.

Figure 6.5.3-2 presents the velocity estimation improvement factor given above.

Simplified Expressions for Low Maneuvering Index

Note that for small λ (up to about 0.1) one has the following simplified expressions of the gains and the velocity estimation improvement factor

$$\alpha \approx \sqrt{2\lambda}$$

(6.5.3-29)

$$\beta \approx \lambda$$

(6.5.3-30)

$$\eta \approx \frac{1}{\sqrt{2}}\lambda^{1.5}$$

(6.5.3-31)

Remarks

A high value of the process noise variance relative to the measurement noise variance, that is, a large maneuvering index λ yields a high position gain α and the filter will give large weight to the latest measurement and consequently little weight to the past data, resulting in less noise reduction.

A small λ yields a lower α and more noise reduction. However, a small α will *not* yield more noise reduction unless it has been *optimally determined* based on λ and *all the modeling assumptions hold.*

The two gains α and β *cannot be chosen independently* — they are both determined by the maneuvering index λ.

Figure 6.5.3-2: Steady-state filter velocity estimation improvement over two-point differencing for the piecewise constant white acceleration model.

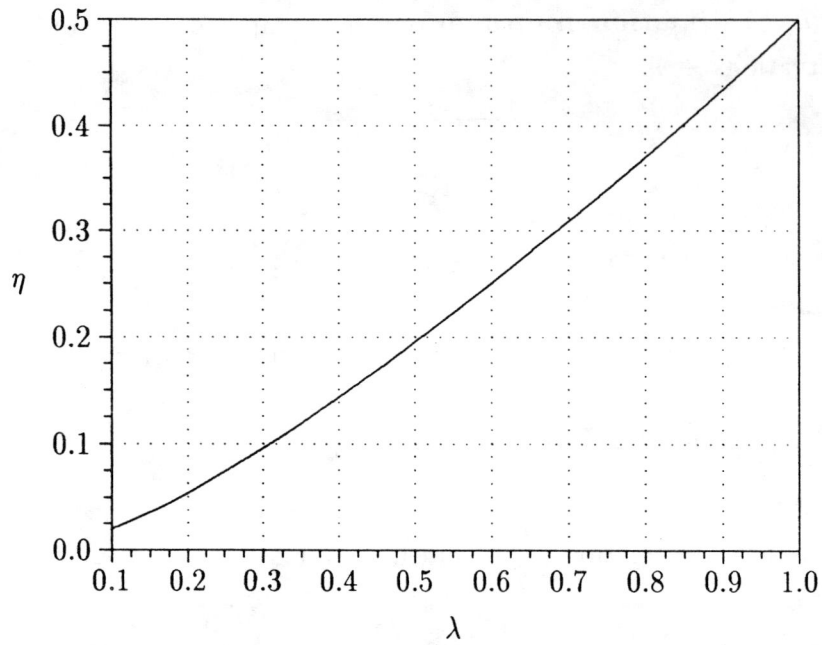

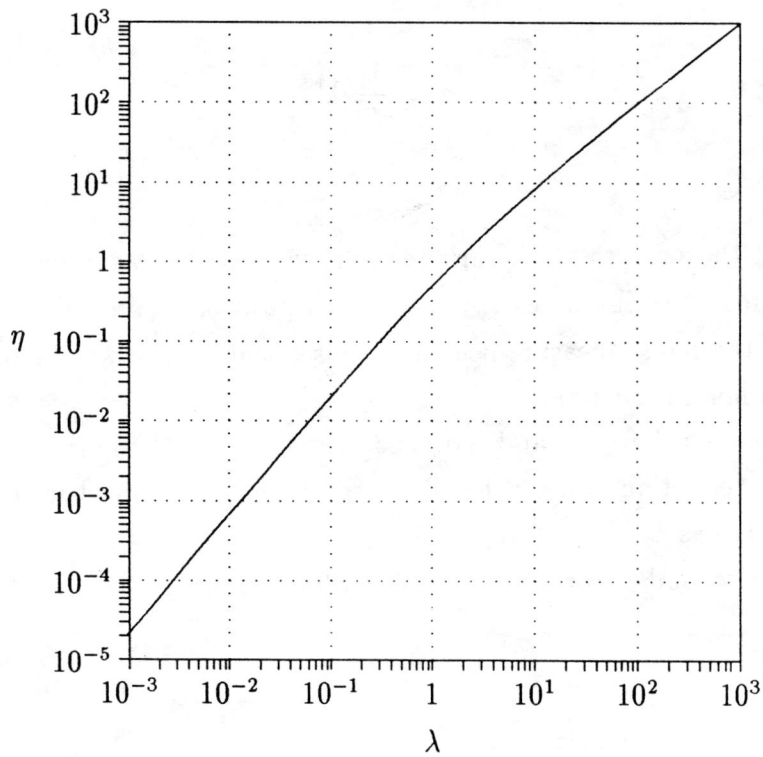

Example

The example of Section 5.3, which dealt with what now is called a white noise acceleration model, is reconsidered. The closed form expressions for the steady-state gain and covariances developed above will be used and compared with the results of the covariance equation iterations that are plotted in Figures 5.3.2-2 and 5.3.2-3.

The two cases of interest are those with nonzero process noise: $q = 1$ and $q = 9$ (the case with $q = 0$ leads to zero variances and gain in steady state). The corresponding maneuvering indices are, using (6.5.3-14) with $T = 1$, $\sigma_v = \sqrt{q}$ and $\sigma_w = 1$, $\lambda = 1$ and $\lambda = 3$, respectively.

Table 6.5.3-1 shows the steady-state values of the gain coefficients as well as the position and velocity variances plotted in Figures 5.3.2-2 and 5.3.2-3, parts (b) and (c), respectively. The innovation variance (6.5.3-27) is also shown.

The updated variances, obtained from (6.5.3-23) and (6.5.3-25), and the predicted variances, obtained from (6.5.3-5) and (6.5.3-7), are seen to match the values plotted in the above figures.

Maneuvering index	Position gain	Velocity gain	Position variance		Velocity variance		Innovation variance
			updated	predicted	updated	predicted	
λ	α	β	p_{11}	m_{11}	p_{22}	m_{22}	s
1	0.75	0.50	0.75	3	1	2	4
3	0.90	0.94	0.90	9.15	4.12	13.12	10.15

Table 6.5.3-1: Steady-state gains and variances for two maneuvering indices.

6.5.4   The Alpha-Beta Filter for the Discretized Continuous Time White Noise Acceleration Model

Next, the **alpha-beta filter** for the **discretized continuous time white noise acceleration model** is derived using the process noise covariance matrix (6.2.2-12) instead of (6.3.2-4). Since almost all the equations stay the same as in Subsection 6.5.3, they will not be repeated and only those which are different will be indicated.

Equating the terms in the updated covariance expressions (6.5.2-11) and (6.5.2-12) with Q given by (6.2.2-12) yields

$$g_1 m_{11} = 2T m_{12} - T^2 m_{22} + \frac{T^3}{3} \tilde{q} \tag{6.5.4-1}$$

$$g_1 m_{12} = T m_{22} - \frac{T^2}{2} \tilde{q} \tag{6.5.4-2}$$

$$g_2 m_{12} = T \tilde{q} \tag{6.5.4-3}$$

To eliminate m_{22}, one has

$$
\begin{aligned}
m_{22} &= \frac{g_1 m_{12}}{T} + \frac{T}{2} \tilde{q} \\
&= \left(\frac{g_1}{T} + \frac{g_2}{2} \right) m_{12}
\end{aligned}
\tag{6.5.4-4}
$$

The counterpart of (6.5.3-8) is

$$\frac{g_1^2}{1 - g_1} \sigma_w^2 = 2T \frac{g_2}{1 - g_1} \sigma_w^2 - T^2 \left(\frac{g_1}{T} + \frac{g_2}{2} \right) \frac{g_2}{1 - g_1} \sigma_w^2 + \frac{T^2}{3} \frac{g_2^2}{1 - g_1} \sigma_w^2 \tag{6.5.4-5}$$

which becomes

$$g_1^2 - 2T g_2 + T g_1 g_2 + \frac{T^2}{6} g_2^2 = 0 \tag{6.5.4-6}$$

In terms of α and β, the above can be written as

$$\alpha^2 - 2\beta + \alpha\beta + \frac{\beta^2}{6} = 0 \tag{6.5.4-7}$$

or

$$\alpha = \sqrt{2\beta + \frac{\beta^2}{12}} - \frac{\beta}{2} \tag{6.5.4-8}$$

The above equation differs from (6.5.3-11) by an extra term.

6.5.4 The Alpha-Beta Filter for the Discretized Continuous Time White Noise Acceleration Model

Equating m_{12} from (6.5.4-3) and (6.5.3-6) yields

$$m_{12} = \frac{T\tilde{q}}{\beta/T}$$
$$= \frac{\beta/T}{1-\alpha}\sigma_w^2 \qquad (6.5.4\text{-}9)$$

which results in

$$\frac{\beta^2}{1-\alpha} = \frac{T^3\tilde{q}}{\sigma_w^2}$$
$$\triangleq \lambda_c^2 \qquad (6.5.4\text{-}10)$$

where λ_c is the **maneuvering index** for this *discretized continuous time system* and has a similar interpretation as λ in (6.5.3-14).

The equation for β becomes

$$\frac{\beta^2}{1-\alpha} = \frac{\beta^2}{1-\sqrt{2\beta+\beta^2/12}+\beta/2}$$
$$= \lambda_c^2 \qquad (6.5.4\text{-}11)$$

which can be solved numerically.

The equations for the updated covariance terms (6.5.3-23) to (6.5.3-25) stay the same.

Remark

This discretized continuous time model is somewhat less common in use than the direct discrete time piecewise constant acceleration model.

6.5.5   The Alpha-Beta-Gamma Filter for the Piecewise Constant Wiener Process Acceleration Model

The **piecewise constant Wiener process acceleration model** is the third order system (6.3.3-1) with the *zero-mean white* process noise, assumed to be the **acceleration increment over a sampling period**, with variance σ_v^2.

The target maneuvering index is defined in the same manner as for the second order model in (6.5.3-14), that is,

$$\lambda = \frac{\sigma_v T^2}{\sigma_w} \tag{6.5.5-1}$$

The steady-state gain for the resulting filter — the **alpha-beta-gamma filter** — is

$$\lim_{k\to\infty} W(k) \triangleq [g_1 \ g_2 \ g_3]' \triangleq \left[\alpha \ \ \frac{\beta}{T} \ \ \frac{\gamma}{2T^2}\right]' \tag{6.5.5-2}$$

It can be shown that the three equations that yield the optimal steady-state filter **gain coefficients** are

$$\frac{\gamma^2}{4(1-\alpha)} = \lambda^2 \tag{6.5.5-3}$$

$$\beta = 2(2-\alpha) - 4\sqrt{1-\alpha} \tag{6.5.5-4}$$

or

$$\alpha = \sqrt{2\beta} - \frac{\beta}{2} \qquad\qquad \gamma = \frac{\beta^2}{\alpha} \tag{6.5.5-5}$$

The relationship between α and β in (6.5.5-4) is the same as (6.5.3-11).

The explicit solution for this system of three nonlinear equations that yields the three gain coefficients from (6.5.5-2) in terms of λ is given at the end of this subsection.

Similarly to the second order system, it can be shown that the corresponding updated state covariance expressions (in steady state) are

$$p_{11} = \alpha\sigma_w^2 \qquad p_{12} = \frac{\beta}{T}\sigma_w^2 \qquad p_{13} = \frac{\gamma}{2T^2}\sigma_w^2 \tag{6.5.5-6}$$

$$p_{22} = \frac{8\alpha\beta + \gamma(\beta - 2\alpha - 4)}{8T^2(1-\alpha)}\sigma_w^2 \tag{6.5.5-7}$$

$$p_{23} = \frac{\beta(2\beta - \gamma)}{4T^3(1-\alpha)}\sigma_w^2 \qquad\qquad p_{33} = \frac{\gamma(2\beta - \gamma)}{4T^4(1-\alpha)}\sigma_w^2 \tag{6.5.5-8}$$

Figure 6.5.5-1 shows the gain coefficients for this filter as a function of the maneuvering index λ in semi-log and log-log scale.

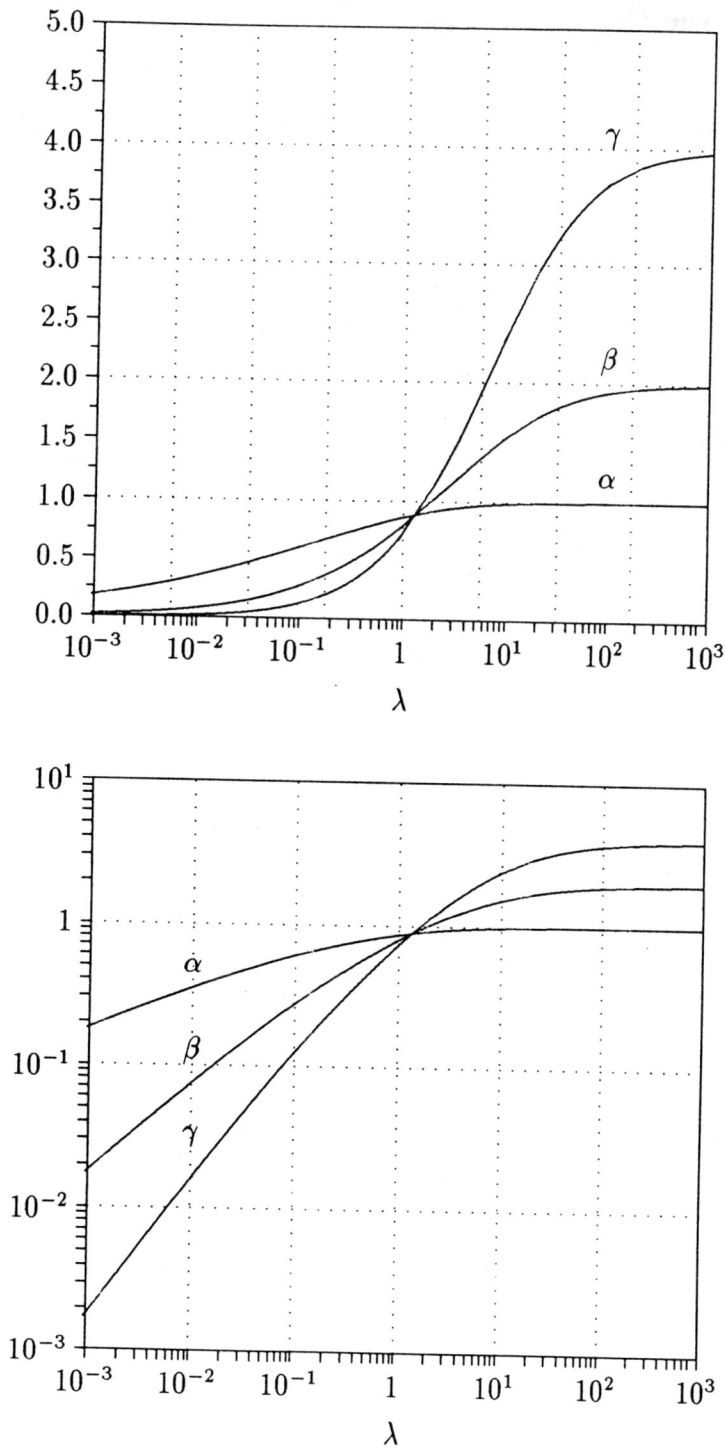

Figure 6.5.5-1: Steady-state filter gain coefficients for the piecewise constant Wiener process acceleration model.

The Solution for the Gain Coefficients

Substituting

$$\alpha = 1 - s^2 \tag{6.5.5-9}$$

in (6.5.5-4) yields

$$\beta = 2(1 - s)^2 \tag{6.5.5-10}$$

Rewriting (6.5.5-3) with (6.5.5-9) yields

$$\gamma = 2\lambda\sqrt{1 - \alpha} = 2\lambda s \tag{6.5.5-11}$$

Equations (6.5.5-9) to (6.5.5-11) provide the *explicit solution for the gain coefficients* in terms of the new variable s for which a cubic equation is obtained next.

Substituting (6.5.5-11) and (6.5.5-9) into the second equation of (6.5.5-5) leads to

$$2\lambda s = \frac{4(1 - s)^4}{1 - s^2} \tag{6.5.5-12}$$

which can be rewritten as

$$s^3 + bs^2 + cs - 1 = 0 \tag{6.5.5-13}$$

where

$$b \triangleq \frac{\lambda}{2} - 3 \qquad\qquad c \triangleq \frac{\lambda}{2} + 3 \tag{6.5.5-14}$$

Substitute again

$$s = y - \frac{b}{3} \tag{6.5.5-15}$$

to obtain

$$y^3 + py + q = 0 \tag{6.5.5-16}$$

where

$$p \triangleq c - \frac{b^2}{3} \qquad\qquad q \triangleq \frac{2b^3}{27} - \frac{bc}{3} - 1 \tag{6.5.5-17}$$

Finally, one more substitution,

$$y = z - \frac{p}{3z} \tag{6.5.5-18}$$

yields

$$z^6 + qz^3 - \frac{p^2}{27} = 0 \tag{6.5.5-19}$$

which has the solution

$$z^3 = \frac{-q \pm \sqrt{q^2 + 4p^3/27}}{2} \tag{6.5.5-20}$$

from which the negative sign should be chosen.

6.5.5 The Alpha-Beta-Gamma Filter for the Piecewise Constant Wiener Process Acceleration Model

Using (6.5.5-15) and (6.5.5-18) yields

$$s = z - \frac{p}{3z} - \frac{b}{3} \qquad (6.5.5\text{-}21)$$

which can be used directly in (6.5.5-9) to (6.5.5-11) to obtain the gain coefficients α, β, and γ.

6.5.6 Alpha-Beta and Alpha-Beta-Gamma Filters — Summary

For discrete time kinematic models

- with zero-mean white process noise that models
  - the acceleration (second order model), or
  - the acceleration increments (third order model — Wiener process acceleration)
- with noisy position measurements,

one has explicit expressions of the *steady-state filter gain and the corresponding covariance*.

These filters, called alpha-beta and alpha-beta-gamma, respectively, are the simplest possible: they use fixed precomputed (steady-state) gains. Consequently, they are *not optimal* during the initial transient period or if the noises are nonstationary.

The gains of these filters depend *only* on the *target maneuvering index*.

The *target maneuvering index* is defined as the ratio between the standard deviations (RMS values) of the following two uncertainties:

- The position displacement over one sampling period due to the process noise (multiplied by 2), and
- The (position) measurement noise.

These filters are usually used independently for each coordinate; however, one can encounter instability under certain extreme circumstances due to the errors introduced by the decoupling [Rog88].

Two classes of models were discussed:

- Discretized continuous time models based on *continuous-time zero-mean white noise,*
- Direct discrete time models based on *piecewise constant zero-mean white noise*, that is, a *zero-mean white sequence.*

These noises model the uncertainties of the motion — acceleration or acceleration increments. These two classes exhibit a different dependence on the sampling period of the effect of the noises on the motion.

None of these assumptions can model exactly target maneuvers, which are neither zero mean nor white — actually they are not even random, but the state models (which have to be Markov processes) require the specification of some randomness.

Nevertheless, they have been used extensively in real-time implementations of target tracking.

In particular, such fixed-gain filters have proven to be useful in implementations where their very modest computational and memory requirements were a major consideration.

One convenient application of these explicit results is to obtain quick (but possibly dirty) evaluations of achievable tracking performance — the quality of estimation, measured by the steady-state error variances.

These kinematic models can also be used as elements in a set of models describing different target behavior modes in the context of multiple model estimation algorithms, to be discussed in Chapter 11.

6.6 NOTES AND PROBLEMS

6.6.1 Bibliographical Notes

The kinematic (polynomial) models for filtering date back to [Skl57, BB62]. They have been extensively discussed in the literature, for example, [WLA70], and several papers presented steady-state filters for them. In [Fri73] analytical expressions of the position and velocity estimation accuracy with position measurements were given. Gain curves as a function of the maneuvering index were presented in [Fit80]. Closed form solutions for the continuous time and discrete time filter with exponentially autocorrelated acceleration were presented in [Fit81]. Tracking accuracies with position and velocity measurements were derived in [Cas81]. Analytical solutions for the steady-state filter gain and covariance with position and velocity measurements were given in [Eks83].

The explicit derivations presented in Section 6.5 using the target maneuvering index are based on [Kal84]. The idea of target maneuvering index has been used in [Fri73] and can be traced back to Sittler. The derivation of the explicit solution for the alpha-beta-gamma filter is based on [GM93].

Track initiation and simple approximations of the gains during the transient for kinematic models have been discussed in [Kal84].

Frequency-domain analysis of alpha-beta filters and the steady-state bias resulting from constant accelerations are discussed in [FS85]. The equivalent bandwidth of polynomial filters has been presented in [NL83]. The response of alpha-beta-gamma filters to step inputs can be found in [Nav77].

A discussion of the possible unbounded errors in decoupled alpha-beta filters is presented in [Rog88].

6.6.2 Problems

6-1 **Simulated kinematic trajectory behavior.** A target is simulated as having a nearly constant velocity motion with white noise acceleration, constant over the sampling period T, as in (6.3.2-1). The noise is $v(k) \sim \mathcal{N}(0, \sigma_v^2)$.

    1. Find the prior pdf of the velocity at time k.

    2. What is the range of the velocity k sampling periods after the initial time?

    3. Assume the initial velocity is 10, the process noise variance and the sampling time are both unity. If, after $k = 25$ samples the velocity became zero, is this a sign that the random number generator is biased?

    4. What conclusion can be drawn from the above about the behavior of the velocity for the third order kinematic model (6.3.3-1) with process noise representing acceleration increments?

6-2 **Process noise rescaling** (for second order direct discrete time kinematic model when sampling period is changed). A target is simulated according to the second order kinematic model (6.3.2-1) with a sampling period T_1 with process noise $v(k)$, which represents the constant acceleration over a sampling period, with variance $\sigma_v^2(T_1)$. Subsequently, the sampling period is changed to T_2 and we want to preserve the statistical properties of the motion. What should be done?

6-3 **Process noise rescaling** (for third order direct discrete time kinematic model when sampling period is changed). A target is simulated according to the third order kinematic model (6.3.3-1) with a sampling period T_1 with process noise $v(k)$, which represents the acceleration increment over a sampling period, with variance $\sigma_v^2(T_1)$. Subsequently, the sampling period is changed to T_2 and we want to preserve the statistical properties of the motion. What should be done?

6-4 **Simulated kinematic trajectory variability when sampling period is changed.** A trajectory is generated as in problem 6-3 with $T_1 = 1$ and $\sigma_v^2(T_1) = 0.1$ for $N_1 = 20$ periods. Then, using the same random number generator seed, the same trajectory is generated with a higher sampling rate, $T_2 = 0.5$ for $N_2 = 40$ samples, that is, the same total time.

1. Can one expect the trajectories to be identical?

2. Find the range of the difference between the two trajectories' accelerations at the common final time t_F.

6-5 **Alpha-beta filter design and evaluation.**

1. Design an α-β filter for a data rate of 40 Hz, with maximum acceleration $|a_M| = 16g$ ($g \approx 10m/s^2$), and measurement noise with $\sigma_w = 10m$.

2. Calculate the (steady-state) error covariance matrix elements and the position and velocity RMS errors.

3. Calculate the RMS position prediction error for a prediction time of $t = 2s$ under the assumptions of the filter.

4. Assume that the target has a constant acceleration of $16g$ during this prediction time. Calculate the position prediction error due to this and compare it with the result from (3).

6-6 **Existence of steady-state filter for a kinematic model.** Given the system

$$x(k+1) = Fx(k) + \Gamma v(k)$$

$$z(k) = Hx(k) + w(k)$$

with

$$F = \begin{bmatrix} 1 & T \\ 0 & 1 \end{bmatrix} \qquad \Gamma = \begin{bmatrix} T^2/2 \\ T \end{bmatrix} \qquad H = [1 \ 0]$$

and the (scalar) process and measurement noise sequences are zero mean white with variances q and r, respectively.

6.6.2 Problems

1. State the condition for stability (existence of steady state) of the Kalman filter for this system in terms of F and H. Prove that this holds for the above system.

2. State the additional condition that the steady-state filter covariance is positive definite and unique. Prove that this also holds.

6-7 **Alpha filter.** Given the scalar system

$$x(k+1) = x(k) + \frac{T^2}{2} v(k)$$

$$z(k) = x(k) + w(k)$$

with the two noise sequences mutually uncorrelated, zero mean, white, and with variances σ_v^2 and σ_w^2, respectively. Let

$$\lambda \triangleq \frac{T^2 \sigma_v}{\sigma_w}$$

Find

1. The steady-state Kalman filter gain α in terms of λ.

2. The noise reduction factor

$$\frac{P_{xx}}{\sigma_w^2}$$

in terms of α, where P_{xx} is the steady-state variance of the estimate of x from the KF.

Chapter 7

COMPUTATIONAL ASPECTS OF ESTIMATION

7.1 INTRODUCTION

7.1.1 Implementation of Linear Estimation

This chapter describes briefly some numerical techniques for the efficient *implementation of the linear estimation techniques* presented earlier.

The techniques to be discussed are for linear systems with state equation

$$x(k+1) = F(k)x(k) + \Gamma(k)v(k) \qquad (7.1.1\text{-}1)$$

and measurement equation

$$z(k) = H(k)x(k) + w(k) \qquad (7.1.1\text{-}2)$$

where $v(k)$ and $w(k)$ are the process and measurement noises, assumed to be zero mean, white, mutually uncorrelated, and with covariances $Q(k)$ and $R(k)$, respectively.

The two properties of a covariance matrix

- symmetry, and
- positive definiteness

can be lost due to **round-off errors** in the course of the calculations of its propagation equations — the covariance prediction and the covariance update.

The covariance propagation equations are discussed next and their propensity for causing **loss of symmetry** and/or **loss of positive definiteness** is examined.

The **covariance prediction equation**

$$P(k+1|k) = F(k)P(k|k)F(k)' + \Gamma(k)Q(k)\Gamma(k)' \qquad (7.1.1\text{-}3)$$

can affect only the symmetry of the resulting matrix. A suitable implementation of the products of three matrices will avoid this problem.

More significant numerical problems arise in the **covariance update equation**, which can be written in the following algebraically equivalent forms:

$$P(k+1|k+1) = \big[I - W(k+1)H(k+1)\big]P(k+1|k) \qquad (7.1.1\text{-}4)$$

$$P(k+1|k+1) = P(k+1|k) - W(k+1)S(k+1)W(k+1)' \qquad (7.1.1\text{-}5)$$

$$P(k+1|k+1) = \big[I - W(k+1)H(k+1)\big]P(k+1|k)\big[I - W(k+1)H(k+1)\big]' + W(k+1)R(k+1)W(k+1)' \qquad (7.1.1\text{-}6)$$

Equation (7.1.1-4) is very sensitive to round-off errors and is bound to lead to loss of symmetry as well as positive definiteness — it is best to avoid it.

Equation (7.1.1-5) will avoid loss of symmetry with a suitable implementation of the last term which is a product of three matrices (as in the covariance prediction) but it can still lead to loss of positive definiteness due to numerical errors in the subtraction.

Equation (7.1.1-6), the **Joseph form covariance update**, while computationally more expensive than (7.1.1-5), is less sensitive to round-off errors. With the proper implementation of the products of three matrices it will preserve symmetry. Furthermore, since the only place it has a subtraction is in the term $I - WH$, which appears "squared," this form of the covariance update has the property of preserving the positive definiteness of the resulting updated covariance.

7.1.2 Outline

The purposes of the techniques to be discussed in this chapter are:

1. Reduction of the computational requirements.
2. Improvement of the numerical accuracy — preservation of the symmetry and the positive definiteness of the state covariance matrix, as well as reduction of its **condition number** (the logarithm of the ratio of its largest to its smallest eigenvalue).

Section 7.2 presents the **information filter**, which carries out the recursive computation of the inverse of the covariance matrix. This is an alternative to (7.1.1-5) and is less demanding computationally for systems with dimension of the measurement vector larger than that of the state.

A technique which carries out the state update with one measurement component at a time — the **sequential updating** technique — rather than the entire measurement vector at once, is described in Section 7.3. This implementation of the Kalman filter is less demanding computationally than the standard implementation.

The technique of **square-root filtering**, which consists of *sequential updating* and **factorization of the covariance matrix**, is described in Section 7.4. This approach avoids the numerical problems that can lead, due to round-off errors, to the loss of symmetry and positive definiteness of the state covariance matrix. It is also less expensive computationally than the best standard implementation (7.1.1-6) while it has double the accuracy.

7.1.3 Computational Aspects — Summary of Objectives

Present techniques for implementation of the Kalman filter that are

- more economical computationally and/or
- more robust numerically.

Typical numerical problems with the state covariance matrix:

- Loss of symmetry.
- Loss of positive definiteness.

The algorithms to be discussed:

- Information filter — calculates recursively the inverse covariance matrix.
- Sequential processing of measurements in updating the state estimate.
- Square-root filter: a combination of sequential updating with covariance factorization that achieves double precision.

7.2 THE INFORMATION FILTER

7.2.1 Recursions for the Information Matrices

The standard version of the Kalman filter calculates the gain in conjunction with a recursive computation of the state covariance. Starting from $P(k-1|k-1)$, the prediction covariance is obtained as

$$P(k|k-1) = F(k-1)P(k-1|k-1)F(k-1)' + \Gamma(k-1)Q(k-1)\Gamma(k-1)' \quad (7.2.1\text{-}1)$$

The gain is

$$W(k) = P(k|k-1)H(k)'\left[H(k)P(k|k-1)H(k)' + R(k)\right]^{-1} \quad (7.2.1\text{-}2)$$

and the updated state covariance $P(k|k)$ is calculated from

$$P(k|k) = P(k|k-1) - P(k|k-1)H(k)'\left[H(k)P(k|k-1)H(k)' + R(k)\right]^{-1}H(k)P(k|k-1) \quad (7.2.1\text{-}3)$$

which completes one cycle of computations.

The **information filter** calculates recursively the *inverses of the covariance matrices*, both for the prediction and the update. The term "information" is used in the sense of the Cramer-Rao lower bound, where the **information matrix** is the *inverse of the covariance matrix*.

A recursion from $P(k-1|k-1)^{-1}$ to $P(k|k-1)^{-1}$ to $P(k|k)^{-1}$ and the expression of the filter gain in terms of (one of) the above inverses is presented next.

The update equation for the information matrix follows immediately from (7.2.1-3) and the matrix inversion lemma (see Subsection 1.3.3) as

$$P(k|k)^{-1} = P(k|k-1)^{-1} + H(k)'R(k)^{-1}H(k) \quad (7.2.1\text{-}4)$$

Denote the information matrix corresponding to the state prediction *without process noise* as

$$A(k-1)^{-1} \triangleq F(k-1)P(k-1|k-1)F(k-1)' \quad (7.2.1\text{-}5)$$

or, since F, as a transition matrix, is invertible,

$$A(k-1) = \left[F(k-1)^{-1}\right]'P(k-1|k-1)^{-1}F(k-1)^{-1} \quad (7.2.1\text{-}6)$$

The prediction information matrix is

$$P(k|k-1)^{-1} = \left[F(k-1)P(k-1|k-1)F(k-1)' + \Gamma(k-1)Q(k-1)\Gamma(k-1)'\right]^{-1}$$
$$= \left[A(k-1)^{-1} + \Gamma(k-1)Q(k-1)\Gamma(k-1)'\right]^{-1} \qquad (7.2.1\text{-}7)$$

and it can be rewritten, again with the matrix inversion lemma, as

$$P(k|k-1)^{-1} = A(k-1) - A(k-1)\Gamma(k-1)$$
$$\cdot\left[\Gamma(k-1)'A(k-1)\Gamma(k-1) + Q(k-1)^{-1}\right]^{-1}\Gamma(k-1)'A(k-1)$$
$$(7.2.1\text{-}8)$$

The expression of the gain (7.2.1-2) is rewritten as

$$W(k) = P(k|k-1)H(k)'\left\{\left[H(k)P(k|k-1)H(k)' + R(k)\right]^{-1} + R(k)^{-1} - R(k)^{-1}\right\}$$
$$= P(k|k-1)H(k)'R(k)^{-1} + P(k|k-1)H(k)'\left[H(k)P(k|k-1)H(k)' + R(k)\right]^{-1}$$
$$\cdot\left\{I - \left[H(k)P(k|k-1)H(k)' + R(k)\right]R(k)^{-1}\right\}$$
$$= \left\{P(k|k-1) - P(k|k-1)H(k)'\left[H(k)P(k|k-1)H(k)' + R(k)\right]^{-1}\right.$$
$$\left.\cdot H(k)P(k|k-1)\right\}H(k)'R(k)^{-1} \qquad (7.2.1\text{-}9)$$

With the matrix inversion lemma, (7.2.1-9) becomes

$$W(k) = \left[P(k|k-1)^{-1} + H(k)'R(k)^{-1}H(k)\right]^{-1}H(k)'R(k)^{-1} \qquad (7.2.1\text{-}10)$$

which is the sought-after expression of the gain. It can be easily shown that the above is equivalent to the alternate form of the gain:

$$W(k) = P(k|k)H(k)'R(k)^{-1} \qquad (7.2.1\text{-}11)$$

Duality Between the Covariance and Information Equations

Note the **duality** [1] between the covariance and the information propagation equations:

Covariance prediction (7.2.1-1) $\longleftrightarrow$ Information update (7.2.1-4)

Covariance update (7.2.1-3) $\longleftrightarrow$ Information prediction (7.2.1-8)

Carrying out in (7.2.1-1), with the time index increased by one, the following replacements

$$P(k+1|k) \rightarrow P(k+1|k+1)^{-1} \qquad (7.2.1\text{-}12)$$

$$A(k) \rightarrow P(k+1|k) \qquad (7.2.1\text{-}13)$$

$$\Gamma(k) \rightarrow H(k)' \qquad (7.2.1\text{-}14)$$

$$Q(k) \rightarrow R(k)^{-1} \qquad (7.2.1\text{-}15)$$

yield (7.2.1-4).

Similarly, carrying out in (7.2.1-3), again with the time index increased by one, the replacements

$$P(k+1|k)^{-1} \leftarrow P(k+1|k+1) \qquad (7.2.1\text{-}16)$$

$$A(k) \leftarrow P(k+1|k) \qquad (7.2.1\text{-}17)$$

$$\Gamma(k) \leftarrow H(k)' \qquad (7.2.1\text{-}18)$$

$$Q(k)^{-1} \leftarrow R(k) \qquad (7.2.1\text{-}19)$$

yield (7.2.1-8).

Since (7.2.1-16) to (7.2.1-19) are exactly the inverses of (7.2.1-12) to (7.2.1-15), this proves the duality.

---

[1]The concept of duality can be illustrated by what the frog said in a restaurant when he got his soup: "Waiter! There is *no* fly in my soup!"

7.2.2 Overview of the Information Filter Algorithm

The sequence of calculations of the information matrices and filter gain for one cycle of the information filter are presented in Figure 7.2.2-1.

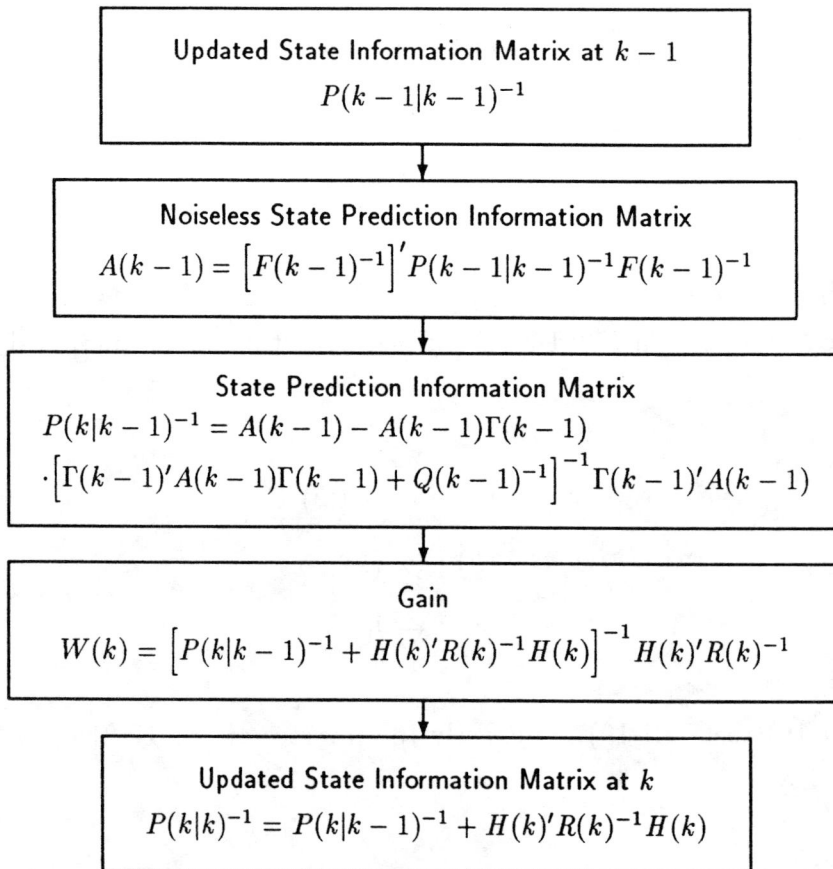

Figure 7.2.2-1: The information matrix and filter gain calculations in the information filter.

This implementation of the Kalman filter is advantageous when the dimension n_z of the measurement vector is larger than the dimension n_x of the state. Note that the inversions in the above sequence of calculations are for $n_x \times n_x$ matrices, while the standard algorithm requires the inversion of the innovation covariance, which is $n_z \times n_z$. The inverse of the measurement noise covariance $R(k)$ is simple to obtain because it is, in general, diagonal.

7.3 SEQUENTIAL PROCESSING OF MEASUREMENTS

7.3.1 Block versus Sequential Processing

The standard implementation of the Kalman filter for a vector measurement is by carrying out the state update simultaneously with the entire measurement vector from a given time, i.e., **block processing**.

If the measurement noise vector components $w_i(k)$, $i = 1, \ldots, n_z$, are uncorrelated, that is,

$$
\begin{aligned}
R(k) &= E[w(k)w(k)'] \\
&= \text{diag}[r_1(k), \ldots, r_{n_z}(k)]
\end{aligned}
\tag{7.3.1-1}
$$

then one can carry out the update of the state with *one component of the measurement at a time*, that is, **sequential processing**, or **scalar updates**.

If the matrix R is not diagonal, one can apply a linear transformation on the measurement to diagonalize its covariance matrix. This can be done using a Cholesky decomposition, to be discussed later in Subsection 7.4.2.

The measurement

$$
\begin{aligned}
z(k) &= \begin{bmatrix} z_1(k) \\ \vdots \\ z_{n_z}(k) \end{bmatrix} = H(k)x(k) + w(k) \\
&= \begin{bmatrix} h_1(k)'x(k) + w_1(k) \\ \vdots \\ h_{n_z}(k)'x(k) + w_{n_z}(k) \end{bmatrix}
\end{aligned}
\tag{7.3.1-2}
$$

will be considered as a *sequence of scalar measurements* $z_i(k)$, $i = 1, \ldots, n_z$.

The uncorrelatedness of the corresponding measurement noises allows the use of the Kalman filter update sequentially for each "scalar measurement" because it implies whiteness for the scalar measurement noise "sequence"

$$
w_1(k), \ldots, w_{n_z}(k), w_1(k+1), \ldots, w_{n_z}(k+1), \ldots
\tag{7.3.1-3}
$$

A natural application of this is in a **multisensor** situation where the measurement noises are independent across sensors.

7.3.2 The Sequential Processing Algorithm

The sequence of *scalar updates* is described below.

Starting from the predicted state to k from $k - 1$, denoted as

$$\hat{x}(k|k, 0) \triangleq \hat{x}(k|k - 1) \tag{7.3.2-1}$$

with associated covariance

$$P(k|k, 0) \triangleq P(k|k - 1) \tag{7.3.2-2}$$

the following sequence of calculations is carried out for $i = 1, \ldots, n_z$:

The scalar innovation corresponding to $z_i(k)$ has variance

$$s(k, i) = h_i(k)'P(k|k, i - 1)h_i(k) + r_i(k) \tag{7.3.2-3}$$

The corresponding gain is

$$W(k, i) = \frac{P(k|k, i - 1)h_i(k)}{s(k, i)} \tag{7.3.2-4}$$

and the updated state

$$\hat{x}(k|k, i) = \hat{x}(k|k, i - 1) + W(k, i)[z_i(k) - h_i(k)'\hat{x}(k|k, i - 1)] \tag{7.3.2-5}$$

with covariance

$$P(k|k, i) = P(k|k, i - 1) - W(k, i)h_i(k)'P(k|k, i - 1) \tag{7.3.2-6}$$

which can be rewritten (and should be implemented) as

$$P(k|k, i) = P(k|k, i - 1) - \frac{P(k|k, i - 1)h_i(k)h_i(k)'P(k|k, i - 1)}{h_i(k)'P(k|k, i - 1)h_i(k) + r_i(k)} \tag{7.3.2-7}$$

Note that, in the i-th update, $P(k|k, i - 1)$ plays the role of a prediction covariance. Finally,

$$\hat{x}(k|k) \triangleq \hat{x}(k|k, n_z) \tag{7.3.2-8}$$

$$P(k|k) \triangleq P(k|k, n_z) \tag{7.3.2-9}$$

Note that this approach eliminates the need for the information filter since only scalar inversions are performed here.

302

If the measurement noise covariance matrix (7.3.1-1) is not diagonal, a transformation that diagonalizes its covariance matrix can be used. An efficient technique to accomplish this is via the Cholesky factorization, to be discussed in detail in Section 7.4. This factorization yields

$$R = LD_RL'$$ (7.3.2-10)

where L is a *lower triangular matrix* and D_R is a *diagonal matrix with positive entries*. Instead of z, one can use the transformed measurement

$$\bar{z} = L^{-1}z$$ (7.3.2-11)

and instead of H one has

$$\bar{H} = L^{-1}H$$ (7.3.2-12)

This then allows the use of the scalar update procedure described above.

A Note on Numerical Accuracy

The covariance update (7.3.2-7) has a subtraction and, therefore, is susceptible to numerical errors, which can cause even the loss of positive definiteness of the covariance matrix. The alternative form, known as the **Joseph form covariance update**,

$$P(k|k,i) = \left[I - W(k,i)h_i(k)'\right]P(k|k,i-1)\left[I - W(k,i)h_i(k)'\right]' + W(k,i)r_i(k)W(k,i)'$$ (7.3.2-13)

avoids this, but is computationally more expensive.

While the above form of the covariance update is better behaved numerically, it is still too sensitive to round-off errors. Further reduction of the effect of numerical errors can be obtained by using the square-root approach, which effectively doubles the numerical precision.

7.4 SQUARE-ROOT FILTERING

7.4.1 The Steps in Square-Root Filtering

The **square-root filtering** implementation of the KF carries out the covariance computations for the square root of the state covariance matrix, symbolically written as $P^{1/2}$, where

$$
\begin{aligned}
P &= P^{1/2}(P^{1/2})' \\
&\triangleq \mathcal{PP}'
\end{aligned}
\tag{7.4.1-1}
$$

The **square root of a matrix** is not unique, so several approaches are possible. The approach described in the sequel uses the **Cholesky factorization** of a (positive definite) matrix

$$
P = LDL' \tag{7.4.1-2}
$$

where L is a **unit lower triangular matrix**, that is,

$$
L_{ii} = 1 \qquad \forall i \tag{7.4.1-3}
$$

$$
L_{ij} = 0 \qquad i < j \tag{7.4.1-4}
$$

and D is a *diagonal matrix with positive elements*.

The matrix square root $\mathcal{P}$ in (7.4.1-1) corresponding to (7.4.1-2) is

$$
\mathcal{P} = LD^{1/2} \tag{7.4.1-5}
$$

which is unique, except for the signs of the (diagonal) elements of $D^{1/2}$, which are taken as positive.

The square-root filtering algorithm consists of the following steps:

1. Factorization of the initial covariance.
2. Computation of the factors of the predicted state covariance.
3. Computation of the factors of the updated state covariance and computation of the filter gain.

These steps are discussed in the next subsections.

7.4.2   The Cholesky Factorization

Given the positive definite $n \times n$ matrix P, its decomposition, called the **Cholesky factorization**,

$$P = [p_{ij}] = LDL' \tag{7.4.2-1}$$

where L is a *unit lower triangular matrix* and D is a *diagonal matrix with positive elements*, is obtained as follows [Bie77]:

For $j = 1, \ldots, n-1$
$\quad d_j = P_{jj}$
$\quad L_{jj} = 1$ (all others are zero)
$\quad$ For $k = j+1, \ldots, n$
$\quad\quad$ For $i = k, \ldots, n$
$\quad\quad\quad P_{ik} := P_{ik} - L_{ij}P_{kj}$
$\quad\quad L_{kj} = \dfrac{P_{kj}}{d_j}$
$d_n = P_{nn}$

Note that since the diagonal terms of P have to be positive (because the matrix was assumed positive definite), the elements of the diagonal matrix D are guaranteed to be positive.

7.4.3   The Predicted State Covariance

The standard equation for the state prediction covariance is

$$P(k|k-1) = F(k-1)P(k-1|k-1)F(k-1)' + \Gamma(k-1)Q(k-1)\Gamma(k-1)' \quad (7.4.3\text{-}1)$$

or, in simpler notation, without time arguments

$$\bar{P} = FPF' + \Gamma Q \Gamma' \quad (7.4.3\text{-}2)$$

where P denotes the previous updated state covariance and $\bar{P}$ denotes state prediction covariance.

Starting from the factorized form

$$P(k-1|k-1) \triangleq P = LDL' \quad (7.4.3\text{-}3)$$

we want to find the new factorized form

$$P(k|k-1) \triangleq \bar{P} = \bar{L}\bar{D}\bar{L}' \quad (7.4.3\text{-}4)$$

The brute force method would be to evaluate (7.4.3-2) and find its Cholesky factorization using the method described in Subsection 7.4.2.

A better technique that takes advantage of the existing factorizations to obtain the **factorized prediction covariance** is as follows.

It is assumed that Q has been factorized as

$$Q = L_Q D_Q L_Q' \quad (7.4.3\text{-}5)$$

It will be shown how the factors of P given in (7.4.3-3) and the factors of Q given in (7.4.3-5) can be used to find the factors of $\bar{P}$.

Then we are looking for $\bar{L}, \bar{D}$ such that

$$\bar{P} = \bar{L}\bar{D}\bar{L}' = FLDL'F' + \Gamma L_Q D_Q L_Q' \Gamma' \quad (7.4.3\text{-}6)$$

Equation (7.4.3-6) can be rewritten as the product

$$\bar{P} = AA' \quad (7.4.3\text{-}7)$$

with the following $n_x \times 2n_x$ matrix

$$A \triangleq \begin{bmatrix} FLD^{1/2} & \Gamma L_Q D_Q^{1/2} \end{bmatrix} \triangleq \begin{bmatrix} a_1' \\ a_2' \\ \vdots \\ a_{n_x}' \end{bmatrix} \triangleq \mathrm{col}(a_i') \quad (7.4.3\text{-}8)$$

where a_i are $2n_x$-vectors.

Using the **Gram-Schmidt orthogonalization** procedure (described in Subsection 7.4.6) on the *rows* of the above matrix, one obtains

$$A = \bar{L}V \tag{7.4.3-9}$$

In the above, $\bar{L}$ is a *unit lower triangular matrix* and V satisfies

$$VV' = \bar{D} \tag{7.4.3-10}$$

where $\bar{D}$ is *diagonal with positive elements.*

Thus

$$\begin{aligned}
\bar{P} &= AA' \\
&= \bar{L}VV'\bar{L}' \\
&= \bar{L}\bar{D}\bar{L}'
\end{aligned} \tag{7.4.3-11}$$

is the sought-after Cholesky factorization of the prediction covariance (7.4.3-4).

Therefore, to obtain the factors $\bar{L}$ and V of the prediction covariance $\bar{P}$, one carries out the Gram-Schmidt orthogonalization of the matrix A in (7.4.3-8), which contains the factors L and D of the previous updated state covariance P and the factors L_Q and D_Q of the process noise covariance Q.

7.4.4 The Filter Gain and the Updated State Covariance

Assuming a scalar measurement update, the implementation of the state update given in (7.3.2-5) requires the calculation of the gain W, which is an n_x-vector. This is obtained together with the factorized form of the updated state covariance as follows.

With the predicted state covariance factorized as $\bar{L}\bar{D}\bar{L}$, we are looking for the factorized updated covariance

$$
\begin{aligned}
P &\triangleq LDL' \\
&= \bar{L}\bar{D}\bar{L}' - \frac{\bar{L}\bar{D}\bar{L}'hh'\bar{L}\bar{D}\bar{L}'}{h'\bar{L}\bar{D}\bar{L}'h + r}
\end{aligned}
\tag{7.4.4-1}
$$

where the time arguments and subscripts have been omitted for simplicity.

Defining

$$
\begin{aligned}
f &= \bar{L}'h \\
&= [f_1 \quad \cdots \quad f_n]'
\end{aligned}
\tag{7.4.4-2}
$$

Equation (7.4.4-1) can be written as

$$
LDL' = \bar{L}\left[\bar{D} - \frac{\bar{D}f(\bar{D}f)'}{f'\bar{D}f + r}\right]\bar{L}'
\tag{7.4.4-3}
$$

With

$$
\bar{D} = \operatorname{diag}[\bar{d}_1, \dots, \bar{d}_n]
\tag{7.4.4-4}
$$

one has

$$
\bar{D}f = [\bar{d}_1 f_1 \quad \cdots \quad \bar{d}_n f_n]'
\tag{7.4.4-5}
$$

Denote the *columns* of $\bar{L}$ as $\bar{\ell}_i$ (which are n_x-vectors), that is,

$$
\bar{L} = [\bar{\ell}_1 \quad \cdots \quad \bar{\ell}_n]
\tag{7.4.4-6}
$$

The **covariance update and gain calculation** algorithm consists of the following [Kle89]:

1. Initialize

$$\alpha_{n+1} = r \tag{7.4.4-7}$$

$$\alpha_n = r + \bar{d}_n f_n^2 \tag{7.4.4-8}$$

$$\xi = [0 \quad \cdots \quad 0 \quad \bar{d}_n f_n]' \tag{7.4.4-9}$$

$$d_n = \frac{\alpha_{n+1}}{\alpha_n}\bar{d}_n \tag{7.4.4-10}$$

2. For $i = n - 1, \ldots, 1$

$$\beta = -\frac{f_i}{\alpha_{i+1}} \tag{7.4.4-11}$$

$$\alpha_i = \alpha_{i+1} + \bar{d}_i f_i^2 \tag{7.4.4-12}$$

$$d_i = \frac{\alpha_{i+1}}{\alpha_i}\bar{d}_i \tag{7.4.4-13}$$

$$\ell_i = \bar{\ell}_i + \beta\xi \tag{7.4.4-14}$$

$$\xi := \xi + \bar{\ell}_i\bar{d}_i f_i \tag{7.4.4-15}$$

3. Gain vector for the state update

$$W = \frac{1}{\alpha_1}\xi \tag{7.4.4-16}$$

In the above d_i, $i = 1, \ldots, n_x$, are the elements of the diagonal matrix D and the n_x-vectors ℓ_i, $i = 1, \ldots, n_x$, are the columns of the unit lower diagonal matrix L. These two matrices define the Cholesky factorization (7.4.4-1) of the updated state covariance matrix P.

7.4.5   Overview of the Square-Root Sequential Scalar Update Algorithm

It is assumed that the measurement noise covariance matrix is diagonal. If this is not the case, a Cholesky factorization is performed as in Subsection 7.4.2 and the transformations (7.3.2-10) to (7.3.2-12) are carried out.

Figure 7.4.5-1 describes the sequence of computations for the square-root sequential scalar update algorithm. The last block is repeated for each component of the n_z-dimensional measurement vector.

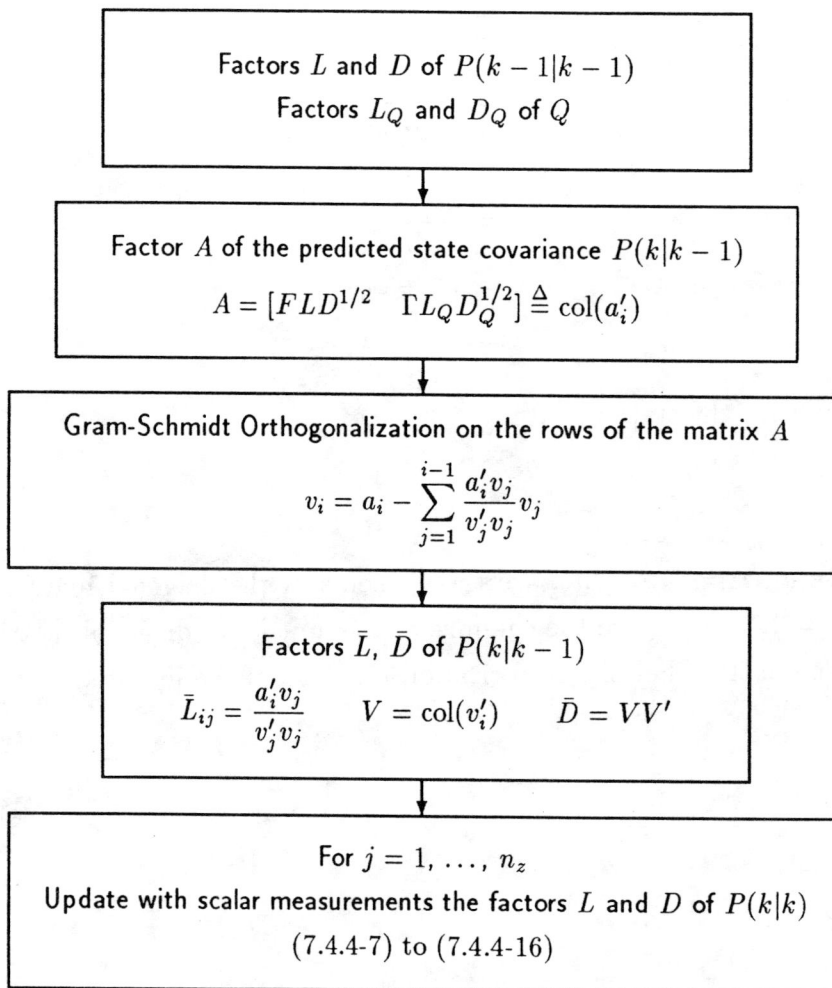

$$
\boxed{
\begin{array}{c}
\text{Factors } L \text{ and } D \text{ of } P(k-1|k-1) \\
\text{Factors } L_Q \text{ and } D_Q \text{ of } Q
\end{array}
}
$$

$$
\boxed{
\begin{array}{c}
\text{Factor } A \text{ of the predicted state covariance } P(k|k-1) \\
A = [FLD^{1/2} \quad \Gamma L_Q D_Q^{1/2}] \triangleq \text{col}(a_i')
\end{array}
}
$$

$$
\boxed{
\begin{array}{c}
\text{Gram-Schmidt Orthogonalization on the rows of the matrix } A \\
v_i = a_i - \sum_{j=1}^{i-1} \frac{a_i' v_j}{v_j' v_j} v_j
\end{array}
}
$$

$$
\boxed{
\begin{array}{c}
\text{Factors } \bar{L}, \bar{D} \text{ of } P(k|k-1) \\
\bar{L}_{ij} = \frac{a_i' v_j}{v_j' v_j} \qquad V = \text{col}(v_i') \qquad \bar{D} = VV'
\end{array}
}
$$

$$
\boxed{
\begin{array}{c}
\text{For } j = 1, \ldots, n_z \\
\text{Update with scalar measurements the factors } L \text{ and } D \text{ of } P(k|k) \\
(7.4.4\text{-}7) \text{ to } (7.4.4\text{-}16)
\end{array}
}
$$

Figure 7.4.5-1: The square-root sequential covariance update algorithm.

7.4.6 The Gram-Schmidt Orthogonalization Procedure

Given N-vectors a_i, $i = 1, \ldots, n$, where $N \geq n$, assumed to be linearly independent, let

$$v_i \triangleq a_i - \sum_{j=1}^{i-1} \frac{a_i' v_j}{v_j' v_j} v_j \qquad i = 1, \ldots, n \qquad (7.4.6\text{-}1)$$

Then the vectors v_i, $i = 1, \ldots, n$, form an orthogonal set. Equation (7.4.6-1) can be rewritten as

$$a_i = v_i + \sum_{j=1}^{i-1} \bar{L}_{ij} v_j \qquad (7.4.6\text{-}2)$$

where

$$\bar{L}_{ij} \triangleq \frac{a_i' v_j}{v_j' v_j} \qquad (7.4.6\text{-}3)$$

Then

$$A = \begin{bmatrix} a_1' \\ a_2' \\ \vdots \\ a_n' \end{bmatrix} = \begin{bmatrix} 1 & 0 & \cdots & 0 \\ \bar{L}_{21} & 1 & \cdots & 0 \\ \vdots & \vdots & \ddots & \vdots \\ \bar{L}_{n1} & \bar{L}_{n2} & \cdots & 1 \end{bmatrix} \begin{bmatrix} v_1' \\ v_2' \\ \vdots \\ v_n' \end{bmatrix} \triangleq \bar{L}V \qquad (7.4.6\text{-}4)$$

where $\bar{L}$ is a unit lower triangular matrix ($n \times n$) and

$$V \triangleq \begin{bmatrix} v_1' \\ v_2' \\ \vdots \\ v_n' \end{bmatrix} \triangleq \mathrm{col}(v_i') \qquad (7.4.6\text{-}5)$$

Note that the transpose of v_i is the i-th *row* of V.

From the orthogonality of the n-vectors v_i, $i = 1, \ldots, n$, it follows that

$$VV' = \bar{D} \qquad (7.4.6\text{-}6)$$

is a diagonal matrix (also $n \times n$).

Equation (7.4.6-1) is called the **Gram-Schmidt orthogonalization procedure**.

The so-called modified Gram-Schmidt orthogonalization procedure (e.g., [Bie77]) has improved numerical properties compared to the standard procedure described above.

Application to the Generation of Correlated Random Variables

This orthogonalization procedure can also be used in the reverse direction to provide a recursion for the *generation of correlated random variables* from a standard random number generator that yields independent random variables.

Similarly, one can obtain a simple recursion for the *generation of a random sequence with exponential autocorrelation* (see problem 7-1).

These techniques are useful in simulations that require the generation of correlated random variables.

7.5 NOTES AND PROBLEMS

7.5.1 Bibliographical Notes

The most comprehensive documentation of numerical algorithms for linear estimation can be found in [Bie77, May79, May82]. The material presented in this chapter is based in part on private communications from D. L. Kleinman and K. R. Pattipati.

The use of parallel computers for the implementation of linear estimation can be found in [OB88]. Parallel implementation of factorized (square-root) estimation is discussed in [IB89].

Several applications of covariance factorization in estimation as well as additional references on this topic can be found in [BL90].

7.5.2 Problems

7-1 **Generation of a random sequence with exponential autocorrelation.** Given a set of independent zero-mean unit-variance random variables, x_i, $i = 1, \ldots$, find a *recursion* that yields the zero-mean sequence y_i, $i = 1, \ldots$, with the autocorrelation function

$$E[y_i y_j] = \rho^{|i-j|} \qquad |\rho| < 1$$

Hint: The form of the recursion should be $y_{i+1} = f(y_i, x_{i+1})$.

Chapter 8

EXTENSIONS OF DISCRETE TIME LINEAR ESTIMATION

8.1 INTRODUCTION

8.1.1 Outline

The major assumptions of the optimal linear estimator for dynamic systems (the Kalman filter) are:

1. The process noise sequence is white.
2. The measurement noise sequence is white.
3. The two noise sequences are uncorrelated.

The next three sections discuss the procedures to reduce the problems where the above assumptions are not satisfied to the standard case.

Section 8.2 presents the *prewhitening* and *state augmentation* technique needed in the case of *autocorrelated process noise*. Section 8.3 deals with the case of *correlated noise sequences* and shows how this correlation can be eliminated. Section 8.4 considers the situation of *autocorrelated measurement noise*, which is solved with the help of the techniques from the previous two sections.

Estimation of the state at times other than the current time is discussed in the following two sections. *Prediction* is treated in Section 8.5 and *smoothing* — estimation of the state at an earlier time than the last data point — is the topic of Section 8.6.

8.1.2 Extensions of Estimation — Summary of Objectives

Reduce the problems of state estimation in the following nonstandard situations:

- autocorrelated process noise,
- cross-correlated process and measurement noise, and
- autocorrelated measurement noise

to the standard problem where both noise sequences are *white and mutually uncorrelated*.

Provide equations for estimation of the state at times other than the time of the latest measurement:

- Prediction — beyond the data interval.
- Smoothing — within the data interval.

8.2 AUTOCORRELATED PROCESS NOISE

8.2.1 The Problem

Consider the dynamic system driven by an **autocorrelated process noise** (also called **colored noise**)

$$x(k+1) = Fx(k) + v_c(k) \qquad (8.2.1\text{-}1)$$

where $v_c(k)$ is a zero-mean stationary but not white sequence

$$E[v_c(k)v_c(j)'] = Q(k-j) \qquad (8.2.1\text{-}2)$$

Then the state $x(k)$ of the above system is *not a Markov sequence*. The goal is to estimate the state from the measurements

$$z(k) = Hx(k) + w(k) \qquad (8.2.1\text{-}3)$$

where

$$E[w(k)w(j)'] = R\delta_{kj} \qquad (8.2.1\text{-}4)$$

$$E[v(k)w(j)'] = 0 \qquad (8.2.1\text{-}5)$$

that is, with white measurement noise uncorrelated from the process noise.

In order to be able to apply the standard estimation results, one has to reformulate the problem into one with a state that is a Markov sequence, that is, one where the process noise driving the dynamic system is white.

This is accomplished by obtaining the **prewhitening system** (or **shaping filter**) for the process noise and appending it to the original system, that is, carrying out a **state augmentation** such that the augmented state becomes a Markov sequence.

This is illustrated in the sequel with an example motivated by maneuvering targets.

8.2.2 An Exponentially Autocorrelated Noise

The Acceleration Model

In Chapter 6 a second order kinematic model for target tracking was obtained modeling the acceleration as *white noise*. In the approach discussed here the target acceleration $a(t)$ is modeled as an **exponentially autocorrelated noise** with mean zero and

$$R(\tau) = E[a(t)a(t+\tau)] = \sigma_m^2 e^{-\alpha|\tau|} \qquad \alpha > 0 \qquad (8.2.2\text{-}1)$$

In the above σ_m^2 is the **instantaneous variance** of the acceleration and $1/\alpha$ is the time constant of the target acceleration autocorrelation, that is, the **decorrelation time** is approximately $2/\alpha$.

The Noise Variance and the Target Physical Parameters

The variance of the (instantaneous) acceleration can be obtained by assuming it to be

1. equal to a maximum value a_M with probability p_M and $-a_M$ with the same probability (chosen by the designer),
2. equal to zero with probability p_0 (also a design parameter),
3. uniformly distributed in $[-a_M, a_M]$ with the remaining probability mass.

This results in the following mixed pmf/pdf (a combination of three point masses and a uniform pdf, with total probability mass unity)

$$p(a) = [\delta(a-a_M)+\delta(a+a_M)]p_M+\delta(a)p_0+[1(a+a_M)-1(a-a_M)]\frac{1-p_0-2p_M}{2a_M} \quad (8.2.2\text{-}2)$$

where $1(\cdot)$ denotes the unit step function.

It can be easily shown that the variance corresponding to (8.2.2-2) is

$$\sigma_m^2 = \frac{a_M^2}{3}(1+4p_M-p_0) \qquad (8.2.2\text{-}3)$$

The dynamic model — the **prewhitening system** — that yields (8.2.2-1) is the first order Markov process

$$\dot{a}(t) = -\alpha a(t) + \tilde{v}(t) \qquad (8.2.2\text{-}4)$$

driven by the white noise $\tilde{v}(t)$ with

$$E[\tilde{v}(t)\tilde{v}(\tau)] = 2\alpha\sigma_m^2\delta(t-\tau) \qquad (8.2.2\text{-}5)$$

This will be used in the next subsection as a subsystem to augment the state equations according to the procedure discussed in Subsection 1.4.21.

Remarks

1. As α increases the process $a(t)$ becomes uncorrelated faster. At the limit (suitably defined — see problem 8-4), as $\alpha \to \infty$ and $\sigma_m \to \infty$, the acceleration becomes white noise, that is,

$$\lim_{\alpha \to \infty,\, \sigma_m \to \infty} R(\tau) = q\delta(\tau) \qquad (8.2.2\text{-}6)$$

The relationship between q in (8.2.2-6) and the parameters σ_m and α is

$$\frac{2}{\alpha}\sigma_m^2 = q \qquad (8.2.2\text{-}7)$$

This case corresponds to the white noise acceleration model (of second order) discussed in Subsection 6.2.2.

2. For $\alpha \to 0$, (8.2.2-4) yields $a(t)$ as the integral of white noise, that is, the acceleration becomes a *Wiener process*, which is not stationary (note that the process defined in (8.2.2-1) is stationary). This is suitable for maneuver modeling for time-varying acceleration with a suitable value for the process noise variance q.

This case corresponds to the Wiener process acceleration (white noise jerk) model, which is of third order, and was presented in Subsection 6.2.3.

8.2.3   The Augmented State Equations

With the acceleration a white noise, the state would consist of position and velocity (per coordinate), as in Subsection 6.2.2. Augmented with the acceleration, the state for the generic coordinate ξ is denoted as

$$x = \begin{bmatrix} \xi & \dot{\xi} & \ddot{\xi} \end{bmatrix}' \tag{8.2.3-1}$$

where $\ddot{\xi} = a$ is the acceleration.

The continuous-time state equations are

$$\dot{x}_1(t) = x_2(t) \tag{8.2.3-2}$$

$$\dot{x}_2(t) = x_3(t) \tag{8.2.3-3}$$

to which one appends the prewhitening subsystem (of dimension 1 in this case — in general it can be of higher dimension)

$$\dot{x}_3(t) = -\alpha x_3(t) + \tilde{v}(t) \tag{8.2.3-4}$$

The **augmented state equation** in vector form is then

$$\dot{x}(t) = Ax(t) + D\tilde{v}(t) \tag{8.2.3-5}$$

with the system matrix

$$A = \begin{bmatrix} 0 & 1 & 0 \\ 0 & 0 & 1 \\ 0 & 0 & -\alpha \end{bmatrix} \tag{8.2.3-6}$$

and $\tilde{v}(t)$ is the (scalar) white process noise from (8.2.2-4), which enters the system through the gain

$$D = \begin{bmatrix} 0 \\ 0 \\ 1 \end{bmatrix} \tag{8.2.3-7}$$

Note that $\alpha = 0$ makes the above identical to (6.2.3-4).

The discrete time equation corresponding to (8.2.3-5) for sampling period T is

$$x(k+1) = Fx(k) + v(k) \tag{8.2.3-8}$$

where

$$F = e^{AT} = \begin{bmatrix} 1 & T & (\alpha T - 1 + e^{-\alpha T})/\alpha^2 \\ 0 & 1 & (1 - e^{-\alpha T})/\alpha \\ 0 & 0 & e^{-\alpha T} \end{bmatrix} \tag{8.2.3-9}$$

320

8.2.3 The Augmented State Equations

The discrete time process noise $v(k)$ has the covariance matrix

$$Q = 2\alpha\sigma_m^2 \begin{bmatrix} T^5/20 & T^4/8 & T^3/6 \\ T^4/8 & T^3/3 & T^2/2 \\ T^3/6 & T^2/2 & T \end{bmatrix} \tag{8.2.3-10}$$

Expression (8.2.3-10) assumes

$$\alpha T \ll 1 \tag{8.2.3-11}$$

that is, that the sampling time T is much less than the time constant $1/\alpha$ of the maneuver autocorrelation. Typical values of $1/\alpha$ for aircraft are $60s$ for a slow turn and $20s$ for an evasive maneuver; atmospheric turbulence has a time constant around $1s$.

The exact expression of Q has the following elements

$$q_{11} = \frac{\sigma_m^2}{\alpha^4}\left[1 - e^{-2\alpha T} + 2\alpha T + \frac{2\alpha^3 T^3}{3} - 2\alpha^2 T^2 - 4\alpha T e^{-\alpha T}\right] \tag{8.2.3-12}$$

$$q_{12} = \frac{\sigma_m^2}{\alpha^3}\left[e^{-2\alpha T} + 1 - 2e^{-\alpha T} + 2\alpha T e^{-\alpha T} - 2\alpha T + \alpha^2 T^2\right] \tag{8.2.3-13}$$

$$q_{13} = \frac{\sigma_m^2}{\alpha^2}\left[1 - e^{-2\alpha T} - 2\alpha T e^{-\alpha T}\right] \quad . \tag{8.2.3-14}$$

$$q_{22} = \frac{\sigma_m^2}{\alpha^2}\left[4e^{-\alpha T} - 3 - e^{-2\alpha T} + 2\alpha T\right] \tag{8.2.3-15}$$

$$q_{23} = \frac{\sigma_m^2}{\alpha}\left[e^{-2\alpha T} + 1 - 2e^{-\alpha T}\right] \tag{8.2.3-16}$$

$$q_{33} = \sigma_m^2[1 - e^{-2\alpha T}] \tag{8.2.3-17}$$

Remark

Assumption (8.2.3-11) implies $T \ll 1/\alpha$, that is, that the sampling period is much smaller than the acceleration decorrelation time. If this is not the case, then the sampling is too slow to "catch" the correlation in the acceleration. Thus, if one uses this model, then (8.2.3-11) should be satisfied and (8.2.3-10) is adequate.

8.2.4 Estimation with Autocorrelated Process Noise — Summary

To estimate the state of a system driven by an *autocorrelated process noise,*

- The prewhitening subsystem for the autocorrelated process noise has to be obtained.
- The original system is augmented with this subsystem such that the resulting system is driven by white noise.

This has been illustrated with an application from maneuvering targets. Target maneuvers, which by their nature have a certain duration in a certain direction, can be modeled by autocorrelated noise.

An exponentially decaying autocorrelation for target maneuvers can model

- The *magnitude of the maneuvers,* and
- The *decorrelation time* of the maneuvers.

The magnitude of the maneuvers is modeled by their instantaneous variance. One possible procedure to obtain it, based on the following:

- the maximum maneuver and the probability that it will take place,
- the probability that no maneuver will take place, and
- a random maneuver of magnitude up to the maximum

has been illustrated.

The time constant of the decorrelation of the maneuvers depends on the target.

This exponentially autocorrelated acceleration is obtained from a subsystem that is a first order Markov process driven by white noise. Higher order prewhitening subsystems can be used in a similar manner.

322

8.3 CROSS-CORRELATED MEASUREMENT AND PROCESS NOISE

Consider the model

$$x(k+1) = Fx(k) + v(k) \tag{8.3.0-1}$$

$$z(k) = Hx(k) + w(k) \tag{8.3.0-2}$$

$$E[v(k)v(j)'] = Q\delta_{kj} \tag{8.3.0-3}$$

$$E[w(k)w(j)'] = R\delta_{kj} \tag{8.3.0-4}$$

$$E[v(k)w(j)'] = U\delta_{kj} \tag{8.3.0-5}$$

where the last equation indicates **cross-correlated noise sequences** — a nonzero cross-correlation between the process and measurement noise sequences, that is, it is *not a standard filtering problem.*

The plant equation can be rewritten such that it has a new process noise uncorrelated with the measurement noise. Using an *arbitrary matrix T*, to be determined later, one can write

$$\begin{aligned} x(k+1) &= Fx(k) + v(k) + T[z(k) - Hx(k) - w(k)] \\ &= (F - TH)x(k) + v(k) - Tw(k) + Tz(k) \end{aligned} \tag{8.3.0-6}$$

Note that (8.3.0-6) is *completely equivalent* to (8.3.0-1) since they differ only by a term that is identically zero.

Denote the new transition matrix

$$F^* \triangleq F - TH \tag{8.3.0-7}$$

new process noise

$$v^*(k) \triangleq v(k) - Tw(k) \tag{8.3.0-8}$$

and the last term in (8.3.0-6), which is a *known input*, is denoted as

$$u^*(k) \triangleq Tz(k) \tag{8.3.0-9}$$

Then one can write the modified state equation (8.3.0-6) as

$$x(k+1) = F^*x(k) + v^*(k) + u^*(k) \tag{8.3.0-10}$$

Setting the cross-correlation between the new process noise and the measurement noise to zero

$$E[v^*(k)w(k)'] = E\Big[[v(k) - Tw(k)]w(k)'\Big] = U - TR = 0 \qquad (8.3.0\text{-}11)$$

yields the unspecified matrix T as

$$T = UR^{-1} \qquad (8.3.0\text{-}12)$$

With the above, the covariance of the new process noise

$$
\begin{aligned}
Q^* &\triangleq E[v^*(k)v^*(k)'] \\
&= E\Big[[v(k) - UR^{-1}w(k)][v(k) - UR^{-1}w(k)]'\Big] \qquad (8.3.0\text{-}13)
\end{aligned}
$$

is obtained as

$$\boxed{Q^* = Q - UR^{-1}U'} \qquad (8.3.0\text{-}14)$$

Using (8.3.0-7), (8.3.0-9), and (8.3.0-12), the modified state equation (8.3.0-10) can be rewritten as

$$\boxed{x(k+1) = (F - UR^{-1}H)x(k) + v^*(k) + UR^{-1}z(k)} \qquad (8.3.0\text{-}15)$$

where the known input — the last term above — is written out explicitly.

Summary

The state estimation problem with cross-correlated process and measurement noise sequences has been transformed into the following equivalent problem defined by

- The modified state equation (8.3.0-15)
  - with covariance of the new process noise (8.3.0-14), and
  - with a known input.
- The original measurement equation (8.3.0-2)
  - with the measurement noise covariance given by (8.3.0-4).

This is now a *standard problem* since the two noise sequences are uncorrelated according to (8.3.0-11).

8.4 AUTOCORRELATED MEASUREMENT NOISE

8.4.1 Whitening of the Measurement Noise

Consider the model

$$x(k+1) = Fx(k) + v(k) \tag{8.4.1-1}$$

$$z(k) = Hx(k) + w_c(k) \tag{8.4.1-2}$$

$$w_c(k+1) = F_c w_c(k) + w_w(k) \tag{8.4.1-3}$$

$$E[w_w(k)w_w(j)'] = R_w \delta_{kj} \tag{8.4.1-4}$$

$$E[v(k)v(j)'] = Q\delta_{kj} \tag{8.4.1-5}$$

$$E[v(k)w_w(j)'] = 0 \tag{8.4.1-6}$$

where (8.4.1-3) indicates a Markov **autocorrelated measurement noise**. The "brute force" solution is to augment the state with w_c. This reduces the problem to a standard one, but with "perfect state observations," which yields a singular state covariance — an undesirable feature.

Introduce the following **differenced measurement**:

$$
\begin{aligned}
y(k) &\triangleq z(k+1) - F_c z(k) = Hx(k+1) + w_c(k+1) - F_c Hx(k) - F_c w_c(k) \\
&= HFx(k) + Hv(k) + w_w(k) - F_c Hx(k)
\end{aligned}
\tag{8.4.1-7}
$$

which can be written as follows

$$y(k) = H^* x(k) + w(k) \tag{8.4.1-8}$$

where the new measurement matrix is

$$H^* \triangleq HF - F_c H \tag{8.4.1-9}$$

The new measurement noise

$$w(k) \triangleq Hv(k) + w_w(k) \tag{8.4.1-10}$$

is *white* but *correlated with the process noise*

$$E[w(k)w(j)'] = (HQH' + R_w)\delta_{kj} \triangleq R\delta_{kj} \tag{8.4.1-11}$$

$$E[v(k)w(k)'] = QH' \tag{8.4.1-12}$$

The technique from the previous section will be used to eliminate the cross-correlation between the two noise sequences. Rewrite the plant equation:

$$
\begin{aligned}
x(k+1) &= Fx(k) + v(k) + T[y(k) - H^*x(k) - w(k)] \\
&= (F - TH^*)x(k) + v(k) - Tw(k) + Ty(k)
\end{aligned}
\tag{8.4.1-13}
$$

Define the new process noise

$$
v^*(k) \triangleq v(k) - Tw(k)
\tag{8.4.1-14}
$$

and new transition matrix

$$
F^* = F - TH^*
\tag{8.4.1-15}
$$

With this, one can write the new state equation as

$$
x(k+1) = F^*x(k) + v^*(k) + Ty(k)
\tag{8.4.1-16}
$$

In the above, T will be chosen such that *the cross-correlation is zero*:

$$
E[v^*(k)w(k)'] = E\Big[[v(k) - Tw(k)]w(k)'\Big] = QH' - TR = 0
\tag{8.4.1-17}
$$

This yields

$$
T = QH'R^{-1} = QH'(HQH' + R_w)^{-1}
\tag{8.4.1-18}
$$

The new transition matrix is then

$$
F^* \triangleq F - TH^* = F - QH'(HQH' + R_w)^{-1}(HF - F_cH)
\tag{8.4.1-19}
$$

The new process noise covariance is

$$
\begin{aligned}
E[v^*(k)v^*(k)'] &= E\Big[[v(k) - QH'R^{-1}w(k)][v(k) - QH'R^{-1}w(k)]'\Big] \\
&= Q - QH'R^{-1}HQ \triangleq Q^*
\end{aligned}
\tag{8.4.1-20}
$$

The estimation can now be done with

- the state equation (8.4.1-16) with white process noise $v^*(k)$ with covariance (8.4.1-20), and

- the measurements (8.4.1-8), with white measurement noise $w(k)$ with covariance (8.4.1-11).

These two noise sequences are mutually uncorrelated according to (8.4.1-17). This is a standard problem except that $y(k)$ contains $z(k+1)$.

326

8.4.2 The Estimation Algorithm with the Whitened Measurement Noise

The estimation algorithm is described next. First note that, in view of (8.4.1-7), Y^k is *equivalent* to Z^{k+1}. Therefore,

$$\hat{x}(k|Y^k) = \hat{x}(k|Z^{k+1}) = \hat{x}(k|k+1); \qquad P(k|Y^k) = P(k|Z^{k+1}) = P(k|k+1) \quad (8.4.2\text{-}1)$$

$$\hat{x}(k+1|Y^k) = \hat{x}(k+1|Z^{k+1}) = \hat{x}(k+1|k+1); \quad P(k+1|Y^k) = P(k+1|Z^{k+1}) = P(k+1|k+1)$$
$$(8.4.2\text{-}2)$$

The filter stages below are labeled according to the indices of x and y.

"*Prediction*" — conditional expectation of (8.4.1-16) given Y^k:

$$
\begin{aligned}
\hat{x}(k+1|Y^k) &= F^* \hat{x}(k|Y^k) + Ty(k) = F^* \hat{x}(k|k+1) + Ty(k) \\
&= \hat{x}(k+1|k+1)
\end{aligned}
\tag{8.4.2-3}
$$

$$
\begin{aligned}
P(k+1|Y^k) &= F^* P(k|Y^k) F^{*'} + Q^* = F^* P(k|k+1) F^{*'} + Q^* \\
&= P(k+1|k+1)
\end{aligned}
\tag{8.4.2-4}
$$

"*Update*":

$$
\begin{aligned}
W(k+1) &= P(k+1|Y^k) H^{*'} S(k+1)^{-1} \\
&= P(k+1|k+1) H^{*'} S(k+1)^{-1}
\end{aligned}
\tag{8.4.2-5}
$$

$$
\begin{aligned}
S(k+1) &= H^* P(k+1|Y^k) H^{*'} + R \\
&= H^* P(k+1|k+1) H^{*'} + R
\end{aligned}
\tag{8.4.2-6}
$$

$$
\begin{aligned}
\hat{x}(k+1|Y^{k+1}) &= \hat{x}(k+1|Y^k) + W(k+1)[y(k+1) - H^* \hat{x}(k+1|Y^k)] \\
&= \hat{x}(k+1|k+1) + W(k+1)[y(k+1) - H^* \hat{x}(k+1|k+1)] \\
&= \hat{x}(k+1|k+2)
\end{aligned}
\tag{8.4.2-7}
$$

$$
\begin{aligned}
P(k+1|Y^{k+1}) &= P(k+1|Y^k) - W(k+1) S(k+1) W(k+1)' \\
&= P(k+1|k+1) - W(k+1) S(k+1) W(k+1)' \\
&= P(k+1|k+2)
\end{aligned}
\tag{8.4.2-8}
$$

The initiation is done starting with the initial estimate $\hat{x}(0|0)$ with covariance $P(0|0)$, and using (8.4.2-7) and (8.4.2-8) with $k = -1$. Following this, the "prediction" stage is carried out and the cycle continues with the "update."

8.5 PREDICTION

8.5.1 Types of Prediction

Prediction is the *estimation of the state* at time j *beyond the data interval*, that is, based on data up to an earlier time $k < j$,

$$\hat{x}(j|k) = E[x(j)|Z^k] \tag{8.5.1-1}$$

There are several types of prediction:

(a) **Fixed point prediction**: this is done to the fixed point in time $j = N$ (e.g., an intercept time) based on the data up to time k, where k is changing.

(b) **Fixed lead prediction**: this is done for a fixed number of L steps ahead, that is, $j = k + L$, where k varies and L is the lead.

(c) **Fixed interval prediction**: this is done based on the data from the fixed interval up to $k = N$ to times beyond it, $j = N + 1, \ldots$.

These are illustrated in Figure 8.5.1-1.

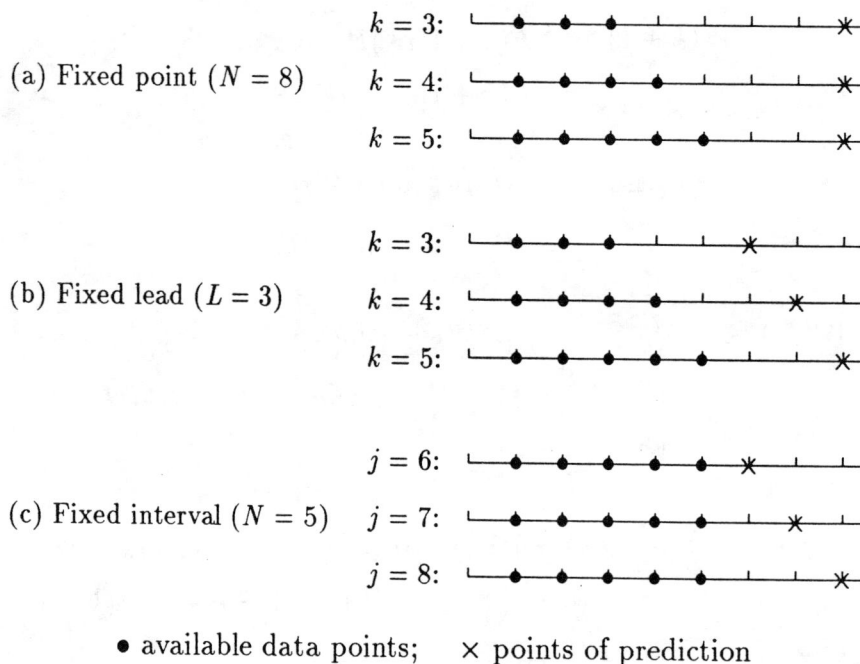

(a) Fixed point ($N = 8$) $k = 3$: $k = 4$: $k = 5$:

(b) Fixed lead ($L = 3$) $k = 3$: $k = 4$: $k = 5$:

(c) Fixed interval ($N = 5$) $j = 6$: $j = 7$: $j = 8$:

● available data points;   × points of prediction

Figure 8.5.1-1: The three types of prediction.

The equations for these three types of prediction are given below.

328

8.5.2   The Algorithms for the Different Types of Prediction

Fixed Point Prediction

Based on (4.3.3-1), the **fixed point prediction** of the state to the fixed time (point) N from time k is

$$
\begin{aligned}
\hat{x}(N|k) &= E\left[\left[\prod_{j=0}^{N-k-1} F(N-1-j)\right]x(k) + \sum_{i=k}^{n-1}\left[\prod_{j=0}^{N-i-2} F(N-1-j)\right]v(i)\Big|Z^k\right] \\
&= \left[\prod_{j=0}^{N-k-1} F(N-1-j)\right]\hat{x}(k|k) \tag{8.5.2-1}
\end{aligned}
$$

since the process noise, being zero mean and white, yields zero after the expectation is applied.

The above can be rewritten as

$$
\hat{x}(N|k) = \left[\prod_{j=0}^{N-k-1} F(N-1-j)\right][F(k-1)\hat{x}(k-1|k-1) + W(k)\nu(k)] \tag{8.5.2-2}
$$

which becomes the recursion

$$
\boxed{\hat{x}(N|k) = \hat{x}(N|k-1) + \left[\prod_{j=0}^{N-k-1} F(N-1-j)\right]W(k)\nu(k)} \tag{8.5.2-3}
$$

It can be shown (see problem 8-1) that the expected value of the prediction to N from i, conditioned on the data through time k, is

$$
E[\hat{x}(N|i)|Z^k] = \hat{x}(N|k) \qquad \forall i > k \tag{8.5.2-4}
$$

This quantity is the *predicted value at k of the prediction to be available at the later time i.*

The covariance of this "prediction of a prediction" is

$$
\boxed{E\left[[\hat{x}(N|i) - \hat{x}(N|k)][\hat{x}(N|i) - \hat{x}(N|k)]'|Z^k\right] = P(N|k) - P(N|i) \quad \forall i > k} \tag{8.5.2-5}
$$

329

Fixed Lead Prediction

Similarly to (8.5.2-1), one has the **fixed lead prediction** for the fixed lead of L steps, given by

$$\hat{x}(k+L|k) = \left[\prod_{j=0}^{L-1} F(k+L-1-j)\right]\hat{x}(k|k)$$

$$= \left[\prod_{j=0}^{L-1} F(k+L-1-j)\right][F(k-1)\hat{x}(k-1|k-1) + W(k)\nu(k)]$$

$$(8.5.2\text{-}6)$$

which becomes the recursion

$$\boxed{\hat{x}(k+L|k) = F(k+L-1)\hat{x}(k+L-1|k-1) + \left[\prod_{j=0}^{L-1} F(k+L-1-j)\right]W(k)\nu(k)}$$

$$(8.5.2\text{-}7)$$

The covariance associated with the above follows from the standard propagation equation

$$\boxed{P(k+l|k) = F(k+l-1)P(k+l-1|k)F(k+l-1)' + Q(k+l-1) \qquad l = 1,\ldots,L}$$

$$(8.5.2\text{-}8)$$

Fixed Interval Prediction

The propagation equation of the **fixed interval prediction** of the state at time j based on the fixed data interval up to N is

$$\hat{x}(j|N) = F(j-1)\hat{x}(j-1|N) \tag{8.5.2-9}$$

or

$$\boxed{\hat{x}(j|N) = F(j-1)\hat{x}(j-1|N) = \left[\prod_{l=0}^{j-N-1} F(j-N-1-l)\right]\hat{x}(N|N) \qquad j > N}$$

$$(8.5.2\text{-}10)$$

The associated covariance follows from the standard state prediction covariance propagation equation

$$\boxed{P(l|N) = F(l-1)P(l-1|N)F(l-1)' + Q(l-1) \qquad l = N+1,\ldots,j} \tag{8.5.2-11}$$

8.6 SMOOTHING

8.6.1 Types of Smoothing

Smoothing or **retrodiction** is the *estimation of the state* at time k *within the data interval*, that is, based on data up to $j > k$,

$$\hat{x}(k|j) = E[x(k)|Z^j] \qquad (8.6.1\text{-}1)$$

There are several types of smoothing:

(a) **Fixed point smoothing**: k is fixed and $j = k+1, k+2, \ldots$.

(b) **Fixed lag smoothing**: k varies and $j = k + L$, where L is the lag.

(c) **Fixed interval smoothing**: $j = N$ (the data interval is up to N) and $k = 0, 1, \ldots, N$.

These are illustrated in Figure 8.6.1-1.

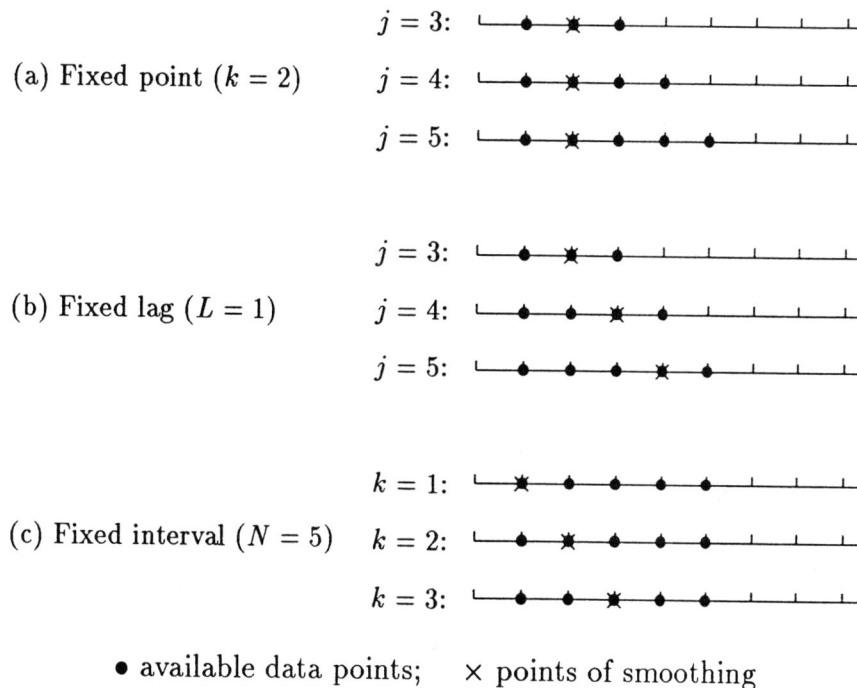

• available data points;   × points of smoothing

Figure 8.6.1-1: The three types of smoothing.

The fixed interval smoothing problem, which is the most common, is discussed in detail next.

8.6.2 Fixed Interval Smoothing

In the following, the **fixed interval smoothing** algorithm will be derived. This is the most commonly encountered situation where smoothing is of interest, for instance, the smoothing of an entire trajectory based on all the available data after an experiment.

The approach consists of

1. setting up the posterior pdf of the states at two consecutive times, conditioned on the measurements in the given interval *under the Gaussian assumption,* and

2. maximizing it to yield the MAP estimates of the states, which are then the sought-after conditional means — the smoothed states.

This will yield a **backward recursion** that calculates the smoothed state at time k from the smoothed state at time $k+1$.

By our definition

$$\hat{x}(k|N) = \arg\max_{x(k)} p[x(k)|Z^N] = \arg\max_{x(k)} p[x(k), Z^N] \qquad (8.6.2\text{-}1)$$

Consider the joint pdf

$$p[x(k), x(k+1), Z^N] = p[x(k), x(k+1), Z^N_{k+1}, Z^k] = p[x(k), x(k+1), Z^N_{k+1}|Z^k]p(Z^k) \qquad (8.6.2\text{-}2)$$

where

$$Z^j_i \triangleq \{z(l)\}^j_{l=i} \qquad (8.6.2\text{-}3)$$

The maximization of (8.6.2-2) with respect to $x(k)$ and $x(k+1)$ will yield the corresponding smoothed states as defined in (8.6.2-1).

Assume that $\hat{x}(k|k)$, $k = 1, \ldots, N$, are available from a ("forward") filtering process that has already been carried out. A backward smoothing recursion will be derived: assuming that $\hat{x}(k+1|N)$ is available, $\hat{x}(k|N)$ will be obtained.

From the above discussion it follows that

$$\hat{x}(k|N) = \arg\max_{x(k)} \left\{ p[x(k), x(k+1), Z^N]\Big|_{x(k+1)=\hat{x}(k+1|N)} \right\} \qquad (8.6.2\text{-}4)$$

The first term on the right hand side of (8.6.2-2) can be rewritten as

$$
\begin{aligned}
p[x(k), x(k+1), Z^N_{k+1}|Z^k] &= p[Z^N_{k+1}|x(k+1), x(k), Z^k]p[x(k+1), x(k)|Z^k] \\
&= p[Z^N_{k+1}|x(k+1)]p[x(k+1)|x(k), Z^k]p[x(k)|Z^k] \\
&= p[Z^N_{k+1}|x(k+1)]p[x(k+1)|x(k)]p[x(k)|Z^k] \quad (8.6.2\text{-}5)
\end{aligned}
$$

The conditioning term dropped in the last line above is irrelevant.

332

In view of (8.6.2-5), (8.6.2-4) can be written as

$$\hat{x}(k|N) = \arg\max_{x(k)} \left\{ p[x(k+1)|x(k)]p[x(k)|Z^k]\Big|_{x(k+1)=\hat{x}(k+1|N)} \right\} \quad (8.6.2\text{-}6)$$

because only the last two terms of (8.6.2-5) contain $x(k)$.

Using now the Gaussian assumption, one has

$$p[x(k+1)|x(k)] = \mathcal{N}[x(k+1); F(k)x(k), Q(k)] \quad (8.6.2\text{-}7)$$

where $F(k)$ is the system transition matrix and $Q(k)$ the process noise covariance,

$$p[x(k)|Z^k] = \mathcal{N}[x(k); \hat{x}(k|k), P(k|k)] \quad (8.6.2\text{-}8)$$

where $P(k|k)$ is the covariance associated with the estimate $\hat{x}(k|k)$.

Using the explicit forms of (8.6.2-7) and (8.6.2-8) in (8.6.2-6), and noting that maximization is equivalent to minimization of the negative of the logarithm, (8.6.2-6) can be replaced by

$$\hat{x}(k|N) = \arg\min_{x(k)} J \quad (8.6.2\text{-}9)$$

where

$$
\begin{aligned}
J \triangleq\ & [\hat{x}(k+1|N) - F(k)x(k)]'Q(k)^{-1}[\hat{x}(k+1|N) - F(k)x(k)] \\
& + [x(k) - \hat{x}(k|k)]'P(k|k)^{-1}[x(k) - \hat{x}(k|k)]
\end{aligned} \quad (8.6.2\text{-}10)
$$

Setting the gradient of J with respect to $x(k)$ to zero

$$\nabla_{x(k)}J = -2F(k)'Q(k)^{-1}[\hat{x}(k+1|N) - F(k)x(k)] + 2P(k|k)^{-1}[x(k) - \hat{x}(k|k)] = 0 \quad (8.6.2\text{-}11)$$

yields the smoothed value $\hat{x}(k|N)$ of $x(k)$ as

$$[F(k)'Q(k)^{-1}F(k) + P(k|k)^{-1}]\hat{x}(k|N) = F(k)'Q(k)^{-1}\hat{x}(k+1|N) + P(k|k)^{-1}\hat{x}(k|k) \quad (8.6.2\text{-}12)$$

Using the matrix inversion lemma

$$(F'Q^{-1}F + P^{-1})^{-1} = P - PF'(FPF' + Q)^{-1}FP \quad (8.6.2\text{-}13)$$

in (8.6.2-12) yields, with the arguments of F, P and Q omitted

$$
\begin{aligned}
\hat{x}(k|N) =\ & [P - PF'(FPF' + Q)^{-1}FP][F'Q^{-1}\hat{x}(k+1|N) + P^{-1}\hat{x}(k|k)] \\
=\ & [PF'Q^{-1} - PF'(FPF' + Q)^{-1}(FPF' + Q - Q)Q^{-1}]\hat{x}(k+1|N) \\
& + \hat{x}(k|k) - PF'(FPF' + Q)^{-1}F\hat{x}(k|k) \\
=\ & [PF'Q^{-1} - PF'Q^{-1} + PF'(FPF' + Q)^{-1}]\hat{x}(k+1|N) \\
& + \hat{x}(k|k) - PF'(FPF' + Q)^{-1}F\hat{x}(k|k) \\
=\ & \hat{x}(k|k) + PF'(FPF' + Q)^{-1}[\hat{x}(k+1|N) - F\hat{x}(k|k)]
\end{aligned} \quad (8.6.2\text{-}14)
$$

From (8.6.2-14), the sought-after backward recursion for the smoothed value of the state — the **smoothed state** — is

$$\hat{x}(k|N) \;=\; \hat{x}(k|k) + C(k)[\hat{x}(k+1|N) - F(k)\hat{x}(k|k)]$$
$$k = N-1,\ldots,0 \qquad\qquad (8.6.2\text{-}15)$$

or

$$\boxed{\hat{x}(k|N) = \hat{x}(k|k) + C(k)[\hat{x}(k+1|N) - \hat{x}(k+1|k)]} \qquad (8.6.2\text{-}16)$$

where the **smoother gain** is

$$C(k) = P(k|k)F(k)'[F(k)P(k|k)F(k)' + Q(k)]^{-1} \qquad (8.6.2\text{-}17)$$

or

$$\boxed{C(k) = P(k|k)F(k)'P(k+1|k)^{-1}} \qquad (8.6.2\text{-}18)$$

The initial index for (8.6.2-16) is $k = N-1$ and $\hat{x}(k|N)$ follows from $\hat{x}(k+1|N)$. This backward recursion starts from the last estimated state $\hat{x}(N|N)$. It can be easily shown that (8.6.2-16) holds in exactly the same form even if there is a known input.

The corresponding recursion for the **smoothed covariance** (the covariance of the smoothed state)

$$P(k|N) \triangleq E\Big[[x(k) - \hat{x}(k|N)][x(k) - \hat{x}(k|N)]'\Big] \qquad (8.6.2\text{-}19)$$

is obtained next.

Equation (8.6.2-16) can be rewritten as

$$\hat{x}(k|N) - C(k)\hat{x}(k+1|N) = \hat{x}(k|k) - C(k)F(k)\hat{x}(k|k) \qquad (8.6.2\text{-}20)$$

Subtracting the above from the identity

$$x(k) \;=\; \hat{x}(k|N) + \tilde{x}(k|N)$$
$$=\; \hat{x}(k|j) + \tilde{x}(k|j) \qquad\qquad (8.6.2\text{-}21)$$

yields

$$\tilde{x}(k|N) + C(k)\hat{x}(k+1|N) = \tilde{x}(k|k) + C(k)F(k)\hat{x}(k|k) \qquad (8.6.2\text{-}22)$$

The following orthogonality properties

$$\tilde{x}(k|N) \perp \hat{x}(k+1|N) \qquad (8.6.2\text{-}23)$$

$$\tilde{x}(k|k) \perp \hat{x}(k|k) \qquad (8.6.2\text{-}24)$$

and a vector version of the fact that the mean of the square of a random variable is equal to the square of its mean plus its variance will be used. The (rearranged) vector versions to be used are

$$
\begin{aligned}
E[\hat{x}(k+1|N)\hat{x}(k+1|N)'] &= E[x(k+1)x(k+1)'] - P(k+1|N) \\
&= F(k)E[x(k)x(k)']F(k)' + Q(k) - P(k+1|N)
\end{aligned}
$$
$$(8.6.2\text{-}25)$$

and

$$E[\hat{x}(k|k)\hat{x}(k|k)'] = E[x(k)x(k)'] - P(k|k) \qquad (8.6.2\text{-}26)$$

Then the expected value of the outer product of each side of (8.6.2-22) with itself yields (with all the cross-terms zero after applying the expectation)

$$
\begin{aligned}
P(k|N) &+ C(k)\big\{F(k)E[x(k)x(k)']F(k)' + Q(k) - P(k+1|N)\big\}C(k)' \\
&= P(k|k) + C(k)F(k)\big\{E[x(k)x(k)'] - P(k|k)\big\}F(k)'C(k)'
\end{aligned}
$$
$$(8.6.2\text{-}27)$$

After cancellations and rearrangements in (8.6.2-27), one obtains

$$P(k|N) = P(k|k) + C(k)[P(k+1|N) - F(k)P(k|k)F(k)' - Q(k)]C(k)' \qquad (8.6.2\text{-}28)$$

or

$$\boxed{P(k|N) = P(k|k) + C(k)[P(k+1|N) - P(k+1|k)]C(k)' \qquad k = N-1,\ldots,0}$$
$$(8.6.2\text{-}29)$$

which is the backward recursion for the **smoothed state covariance**. It can be easily shown that the term in the brackets in (8.6.2-29) is negative semidefinite.

Equations (8.6.2-16) and (8.6.2-29) are the backward smoothing recursions.

Remark

The quantity multiplied by the smoothing gain in (8.6.2-16)

$$\mu(k|N) \triangleq \hat{x}(k+1|N) - \hat{x}(k+1|k) \qquad (8.6.2\text{-}30)$$

can be interpreted as the **smoothing residual**.

8.7 NOTES AND PROBLEMS

8.7.1 Bibliographical Notes

The use of autocorrelated noise to model maneuvers, presented in Section 8.2 is based on [Sin70]. The steady-state solution for the filter gains in explicit form, similar to the alpha-beta-gamma filter, is available in [Fit81].

The situation of cross-correlated noise sequences and colored measurement noise has been treated in [BH75, SM71, AM79, May82]. These references also discuss prediction and smoothing. The fixed-interval smoothing problem discussed in Section 8.6 follows the development of [RTS65].

8.7.2 Problems

8-1 **Fixed point prediction — the mean and covariance.**

    1. Prove (8.5.2-4).

    2. Prove (8.5.2-5). Hint: Add and subtract $x(N)$ and use the fact that

$$\hat{x}(N|k) \perp \tilde{x}(N|i) \qquad \forall i > k$$

8-2 **Updated state covariance conditioned on prior data.**

    Find, for $k < N$, $\text{cov}[\hat{x}(N|N)|Z^k]$.

8-3 **Acceleration variance in the mixed pmf/pdf model.** Show how (8.2.2-3) follows from (8.2.2-2).

8-4 **Limiting process of an autocorrelated noise to white noise.** Prove (8.2.2-7). Hint: define the limit such that the power spectral density at frequency zero stays the same.

8-5 **Prewhitening of a colored noise — the shaping filter.** Given the zero-mean random sequence with autocorrelation function

$$E[w(i)w(j)] = \delta_{ij} + \alpha(\delta_{i+1,j} + \delta_{i-1,j})$$

find its prewhitening system, that is, the linear system driven by white noise whose output has this autocorrelation.

8-6 **Smoothing compared with estimation.**

    1. For problem 5-1(4), with the same noise sequence as used in the forward estimation, calculate and list:

$$\hat{x}(k|50), \tilde{x}(k|50), P(k|50), \tilde{x}(k|50)/\sqrt{P(k|50)}, \qquad k = 1,\ldots,50$$

    2. Compare $P(k|k)$ with $P(k|50)$.

8-7 **Dyad of a state estimate — the expected value.** Prove the identities (8.6.2-25) and (8.6.2-26).

8-8 **Mixed pmf/pdf of the acceleration model.** Derive (8.2.2-2) from the three assumptions right above it.

Chapter 9

CONTINUOUS TIME LINEAR STATE ESTIMATION

9.1 INTRODUCTION

9.1.1 Outline

This chapter deals with the estimation of the state of continuous time linear dynamic systems based on observations which are made also in continuous time.

The linear minimum mean square error (LMMSE) filter for this *continuous time* problem, known as the **Kalman-Bucy filter**, is obtained in Section 9.2 from a limiting process of the discrete time problem. The properties of the innovation process are derived. The asymptotic properties of the filtered state covariance equation — the continuous time Riccati equation — which yield a steady-state filter, are also presented.

Section 9.3 deals with the prediction of the state of a continuous time system.

The *duality* [1] of the LMMSE estimation with the linear-quadratic (LQ) control problem is discussed in Section 9.4. These two problems have their solutions determined by the same Riccati equation.

The Wiener-Hopf problem, which consists of the estimation of a stochastic process based on another observed process using their auto- and cross-covariances is the topic of a brief discussion in Section 9.5.

---

[1] This duality is conceptually similar to the one discussed on page 299.

9.1.2   Continuous Time Estimation — Summary of Objectives

- Derive the LMMSE filter for continuous time dynamic systems — the Kalman-Bucy filter — from a limiting process of the discrete time filter.

- Present the asymptotic properties of the continuous time Riccati matrix differential equation and the convergence of the filter to steady state.

- Discuss the state prediction.

- Show the duality between the following problems:

  - LMMSE estimation, and
  - LQ control.

- Briefly review the Wiener-Hopf problem — LMMSE estimation for stationary stochastic processes — and indicate its relationship with the Kalman-Bucy filter.

9.2 THE CONTINUOUS TIME LINEAR STATE ESTIMATION FILTER

9.2.1 The Continuous Time Estimation Problem

The state equation is (without input, for simplicity)

$$\dot{x}(t) = A(t)x(t) + D(t)\tilde{v}(t) \qquad (9.2.1\text{-}1)$$

with $\tilde{v}(t)$ the process noise, zero mean and white

$$E[\tilde{v}(t)] = 0 \qquad (9.2.1\text{-}2)$$

$$E[\tilde{v}(t)\tilde{v}(\tau)'] = \tilde{q}(t)\delta(t - \tau) \qquad (9.2.1\text{-}3)$$

where $\tilde{q}(t)$ is the (possibly time-varying) **intensity of the white noise**. This becomes the power spectral density if it is time invariant.

The measurement equation is

$$z(t) = C(t)x(t) + \tilde{w}(t) \qquad (9.2.1\text{-}4)$$

with the process noise $\tilde{w}(t)$ zero mean and white

$$E[\tilde{w}(t)] = 0 \qquad (9.2.1\text{-}5)$$

$$E[\tilde{w}(t)\tilde{w}(\tau)'] = \tilde{R}(t)\delta(t - \tau) \qquad (9.2.1\text{-}6)$$

The process noise, the measurement noise and the initial state are mutually uncorrelated. Note that the noises can be *nonstationary*.

The continuous time filter will be derived by taking the limit of the discrete time filter as $\Delta \triangleq (t_{k+1} - t_k) \to 0$.

Ito Differential Equation Form

The state equation is sometimes written in **Ito differential equation** form

$$dx(t) = A(t)x(t)dt + D(t)d\mathbf{w}_v(t) \tag{9.2.1-7}$$

where $d\mathbf{w}_v(t)$ is the infinitesimal increment of the Wiener process $\mathbf{w}_v(t)$ with moments

$$E[d\mathbf{w}_v(t)] = 0 \tag{9.2.1-8}$$

and

$$E[d\mathbf{w}_v(t)d\mathbf{w}_v(t)'] = \tilde{q}(t)dt \tag{9.2.1-9}$$

The incremental Wiener process in (9.2.1-7) is

$$d\mathbf{w}_v(t) = \tilde{v}(t)dt \tag{9.2.1-10}$$

Equation (9.2.1-7) is more rigorous than (9.2.1-1), since the white noise exists only in the sense of generalized functions (the Wiener process is nondifferentiable).

Similarly, the measurement equation (9.2.1-4) can be written as

$$d\zeta(t) \triangleq z(t)dt = C(t)x(t)dt + d\mathbf{w}_w(t) \tag{9.2.1-11}$$

where

$$d\mathbf{w}_w(t) = \tilde{w}(t)dt \tag{9.2.1-12}$$

and

$$E[d\mathbf{w}_w(t)d\mathbf{w}_w(t)'] = \tilde{R}(t)dt \tag{9.2.1-13}$$

9.2.2 Connection between Continuous and Discrete Time Representations and Noise Statistics

The discrete time state equation is, with F denoting the transition matrix,

$$x(t_{k+1}) = F(t_{k+1}, t_k)x(t_k) + v(t_k) \qquad (9.2.2\text{-}1)$$

The connection between the continuous and discrete time representations is, taking a small $\Delta \triangleq t_{k+1} - t_k$

$$
\begin{aligned}
\dot{x}(t_k) &= \frac{1}{\Delta}[x(t_k + \Delta) - x(t_k)] = \frac{1}{\Delta}[F(t_k + \Delta, t_k)x(t_k) + v(t_k) - x(t_k)] \\
&= \frac{1}{\Delta}[F(t_k + \Delta, t_k) - I]x(t_k) + \frac{1}{\Delta}v(t_k) \qquad (9.2.2\text{-}2)
\end{aligned}
$$

Using the following two properties of the transition matrix

$$F(t, t) = I \qquad (9.2.2\text{-}3)$$

$$\frac{\partial}{\partial \tau}F(\tau, t) = A(\tau)F(\tau, t) \qquad (9.2.2\text{-}4)$$

one has, for $\Delta \to 0$,

$$\frac{1}{\Delta}[F(t + \Delta, t) - I] = A(t) \qquad (9.2.2\text{-}5)$$

The expression of the discrete time process noise in (9.2.2-1) is

$$v(t_k) = \int_{t_k}^{t_{k+1}} F(t_{k+1}, \tau)D(\tau)\tilde{v}(\tau)d\tau \qquad (9.2.2\text{-}6)$$

Using the fact that, according to (9.2.2-3), the transition matrix over a very short interval tends to the identity matrix, one has from (9.2.2-6)

$$D(t_k)\tilde{v}(t_k) = \frac{1}{\Delta}v(t_k) \qquad (9.2.2\text{-}7)$$

which then completes the proof of equivalence between (9.2.2-2) and (9.2.1-1).

The covariance of the discrete time process noise is

$$Q(t_k) = E[v(t_k)v(t_k)'] = E\left[\left[\int_{t_k}^{t_k+\Delta} D(\tau_1)\tilde{v}(\tau_1)d\tau_1\right]\left[\int_{t_k}^{t_k+\Delta} D(\tau_2)\tilde{v}(\tau_2)d\tau_2\right]'\right] \qquad (9.2.2\text{-}8)$$

Rewriting the above as a double integral (with the limits omitted) and applying the expectation on the integrand, yields, using (9.2.1-2)

$$
\begin{aligned}
Q(t_k) &= \int\int D(\tau_1)\tilde{q}(\tau_1)D(\tau_2)'\delta(\tau_1 - \tau_2)d\tau_1 d\tau_2 = \int D(\tau_1)\tilde{q}(\tau_1)D(\tau_1)'d\tau_1 \\
&= D(t_k)\tilde{q}(t_k)D(t_k)'\Delta \qquad (9.2.2\text{-}9)
\end{aligned}
$$

and thus

$$D(t_k)\tilde{q}(t_k)D(t_k)' = \frac{1}{\Delta}Q(t_k) \qquad (9.2.2\text{-}10)$$

Remark

It follows from (9.2.1-3) that the **physical dimension** of the continuous time process noise intensity $\tilde{q}$ is the physical dimension of $\tilde{v}$ squared *multiplied* by time, since the physical dimension of $\delta(t)$ is t^{-1}. This is consistent with $\tilde{q}$ being a power spectral density with the physical dimension of $\tilde{v}$ squared divided by frequency.

On the other hand, the physical dimension of the discrete time intensity (covariance) Q is the physical dimension of v squared.

This is the reason for the similarity of (9.2.2-10) and (9.2.2-7): both the discrete time noise and its covariance are divided by the time interval Δ to yield their continuous time counterparts.

The **discrete time measurement** can be viewed as a **short-term average** of the continuous time measurement

$$z(t_k) = \frac{1}{\Delta} \int_{t_k}^{t_k+\Delta} z(\tau)d\tau = C(t_k)x(t_k) + \frac{1}{\Delta} \int_{t_k}^{t_k+\Delta} \tilde{w}(\tau)d\tau \qquad (9.2.2\text{-}11)$$

Comparing the above with the discrete time measurement equation

$$z(t_k) = H(t_k)x(t_k) + w(t_k) \qquad (9.2.2\text{-}12)$$

yields

$$H(t_k) = C(t_k) \qquad (9.2.2\text{-}13)$$

and the relationship between the discrete time and continuous time measurement noises is

$$w(t_k) = \frac{1}{\Delta} \int_{t_k}^{t_k+\Delta} \tilde{w}(\tau)d\tau \qquad (9.2.2\text{-}14)$$

The relationship between the covariance of the discrete time measurement noise and the intensity of the continuous time measurement noise is obtained, similarly to (9.2.2-8), as

$$
\begin{aligned}
R(t_k) &= E[w(t_k)w(t_k)'] = E\left[\frac{1}{\Delta^2} \int\int \tilde{w}(\tau_1)\tilde{w}(\tau_2)'d\tau_1 d\tau_2\right] \\
&= \frac{1}{\Delta^2} \int\int \tilde{R}(\tau_1)\delta(\tau_1 - \tau_2)d\tau_1 d\tau_2 \\
&= \frac{1}{\Delta^2} \int \tilde{R}(\tau_1)d\tau_1 \\
&= \frac{1}{\Delta}\tilde{R}(t_k) \qquad\qquad\qquad\qquad\qquad (9.2.2\text{-}15)
\end{aligned}
$$

or

$$\tilde{R}(t_k) = R(t_k)\Delta \qquad (9.2.2\text{-}16)$$

344

9.2.3   The Continuous Time Filter Equations

Differencing the discrete time filter (5.2.3-12), one has

$$
\begin{aligned}
\hat{x}(k+1|k+1) - \hat{x}(k|k) &= F(k)\hat{x}(k|k) - \hat{x}(k|k) \\
&\quad + W(k+1)[z(k+1) - H(k+1)\hat{x}(k+1|k)]
\end{aligned}
$$

$$(9.2.3\text{-}1)$$

Replacing $F(k)$ by $F(t_{k+1}, t_k)$ with the explicit full notation, using expression (5.2.3-17) for the filter gain and identity (9.2.2-13), yields, after division by Δ,

$$
\begin{aligned}
\frac{1}{\Delta}[\hat{x}(t_{k+1}|t_{k+1}) - \hat{x}(t_k|t_k)] &= \frac{1}{\Delta}[F(t_{k+1}, t_k) - I]\hat{x}(t_k|t_k) \\
&\quad + P(t_{k+1}|t_{k+1})C(t_{k+1})'[R(t_{k+1})\Delta]^{-1} \\
&\quad \cdot [z(t_{k+1}) - C(t_k)\hat{x}(t_{k+1}|t_k)]
\end{aligned}
$$

$$(9.2.3\text{-}2)$$

For $\Delta \to 0$, using (9.2.2-5) and (9.2.2-15) and the fact that

$$
\hat{x}(t_{k+1}|t_k) - \hat{x}(t_k|t_k) \doteq O(\Delta)
$$

$$(9.2.3\text{-}3)$$

that is, this difference is of the order of Δ and

$$
\hat{x}(t_{k+1}|t_k) \xrightarrow{\Delta \to 0} \hat{x}(t_k|t_k) \triangleq \hat{x}(t_k)
$$

$$(9.2.3\text{-}4)$$

one obtains, after dropping the subscript of t,

$$
\boxed{\dot{\hat{x}}(t) = A(t)\hat{x}(t) + L(t)[z(t) - C(t)\hat{x}(t)]}
$$

$$(9.2.3\text{-}5)$$

where the (optimal) *continuous time* **filter gain** is

$$
L(t) \triangleq P(t)C(t)\tilde{R}(t)^{-1}
$$

$$(9.2.3\text{-}6)$$

This is the *differential equation* of the **continuous time state estimate**, namely, the **continuous time state estimation filter**, called the **Kalman-Bucy filter**.

345

The discrete time covariance update equation (5.2.3-14) can be written as

$$
\begin{aligned}
P(t_{k+1}|t_{k+1}) &= P(t_{k+1}|t_k) - P(t_{k+1}|t_{k+1})H(t_{k+1})'R(t_{k+1})^{-1}H(t_{k+1})P(t_{k+1}|t_k) \\
&= F(t_{k+1}, t_k)P(t_k|t_k)F(t_{k+1}, t_k)' + Q(t_k) \\
&\quad - P(t_{k+1}|t_{k+1})H(t_{k+1})'R(t_{k+1})^{-1}H(t_{k+1})P(t_{k+1}|t_k) \\
&= [I + A(t_k)\Delta]P(t_k|t_k)[I + A(t_k)\Delta]' + Q(t_k) \\
&\quad - P(t_{k+1}|t_{k+1})H(t_{k+1})'R(t_{k+1})^{-1}H(t_{k+1})P(t_{k+1}|t_k) \\
&= P(t_k|t_k) + A(t_k)P(t_k|t_k)\Delta + P(t_k|t_k)A(t_k)'\Delta + Q(t_k) \\
&\quad - P(t_{k+1}|t_{k+1})H(t_{k+1})'R(t_{k+1})^{-1}H(t_{k+1})P(t_{k+1}|t_k) \qquad (9.2.3\text{-}7)
\end{aligned}
$$

where the term in Δ^2 has been neglected. In difference form, one has

$$
\begin{aligned}
\frac{1}{\Delta}[P(t_{k+1}|t_{k+1}) - P(t_k|t_k)] &= A(t_k)P(t_k|t_k) + P(t_k|t_k)A(t_k)' + \frac{1}{\Delta}Q(t_k) \\
&\quad - P(t_{k+1}|t_{k+1})H(t_{k+1})'[R(t_{k+1})\Delta]^{-1} \\
&\quad \cdot H(t_{k+1})P(t_{k+1}|t_k) \qquad (9.2.3\text{-}8)
\end{aligned}
$$

Using (9.2.2-9), (9.2.2-12) and (9.2.2-15) and the fact that, under the assumption of continuity,

$$
P(t_{k+1}|t_k) \xrightarrow{\Delta \to 0} P(t_{k+1}|t_{k+1}) \triangleq P(t_{k+1}) \qquad (9.2.3\text{-}9)
$$

yields, after dropping the subscript of t,

$$
\boxed{\dot{P}(t) = A(t)P(t) + P(t)A(t)' + D(t)\tilde{q}(t)D(t)' - P(t)C(t)'\tilde{R}(t)^{-1}C(t)P(t)} \qquad (9.2.3\text{-}10)
$$

The above is the (continuous time) **matrix Riccati differential equation** for the state estimate covariance. Its solution can be obtained using numerical techniques.

Remarks

The structure of the Kalman-Bucy filter, as given in (9.2.3-5), is the same as that of the Luenberger observer [Che84].

Note that there is no prediction stage — this is a direct consequence of the continuous time formulation.

The measurement noise covariance $\tilde{R}$ has to be invertible — no "perfect" measurements are allowed in the Kalman-Bucy filter. If some of the components of the measurement are noiseless (the measurement noise covariance has some zero eigenvalues), then one can use a so-called observer-estimator or assume a "small" noise to be present in those components.

In the absence of measurements, which can be modeled by letting $\tilde{R}^{-1} = 0$, the state estimate and its covariance evolve according to the following "open-loop" equations

$$\dot{\hat{x}}(t) = A(t)\hat{x}(t) \tag{9.2.3-11}$$

$$\dot{P}(t) = A(t)P(t) + P(t)A(t)' + D(t)\tilde{q}(t)D(t)' \tag{9.2.3-12}$$

obtained from (9.2.3-5) and (9.2.3-10), respectively, by dropping the last term in each. Equation (9.2.3-12) is known as the **Lyapunov differential equation**.

The above two equations are used later in the **continuous-discrete filter**, for the case of continuous time state equation and discrete time measurements.

The **Ito differential equation** form of the filter (9.2.3-5) is

$$\begin{aligned} d\hat{x}(t) &= A(t)\hat{x}(t)dt + L(t)[z(t) - C(t)\hat{x}(t)]dt \\ &= [A(t) - L(t)C(t)]\hat{x}(t)dt + L(t)d\zeta(t) \end{aligned} \tag{9.2.3-13}$$

where the last form uses the notation (9.2.1-11).

9.2.4   The Continuous Time Innovation

The **continuous time innovation** is, from (9.2.3-4)

$$\nu(t) \triangleq z(t) - C(t)\hat{x}(t) = C(t)x(t) + \tilde{w}(t) - C(t)\hat{x}(t) = C(t)\tilde{x}(t) + \tilde{w}(t) \qquad (9.2.4\text{-}1)$$

where

$$\tilde{x}(t) \triangleq x(t) - \hat{x}(t) \qquad (9.2.4\text{-}2)$$

is the **state estimation error**.

The fact that the continuous time innovation is *zero mean* follows immediately from the limiting process used to derive the continuous time filter.

The continuous time innovation is also *white* as it is shown next. The autocorrelation of the innovation is

$$
\begin{aligned}
E[\nu(t)\nu(\tau)'] &= E\Big[[C(t)\tilde{x}(t) + \tilde{w}(t)][C(\tau)\tilde{x}(\tau) + \tilde{w}(\tau)]'\Big] \\
&= E[\tilde{x}(\tau)\tilde{w}(t)']'C(\tau)' + C(t)E[\tilde{x}(t)\tilde{w}(\tau)'] \\
&\quad + C(t)E[\tilde{x}(t)\tilde{x}(\tau)']C(\tau)' + E[\tilde{w}(t)\tilde{w}(\tau)']
\end{aligned}
\qquad (9.2.4\text{-}3)
$$

Assume $t > \tau$. The first term after the last equal sign above is zero since $\tilde{w}(t)$ is white and zero mean, and thus orthogonal to $\tilde{x}(\tau)$, $\forall \tau < t$. Also note that

$$E[\tilde{x}(t)\tilde{x}(\tau)'] = E\Big[\tilde{x}(t)[x(\tau) - \hat{x}(\tau)]'\Big] = E[\tilde{x}(t)x(\tau)'] \qquad (9.2.4\text{-}4)$$

since $\tilde{x}(t)$ is orthogonal to any function of the measurements up to τ, $\forall \tau \leq t$, and thus on $\hat{x}(\tau)$.

Using these results and the fact that

$$E[\tilde{w}(t)\tilde{w}(\tau)'] = \tilde{R}(t)\delta(t - \tau) \qquad (9.2.4\text{-}5)$$

in (9.2.4-3) yields

$$
\begin{aligned}
E[\nu(t)\nu(\tau)'] &= C(t)E[\tilde{x}(t)\tilde{w}(\tau)'] + C(t)E[\tilde{x}(t)x(\tau)']C(\tau)' + \tilde{R}(t)\delta(t - \tau) \\
&= C(t)E\Big[\tilde{x}(t)[C(\tau)x(\tau) + \tilde{w}(\tau)]'\Big] + \tilde{R}(t)\delta(t - \tau) \\
&= \tilde{R}(t)\delta(t - \tau) + C(t)E[\tilde{x}(t)z(\tau)]
\end{aligned}
\qquad (9.2.4\text{-}6)
$$

Again, since

$$\tilde{x}(t) \perp z(\tau) \qquad \forall \tau \leq t \qquad (9.2.4\text{-}7)$$

the last term in (9.2.4-6) vanishes.

9.2.4 The Continuous Time Innovation

Thus the autocorrelation of the continuous time innovation is

$$\boxed{E[\nu(t)\nu(\tau)'] = \tilde{R}(t)\delta(t - \tau)}$$

(9.2.4-8)

that is, the innovation is white. Furthermore, the intensity of the continuous time innovation is *equal* to the intensity of the measurement noise.

Remark

Note the difference from the discrete time case where the innovation covariance is

$$
\begin{aligned}
S(k) &= E[\nu(k)\nu(k)'] \\
&= H(k)P(k|k - 1)H(k)' + R(k)
\end{aligned}
$$

(9.2.4-9)

that is, *larger* than the measurement noise covariance $R(k)$, in the sense that the difference between them is positive semidefinite. This difference is due to the fact that in discrete time there is a predicted state distinct from the (current or updated) state estimate, while in continuous time there is no prediction since the measurements arrive continuously.

9.2.5   Asymptotic Properties of the Continuous Time Riccati Equation

Time-Varying Systems

The continuous time **matrix Riccati differential equation** (9.2.3-9) is, for a time-varying system

$$\dot{P}(t) = A(t)P(t) + P(t)A(t)' + D(t)\tilde{q}(t)D(t)' - P(t)C(t)'\tilde{R}(t)^{-1}C(t)P(t) \qquad (9.2.5\text{-}1)$$

Consider (9.2.5-1) for a system *without process noise*. Then, using the identity

$$\frac{d}{dt}[P(t)^{-1}] = -P(t)^{-1}\dot{P}(t)P(t)^{-1} \qquad (9.2.5\text{-}2)$$

one can rewrite (9.2.5-1), with $\tilde{q} = 0$ and without the time arguments for simplicity, as (see problem 9-1)

$$\frac{d}{dt}(P^{-1}) = -A'P^{-1} - P^{-1}A + C'\tilde{R}^{-1}C \qquad (9.2.5\text{-}3)$$

The above equation for the inverse covariance or **information matrix** has the following explicit solution

$$P(t)^{-1} = F(t_0,t)'P(t_0)^{-1}F(t_0,t) + \int_{t_0}^{t} F(\tau,t)'C(\tau)'R(\tau)^{-1}C(\tau)F(\tau,t)d\tau \qquad (9.2.5\text{-}4)$$

where F is the transition matrix corresponding to $A(t)$ (see problem 9-2).

Equation (9.2.5-4) indicates that, even if $P(t_0)^{-1}$ is singular (i.e., no prior information is available about some of the state components), such information will become available from the observations if the last term in (9.2.5-4), called the **observability Gramian**

$$\mathbf{G} \triangleq \int_{t_0}^{t} F(\tau,t)'C(\tau)'R(\tau)^{-1}C(\tau)F(\tau,t)d\tau \qquad (9.2.5\text{-}5)$$

is *positive definite*.

It can be observed from (9.2.5-4), which pertains to the situation without process noise, that as $t \to \infty$ the information matrix can also tend to infinity, in which case the covariance will tend to zero.

Time-Invariant Systems

For a *time-invariant system*, the positive definiteness of the Gramian (9.2.5-5) is equivalent to the condition that the **observability matrix** of the pair (A, C) given by

$$\mathcal{Q}_O \triangleq \begin{bmatrix} C' & A'C' & \cdots & (A')^{n_x-1}C' \end{bmatrix}' \tag{9.2.5-6}$$

where n_x is the dimension of the state, is of full rank (i.e., n_x).

Observability of the system guarantees the positive definiteness of the inverse covariance matrix, which makes it invertible, that is, it guarantees the existence (boundedness) of the covariance matrix.

The asymptotic properties of the solution of the Riccati equation for a time-invariant system *with process noise* are:

1. Under the observability condition of the pair (A, C), the covariance converges to a *finite **steady-state covariance***;
2. If, in addition, with $\tilde{D}$ being the square root of the process noise intensity matrix premultiplied by D and postmultiplied by D', the pair $(A, \tilde{D})$ is controllable, then the resulting steady-state covariance is *unique and positive definite.*

The matrix square root $\tilde{D}$ is defined in this case by

$$\tilde{D}\tilde{D}' \triangleq D\tilde{q}D' \tag{9.2.5-7}$$

The controllability condition for the pair $(A, \tilde{D})$ is that the matrix

$$\mathcal{Q}_C \triangleq \begin{bmatrix} \tilde{D} & A\tilde{D} & \cdots & A^{n_x-1}\tilde{D} \end{bmatrix} \tag{9.2.5-8}$$

is of full rank (i.e., n_x).

The steady-state covariance is the solution of the **algebraic matrix Riccati equation**

$$AP + PA' + D\tilde{q}D' - PC'\tilde{R}^{-1}CP = 0 \tag{9.2.5-9}$$

which, under conditions (1) and (2), has a *unique positive definite solution.* To this corresponds a **steady-state filter.**

Stability of the Filter

Consider the filter equation

$$\dot{\hat{x}}(t) \;=\; A\hat{x}(t) + P(t)C'\tilde{R}^{-1}[z(t) - C\hat{x}(t)]$$
$$=\; A\hat{x}(t) + P(t)C'\tilde{R}^{-1}[Cx(t) + \tilde{w}(t) - C\hat{x}(t)] \qquad (9.2.5\text{-}10)$$

and the system equation

$$\dot{x}(t) = Ax(t) + D\tilde{v}(t) \qquad (9.2.5\text{-}11)$$

Subtracting (9.2.5-10) from (9.2.5-11) and using the notation

$$\tilde{x} \overset{\Delta}{=} x - \hat{x} \qquad (9.2.5\text{-}12)$$

for the state estimation error, it follows that this error obeys the following differential equation

$$\dot{\tilde{x}}(t) \;=\; A\tilde{x}(t) - P(t)C'\tilde{R}^{-1}\tilde{x}(t) - P(t)C'\tilde{R}^{-1}\tilde{w}(t) + D\tilde{v}(t)$$
$$=\; [A - P(t)C'\tilde{R}^{-1}]\tilde{x}(t) - P(t)C'\tilde{R}^{-1}\tilde{w}(t) + D\tilde{v}(t) \qquad (9.2.5\text{-}13)$$

Under conditions (1) and (2) — which guarantee that (9.2.5-13) has a stationary solution (the expected value of $\tilde{x}\tilde{x}'$, which is the error covariance, is bounded) — the system matrix for the error equation

$$\tilde{A} \overset{\Delta}{=} A - P(t)C\tilde{R}^{-1} \qquad (9.2.5\text{-}14)$$

is **stable** (i.e., all its eigenvalues are in the left-half plane). Note that there is no stability requirement for the dynamic system (9.2.5-11).

9.2.6   Example of a Continuous Time Filter

Consider the scalar system

$$\dot{x} = -ax(t) + \tilde{v}(t) \tag{9.2.6-1}$$

with

$$E[\tilde{v}(t)] = 0 \tag{9.2.6-2}$$

$$E[\tilde{v}(t)\tilde{v}(\tau)] = q\delta(t - \tau) \tag{9.2.6-3}$$

and the measurement

$$z(t) = x(t) + \tilde{w}(t) \tag{9.2.6-4}$$

with

$$E[\tilde{w}(t)] = 0 \tag{9.2.6-5}$$

$$E[\tilde{w}(t)\tilde{w}(\tau)] = r\delta(t - \tau) \tag{9.2.6-6}$$

and the two noises and the initial state uncorrelated from each other.

The Riccati equation for the variance $p(t)$ of the filtered state is

$$\dot{p} = -2ap + q - \frac{p^2}{r} \tag{9.2.6-7}$$

with initial condition p_0, the initial state variance.

The scalar Riccati equation (9.2.6-7) has the following explicit solution:

$$p(t) = p_1 + \frac{p_1 + p_2}{\dfrac{p_0 + p_2}{p_0 - p_1}e^{2\alpha t} - 1} \tag{9.2.6-8}$$

where

$$\alpha = \sqrt{a^2 + \frac{q}{r}} \tag{9.2.6-9}$$

$$p_1 = r(\alpha - a) \tag{9.2.6-10}$$

$$p_2 = r(\alpha + a) \tag{9.2.6-11}$$

The resulting Kalman-Bucy filter is

$$\dot{\hat{x}}(t) = -a\hat{x}(t) + \frac{p(t)}{r}[z(t) - \hat{x}(t)] \tag{9.2.6-12}$$

The Steady-State Filter

Since this is a time-invariant system with stationary noises and the conditions for the existence and positiveness of the steady-state solution are satisfied, the filter itself will become time invariant after the transient — the **steady-state filter**.

The solution of the Riccati equation, which is the optimal state estimate variance, has the asymptotic value

$$\lim_{t\to\infty} p(t) = p_1 \qquad (9.2.6\text{-}13)$$

It can be easily shown that this value is also the only positive solution of the algebraic Riccati equation

$$0 = -2ap + q - \frac{p^2}{r} \qquad (9.2.6\text{-}14)$$

This yields the steady-state filter with the fixed gain p_1/r. The differential equation of the filter is

$$\dot{\hat{x}}(t) = -a\hat{x}(t) + \frac{p_1}{r}[z(t) - \hat{x}(t)] \qquad (9.2.6\text{-}15)$$

Since this is time invariant, one can obtain its Laplace transfer function, which is

$$\frac{\hat{x}(s)}{Z(s)} = \frac{\alpha - a}{s + \alpha} \qquad (9.2.6\text{-}16)$$

On the Solution of the Riccati Equation

For dimension greater than one, the Riccati equation cannot be solved, in general, explicitly. There are some techniques that transform the n_x-dimensional nonlinear matrix Riccati equation into a $2n_x$-dimensional linear equation, but even that requires, in general, numerical solution. The most common method of solving the continuous time Riccati equation is via numerical integration.

The steady-state solution of the Riccati equation can be obtained for low dimension kinematic models, similarly to the discrete time case, in closed form.

9.2.7 Overview of the Kalman-Bucy Filter

Figure 9.2.7-1 presents the block diagram of the Kalman-Bucy filter.

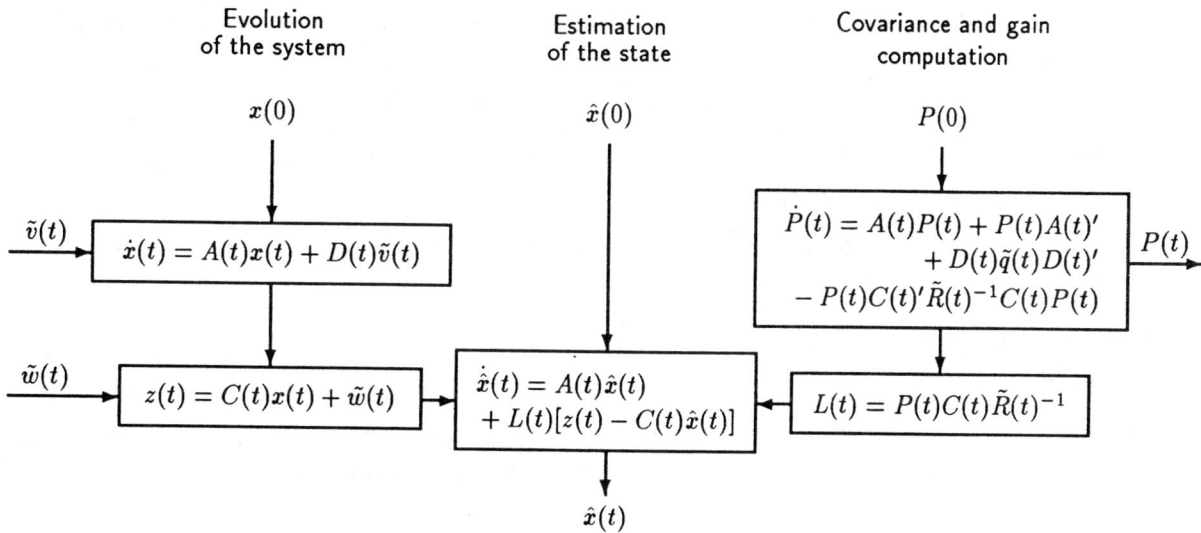

Figure 9.2.7-1: Block diagram of the Kalman-Bucy filter.

The statistical assumptions about the initial state and the noises are:

$$E[x(0)] = \hat{x}(0) \tag{9.2.7-1}$$

$$\text{cov}[x(0)] = P(0) \tag{9.2.7-2}$$

$$E[\tilde{v}(t)] = 0 \tag{9.2.7-3}$$

$$E[\tilde{v}(t)\tilde{v}(\tau)'] = \tilde{q}(t)\delta(t-\tau) \tag{9.2.7-4}$$

$$E[\tilde{w}(t)] = 0 \tag{9.2.7-5}$$

$$E[\tilde{w}(t)\tilde{w}(\tau)'] = \tilde{R}(t)\delta(t-\tau) \tag{9.2.7-6}$$

and that these three sources of uncertainty are mutually uncorrelated.

9.2.8 Continuous Time State Estimation — Summary

The LMMSE state estimator for a continuous time linear dynamic system driven by white process noise, and whose state is observed in the presence of white measurement noise, has been derived.

This estimator, called the Kalman-Bucy filter, consists of a linear differential equation, similar to the state equation but driven by the innovation multiplied by the filter gain.

The filter gain is given in terms of the covariance matrix of the state estimate, which is the same as the covariance of the estimation error. This is obtained as the solution of a Riccati matrix differential equation.

This filter has been derived for a system which can be *time varying* and with noises that can be *nonstationary*.

Conditions, under which for a time-invariant system with stationary noises the Riccati equation for the error covariance has a unique positive definite solution, have been given. Under these conditions the filter gain converges to a steady-state value and the resulting steady-state filter is stable — the estimation error becomes stationary.

Similarly to the discrete time case, the continuous time LMMSE estimator is the best among the class of linear estimators. If

- the initial state,
- the process noise stochastic process, and
- the measurement noise stochastic process

are all Gaussian, then this estimator yields the optimal MMSE estimate — the conditional mean of the state.

9.3 PREDICTION AND THE CONTINUOUS-DISCRETE FILTER

9.3.1 Prediction of the Mean and Covariance

Prediction is the estimation of the state at time t_2 beyond the data interval, that is, based on data up to an earlier time $t_1 < t_2$,

$$\hat{x}(t_2|t_1) = E[x(t_2)|Z^{t_1}] \tag{9.3.1-1}$$

where

$$Z^{t_1} \triangleq \{z(t), t \in [t_0, t_1]\} \tag{9.3.1-2}$$

and t_0 denotes the initial time.

With the system

$$\dot{x}(t) = A(t)x(t) + D(t)\tilde{v}(t) \tag{9.3.1-3}$$

one has

$$x(t_2) = F(t_2, t_1)x(t_1) + \int_{t_1}^{t_2} F(t_2, \tau)D(\tau)\tilde{v}(\tau)d\tau \tag{9.3.1-4}$$

Applying the conditional expectation (9.3.1-1) on (9.3.1-4) yields the **predicted state**

$$\hat{x}(t_2|t_1) = F(t_2, t_1)\hat{x}(t_1) \tag{9.3.1-5}$$

in view of the fact that the process noise $\tilde{v}(t)$ is zero mean and white.

The above is actually the solution of the differential equation (9.2.3-11) repeated below

$$\dot{\hat{x}}(t) = A(t)\hat{x}(t) \tag{9.3.1-6}$$

at $t = t_2$ with initial condition $\hat{x}(t_1)$. If the transition matrix $F(t_2, t_1)$ is not available in explicit form, then (9.3.1-5) cannot be used — instead one has to carry out numerical integration of (9.3.1-6) from t_1 to t_2 to obtain $\hat{x}(t_2|t_1)$.

The covariance associated with the prediction (9.3.1-5) — the **prediction covariance** — is

$$P(t_2|t_1) \triangleq \text{cov}[x(t_2)|Z^{t_1}] = F(t_2, t_1)P(t_1)F(t_2, t_1)' + \int_{t_1}^{t_2} F(t_2, \tau)D(\tau)\tilde{q}(\tau)D(\tau)'F(t_2, \tau)'d\tau \tag{9.3.1-7}$$

where $P(t_1)$ is the covariance of $x(t_1)$ and $\tilde{q}$ is the covariance of the process noise.

Similarly to the situation for the predicted state, which is given explicitly by (9.3.1-5) or is the solution of (9.3.1-6), (9.3.1-7) is the solution of (9.2.3-12), repeated below

$$\dot{P}(t) = A(t)P(t) + P(t)A(t)' + D(t)\tilde{q}(t)D(t)' \tag{9.3.1-8}$$

at t_2 starting from t_1 with initial condition $P(t_1)$.

9.3.2   The Various Types of Prediction

As in the discrete time case, there are several types of prediction:

(a) **Fixed point prediction**: prediction is done to the fixed point in time τ (e.g., an intercept time) based on the data up to time t, where t is changing.

(b) **Fixed lead prediction**: prediction is done for a fixed time T ahead (i.e., to $t + T$, where t varies and T is the lead).

(c) **Fixed interval prediction**: prediction is done based on the data from the fixed interval up to t_1 to times beyond it, $t > t_1$.

The equations for these three types of prediction are given below.

Fixed Point Prediction

The predicted state is

$$\hat{x}(\tau | t) = F(\tau, t)\hat{x}(t) \tag{9.3.2-1}$$

It can be shown that the differential equation obeyed by this prediction is (see problem 9-3)

$$
\begin{aligned}
\frac{\partial}{\partial t}\hat{x}(\tau | t) &= -F(\tau, t)A(t)\hat{x}(t) + F(\tau, t)\dot{\hat{x}}(t) \\
&= -F(\tau, t)A(t)\hat{x}(t) + F(\tau, t)[A(t)\hat{x}(t) + L(t)\nu(t)] \\
&= F(\tau, t)L(t)\nu(t) \tag{9.3.2-2}
\end{aligned}
$$

(See also problem 9-4.)

Fixed Lead Prediction

This prediction for lead time T is given by the expression

$$\hat{x}(t + T | t) = F(t + T, t)\hat{x}(t) \tag{9.3.2-3}$$

The differential equation for this prediction is (see problem 9-5)

$$\frac{d}{dt}\hat{x}(t + T | t) = A(t + T)\hat{x}(t + T | t) + F(t + T, t)W(t)\nu(t) \tag{9.3.2-4}$$

Fixed Interval Prediction

The state prediction to t based on data up to $t_1 < t$ is

$$\hat{x}(t|t_1) = F(t, t_1)\hat{x}(t_1) \tag{9.3.2-5}$$

and its propagation equation is (see problem 9-3)

$$
\begin{aligned}
\dot{\hat{x}}(t|t_1) &= A(t)F(t, t_1)\hat{x}(t_1) \\
&= A(t)\hat{x}(t|t_1)
\end{aligned} \tag{9.3.2-6}
$$

It can be easily shown that the covariance associated with (9.3.2-5)

$$P(t|t_1) \triangleq \text{cov}[x(t)|Z^{t_1}] \tag{9.3.2-7}$$

obeys the differential equation

$$\frac{d}{dt}P(t|t_1) = A(t)P(t|t_1) + P(t|t_1)A(t)' + D(t)\tilde{q}(t)D(t)' \tag{9.3.2-8}$$

Note that this is the same as (9.2.3-12).

9.3.3 The Continuous-Discrete Filter

The state equation is in *continuous time* (as before, without input for simplicity)

$$\dot{x}(t) = A(t)x(t) + D(t)\tilde{v}(t) \qquad (9.3.3\text{-}1)$$

with $\tilde{v}(t)$ the process noise, zero mean and white

$$E[\tilde{v}(t)] = 0 \qquad (9.3.3\text{-}2)$$

$$E[\tilde{v}(t)\tilde{v}(\tau)'] = \tilde{q}(t)\delta(t - \tau) \qquad (9.3.3\text{-}3)$$

where $\tilde{q}(t)$ is the possibly time-varying intensity of the white noise.

The measurement equation is in *discrete time*

$$
\begin{aligned}
z(k) &\triangleq z(t_k) \\
&= C(t_k)x(t_k) + w(t_k) \\
&\triangleq H(k)x(k) + w(k)
\end{aligned}
\qquad (9.3.3\text{-}4)
$$

with the process noise $w(t_k)$ a zero-mean white sequence

$$E[w(t_k)] = 0 \qquad (9.3.3\text{-}5)$$

$$E[w(t_k)w(t_j)'] = R(k)\delta_{kj} \qquad (9.3.3\text{-}6)$$

The process noise, the measurement noise and the initial state are mutually uncorrelated. The noises can be *nonstationary*.

The Algorithm

The state estimation filter for this problem of continuous time state equation and discrete time measurements, called the **continuous-discrete filter**, consists of the following:

1. Prediction stage: between the times at which measurements are obtained the state estimate and its covariance are propagated according to (9.2.3-11) and (9.2.3-12), yielding the corresponding predictions $\hat{x}(k+1|k) \triangleq \hat{x}(t_{k+1}|t_k)$ and $P(k+1|k) \triangleq P(t_{k+1}|t_k)$;

2. Update stage: at the discrete times when the measurements are obtained, the state estimate and its covariance are updated according to the standard discrete time Kalman filter update equations from Subsection 5.3.1.

360

9.4 DUALITY OF ESTIMATION AND CONTROL

9.4.1   The Two Problems

The Estimation Problem

Consider the LMMSE estimation problem for the following time-invariant system

$$\mathcal{S} : \begin{cases} \dot{x}(t) = Ax(t) + D\tilde{v}(t) \\ z(t) = Cx(t) + \tilde{w}(t) \end{cases} \tag{9.4.1-1}$$

with the following statistics

$$\mathrm{cov}[x(0)] = P_0 \tag{9.4.1-2}$$

$$E[\tilde{v}(t)\tilde{v}(\tau)'] = Q\delta(t - \tau) \qquad Q > 0 \tag{9.4.1-3}$$

$$E[\tilde{w}(t)\tilde{w}(\tau)'] = R\delta(t - \tau) \qquad R > 0 \tag{9.4.1-4}$$

The initial state and the noises are mutually uncorrelated. The noise intensities Q and R are denoted in this section without a tilde.

The Control Problem

Consider the **linear-quadratic control problem (LQ)** for the *linear system*

$$\mathcal{S}^* : \begin{cases} \dot{y}(t) = -A'y(t) - C'u(t) \\ \psi(t) = D'y(t) \end{cases} \tag{9.4.1-5}$$

with the following *quadratic* **cost functional** (which maps the state and control time-functions into a scalar "cost")

$$J(u, t_0, t_1) = y(t_1)'P_0\, y(t_1) + \int_{t_0}^{t_1} [\psi(t)'Q\psi(t) + u(t)'Ru(t)]dt \tag{9.4.1-6}$$

This cost is to be *minimized* by the function $u(t)$, $t \in [t_0, t_1]$.

The **cost weighting matrices** Q and R in (9.4.1-6) are the same as the noise intensity matrices in (9.4.1-3) and (9.4.1-4), that is, positive definite (relaxation of the requirement on Q to positive semidefinite is possible).

The system $\mathcal{S}^*$ is called the **adjoint** of $\mathcal{S}$, since $\mathcal{S}^*$ propagates backwards in time according to the system matrix of $\mathcal{S}$ transposed.

9.4.2   The Solutions to the Estimation and the Control Problems

The Estimation Solution

The optimal solution to the LMMSE estimation problem for system $\mathcal{S}$ is

$$\begin{aligned}
\dot{\hat{x}}(t) &= A\hat{x}(t) + L_e(t)[z(t) - C\hat{x}(t)] \\
&= [A - L_e(t)C]\hat{x}(t) + L_e(t)z(t)
\end{aligned} \tag{9.4.2-1}$$

where the estimator (filter) gain is denoted as

$$L_e(t) = P(t)C'R^{-1} \tag{9.4.2-2}$$

The state estimation covariance follows from the Riccati equation

$$\dot{P}(t) = AP(t) + P(t)A' - P(t)C'R^{-1}CP(t) + DQD' \qquad t > t_0 \tag{9.4.2-3}$$

with boundary condition at the *initial time*

$$P(t_0) = P_0 \tag{9.4.2-4}$$

The Control Solution

The optimal solution to the LQ control problem for the adjoint system $\mathcal{S}^*$ is the optimal feedback control

$$u^o(t) = -L_c(t)'y^o(t) \tag{9.4.2-5}$$

where the optimal state $y^o(t)$ evolves according to the closed loop equation

$$\dot{y}^o = -A'y^o(t) - C'u^o(t) = -[A - L_c(t)C]'y^o(t) \tag{9.4.2-6}$$

The control feedback gain in (9.4.2-5) is given by

$$L_c(t) = K(t)C'R^{-1} \tag{9.4.2-7}$$

and the **optimal cost matrix** $K(t)$ follows from the Riccati equation

$$\dot{K}(t) = AK(t) + K(t)A' - K(t)C'R^{-1}CK(t) + DQD' \qquad t < t_1 \tag{9.4.2-8}$$

with boundary condition at the *terminal time* t_1

$$K(t_1) = P_0 \tag{9.4.2-9}$$

The optimized (minimized) value of the cost (9.4.1-6) is the **optimal cost**

$$J^o(t_0, t_1) \triangleq \min_u J(u)$$
$$= y(t_0)'K(t_0)y(t_0) \qquad (9.4.2\text{-}10)$$

The proof of the solution of the LQ problem can be found in, for example, [BH75].

Duality of the Solutions

The optimal performance of the LMMSE state estimation problem starting from time t_0 is given at time $t > t_0$ by the *(optimal) state covariance matrix* $P(t)$.

The optimal performance of the LQ control problem starting at the arbitrary time $t < t_1$ and running up to time t_1 is given, similarly to (9.4.2-10), by

$$J^o(t, t_1) = y(t)'K(t)y(t) \qquad (9.4.2\text{-}11)$$

that is, it is determined by the *optimal cost matrix* $K(t)$.

The following similarities between the solutions of the LMMSE estimation and the LQ control problems can be observed:

1. The covariance equation (9.4.2-3) of the estimation problem is identical to the cost matrix equation (9.4.2-8) of the control problem. The only difference is that, while the covariance equation propagates forward from the initial condition (9.4.2-4), the cost matrix equation propagates backwards from the terminal condition (9.4.2-9), which is, however, the same as (9.4.2-4).

2. The filter gain (9.4.2-2) has the same expression as the control feedback gain (9.4.2-7).

3. The closed loop systems (9.4.2-1) [without the driving term] and (9.4.2-6) are the adjoint of each other.

In view of the fact that their basic equations are the same, these two problems are **dual problems**, and, as it will be seen in the next subsection, their solutions exhibit similar properties under the same conditions.

9.4.3 Properties of the Solutions

The controllability matrix of system $\mathcal{S}$ is

$$\mathcal{Q}_C(\mathcal{S}) \triangleq \begin{bmatrix} D & AD & \cdots & A^{n_x-1}D \end{bmatrix} \tag{9.4.3-1}$$

and this is the same as the observability matrix of system $\mathcal{S}^*$

$$\mathcal{Q}_C(\mathcal{S}) = Q_O(\mathcal{S}^*)' \tag{9.4.3-2}$$

The observability matrix of system $\mathcal{S}$ is

$$\mathcal{Q}_O(\mathcal{S}) \triangleq \begin{bmatrix} C' & A'C' & \cdots & (A')^{n_x-1}C' \end{bmatrix}' \tag{9.4.3-3}$$

and this is the same as the controllability matrix of system $\mathcal{S}^*$

$$\mathcal{Q}_O(\mathcal{S}) = \mathcal{Q}_C(\mathcal{S}^*)' \tag{9.4.3-4}$$

In view of the above equalities (9.4.3-2) and (9.4.3-4), the **dual problems** of LMMSE estimation and LQ control exhibit *similar properties* under the *same conditions*. These properties will be detailed next.

The solution to the estimation Riccati equation (9.4.2-3) with initial time t_0 will be denoted as $P_{t_0}(t)$. Similarly, the solution to the control Riccati equation (9.4.2-8) with terminal time t_1 will be denoted as $K_{t_1}(t)$.

With these notations one has

- $P_{t_0}(\infty)$ — the steady-state covariance in the LMMSE estimation problem (i.e., with "infinite horizon");
- $K_{t_1}(-\infty)$ — the LQ problem cost starting at $t \to -\infty$ and going up to t_1, that is, the control problem over an infinite period of time (with "infinite horizon").

The dual estimation and control problems have the following properties:

1. Under the equivalent conditions

$$\{\mathcal{S} \text{ is completely observable}\} \equiv \{\mathcal{S}^* \text{ is completely controllable}\}$$

one has

$$\lim_{t_0 \to -\infty} P_{t_0}(t) = \lim_{t \to \infty} P_{t_0}(t) \triangleq P(\infty) = K(-\infty) \triangleq \lim_{t \to -\infty} K_{t_1}(t) = \lim_{t_1 \to \infty} K_{t_1}(t) \tag{9.4.3-5}$$

where the common limit above is the not necessarily unique solution of the algebraic Riccati equation

$$AP + PA' + DQD' - PC'R^{-1}CP = 0 \tag{9.4.3-6}$$

2. If, in addition,

$$\{\mathcal{S} \text{ is completely controllable}\} \equiv \{\mathcal{S}^* \text{ is completely observable}\}$$

then

$$P_{t_0}(t) > 0 \qquad \forall t > t_0 \qquad (9.4.3\text{-}7)$$

$$K_{t_1}(t) > 0 \qquad \forall t < t_1 \qquad (9.4.3\text{-}8)$$

and the limit in (9.4.3-5) is the unique positive definite solution of (9.4.3-6).

3. Under the complete controllability and complete observability conditions the eigenvalues of the **closed-loop system matrix**

$$A_c \triangleq A - P(\infty)C'R^{-1}C \qquad (9.4.3\text{-}9)$$

are in the left-half plane and thus the filter and the optimal feedback control system are asymptotically stable. This holds regardless of the eigenvalues of the open-loop system matrix A.

Remark

In view of (9.4.2-10) a *finite* $K_{t_1}(-\infty)$ guarantees a *finite cost for the infinite horizon* problem and, hence, the state and control vectors *converge to zero* since the integral of the sum of their norms is finite.

9.5 THE WIENER-HOPF PROBLEM

9.5.1 Formulation of the Problem

The **Wiener-Hopf problem** deals with the LMMSE estimation of a vector-valued stochastic process $x(t)$ with mean and autocovariance

$$E[x(t)] = \bar{x}(t) \tag{9.5.1-1}$$

$$\text{cov}[x(t), x(\tau)] = V_{xx}(t, \tau) \tag{9.5.1-2}$$

based on the observed process $z(t)$ with statistics

$$E[z(t)] = \bar{z}(t) \tag{9.5.1-3}$$

$$\text{cov}[z(t), z(\tau)] = V_{zz}(t, \tau) \tag{9.5.1-4}$$

The **cross-covariance** between the two processes is denoted as

$$\text{cov}[x(t), z(\tau)] = V_{xz}(t, \tau) \tag{9.5.1-5}$$

The estimator is of the form

$$\hat{x}(t) = \xi(t) + \int_0^t H(t, \tau) z(\tau) d\tau \tag{9.5.1-6}$$

and it should minimize the mean-square value of the norm of the error

$$E\left[\|\tilde{x}(t)\|^2\right] = \text{tr}\left\{\text{cov}[\tilde{x}(t)]\right\} \tag{9.5.1-7}$$

where the estimation error is defined as

$$\tilde{x}(t) \triangleq x(t) - \hat{x}(t) \tag{9.5.1-8}$$

Equation (9.5.1-6) has the form of an **affine transformation** of the function $z(\tau)$ — a linear operator on this function plus another function.

9.5.2 The Wiener-Hopf Equation

The LMMSE estimator (9.5.1-6) is obtained by setting the following conditions on the estimation error (9.5.1-8):

1. the error should be zero mean (i.e., the estimator should be unbiased), and
2. the error should be orthogonal to the observations.

The **unbiasedness** condition is

$$E[\hat{x}(t)] = \xi(t) + \int_0^t H(t,\tau)\bar{z}(\tau)d\tau = \bar{x}(t) \tag{9.5.2-1}$$

and it yields

$$\xi(t) = \bar{x}(t) - \int_0^t H(t,\tau)\bar{z}(\tau)d\tau \tag{9.5.2-2}$$

which leads to

$$\hat{x}(t) = \bar{x}(t) + \int_0^t H(t,\tau)[z(\tau) - \bar{z}(\tau)]d\tau \tag{9.5.2-3}$$

The **orthogonality** condition extended from random vectors to vector-valued random processes is

$$E[\tilde{x}(t)z(\theta)'] = 0 \qquad \forall \theta \leq t \tag{9.5.2-4}$$

or

$$E\Big[[x(t) - \hat{x}(t)][z(\theta) - \bar{z}(\theta)]'\Big] = 0 \qquad \forall \theta \leq t \tag{9.5.2-5}$$

Inserting (9.5.2-3) into the above, one has

$$E\left[\left\{x(t) - \bar{x}(t) - \int_0^t H(t,\tau)[z(\tau) - \bar{z}(\tau)]d\tau\right\}\{z(\theta) - \bar{z}(\theta)\}'\right] = 0 \tag{9.5.2-6}$$

which yields the **Wiener-Hopf equation**

$$\boxed{V_{xz}(t,\theta) - \int_0^t H(t,\tau)V_{zz}(\tau,\theta)d\tau = 0} \tag{9.5.2-7}$$

Remarks

This equation does not assume an explicit model that relates x to z — the modeling is implicit in their moments (means, autocovariances and the cross-covariance).

Equation (9.5.2-7) is an *integral equation* for the *impulse response* $H(t,\tau)$ of the filter that is the optimal estimator for this problem. This equation cannot be solved for the general nonstationary case discussed above.

In the *stationary case*, the Wiener-Hopf equation becomes

$$V_{xz}(t - \theta) - \int_0^t H(t - \tau)V_{zz}(\tau - \theta)d\tau = 0 \qquad \forall \theta \leq t \tag{9.5.2-8}$$

Outline of the Solution

Using Fourier transforms, a modified version of this equation can be solved for *rational spectra* (power spectral densities of the processes involved), yielding the **Wiener filter**.

Assuming that the observation process starts at $t = -\infty$ rather than at $t = 0$, (9.5.2-8) can be replaced by

$$V_{xz}(t - \theta) = \int_{-\infty}^{t} H(t - \tau)V_{zz}(\tau - \theta)d\tau \qquad \forall \theta \leq t \qquad (9.5.2\text{-}9)$$

Denoting

$$\alpha \overset{\Delta}{=} t - \tau \qquad (9.5.2\text{-}10)$$

$$\beta \overset{\Delta}{=} t - \theta \qquad (9.5.2\text{-}11)$$

one has

$$\tau - \theta = -\alpha + \beta \qquad (9.5.2\text{-}12)$$

and (9.5.2-9) becomes

$$V_{xz}(\beta) = \int_{0}^{\infty} H(\alpha)V_{zz}(\beta - \alpha)d\alpha \qquad \forall \theta \leq t \qquad (9.5.2\text{-}13)$$

If the lower limit of the integral in (9.5.2-13) is changed to $-\infty$, this equation can be solved via Fourier transforms for the impulse response $H(\alpha)$; however, the resulting system will have poles in both the left and right half plane, that is, it will be **noncausal**. Using spectral factorization one can obtain a *causal system* which is the desired filter, known as the Wiener filter.

Wiener Filter versus Kalman Filter

It can be shown that the Wiener filter is the *steady-state version* of the Kalman-Bucy filter. In view of this, and the fact that the Kalman-Bucy filter is for the more general *nonstationary* case, the Wiener filter will not be pursued further.

9.6 NOTES AND PROBLEMS

9.6.1 Bibliographical Notes

More details on continuous time filtering, including Ito differential equations, can be found in [SM71, Med69, May79, May82].

The linear-quadratic control problem is covered in many texts, for instance, [BH75, Med69, May79, May82].

The duality between estimation and control is treated in [Rho71].

More details on the Wiener filter can be found in [SM71, Kai81, Lew86].

9.6.2 Problems

9-1 Derivative of the inverse of a matrix.

1. Prove (9.2.5-2). Hint: Use the fact that $P(t)^{-1}P(t) = I$.

2. Prove (9.2.5-3).

9-2 Riccati differential equation solution for the inverse covariance.

1. Show that if

$$\dot{x}(t) = A(t)x(t) \quad \text{and} \quad x(t)'x_a(t) = c$$

then $x_a(t)$ is the solution of the *adjoint system*

$$\dot{x}_a(t) = -A(t)'x_a(t) \triangleq A_a(t)x_a(t)$$

and that the transition matrix of the adjoint system from t_0 to t relates to the one of the original system as follows:

$$F_a(t, t_0) = F(t_0, t)' = [F(t, t_0)']^{-1}$$

2. Rewrite (9.2.5-3) in terms of the system matrix $A_a(t)$ of the adjoint system and show that it has the solution

$$P(t)^{-1} = F_a(t, t_0)P(t_0)^{-1}F_a(t, t_0)' + \int_{t_0}^{t} F_a(t, \tau)C(\tau)'R(\tau)^{-1}C(\tau)F_a(t, \tau)'d\tau$$

3. Show that the above yields (9.2.5-4).

9-3 Fixed-point prediction evolution in continuous time. Let $F(t_2, t_1)$ be the transition matrix corresponding to the system matrix $A(t)$, that is, the equation

$$\dot{x}(t) = A(t)x(t)$$

has the solution

$$x(t_2) = F(t_2, t_1)x(t_1)$$

1. Show that

$$\frac{\partial}{\partial t_2} F(t_2, t_1) = A(t_2) F(t_2, t_1)$$

$$\frac{\partial}{\partial t_1} F(t_2, t_1) = -F(t_2, t_1) A(t_1)$$

2. Prove (9.3.2-2).

9-4 **Fixed-point prediction prior moments.** Show that, for $t_2 > t_1$,

$$E[\hat{x}(\tau|t_2)|Z^{t_1}] = \hat{x}(\tau|t_1)$$

$$\text{cov}[\hat{x}(\tau|t_2)|Z^{t_1}] = P(\tau|t_1) - P(\tau|t_2)$$

9-5 **Fixed-lead prediction evolution in continuous time.** Prove (9.3.2-4).

9-6 **State MSE matrix for continuous time filter with arbitrary gain.** Show that, similarly to (5.2.3-18), the continuous time Riccati equation (9.2.3-10) has the following alternative expression

$$\dot{P}(t) = [A(t) - L(t)C(t)]P(t) + P(t)[A(t) - L(t)C(t)]' + L(t)\tilde{R}(t)L(t)' + D(t)\tilde{q}(t)D(t)'$$

which holds for an *arbitrary filter gain* $L(t)$, that is, $P(t)$ is the state estimation MSE matrix. Hint: utilize the limiting form of the derivative of $P(t) \triangleq E[\tilde{x}(t)\tilde{x}(t)']$ and use the differential forms (9.2.1-7) and (9.2.3-13).

Chapter 10

STATE ESTIMATION FOR NONLINEAR DYNAMIC SYSTEMS

10.1 INTRODUCTION

10.1.1 Outline

This chapter deals with the estimation of the state of discrete time nonlinear dynamic systems observed via nonlinear measurements.

The optimal estimator for this problem is presented in Section 10.2 and the difficulty in its implementation is discussed. Also a comparison is made between the optimal nonlinear estimator and the best linear estimator for a simple problem.

The suboptimal filter known as the **extended Kalman filter (EKF)** is derived in Section 10.3 and some of the problems encountered in its implementation are illustrated in an example. Methods of compensation for linearized filters are discussed in Sections 10.4 and 10.5.

Section 10.6 describes a numerical procedure that relies on the technique of dynamic programming to obtain the maximum a posteriori (MAP) estimate of the sequence of states of a nonlinear dynamic system — the **modal trajectory**.

10.1.2 Nonlinear Estimation — Summary of Objectives

- Present the optimal estimator.
- Discuss the difficulty in implementing it in practice due to

    - memory requirements, and
    - computational requirements.

- Compare the optimal nonlinear estimator to the best linear estimator for a simple problem.
- Derive the suboptimal filter known as the extended Kalman filter.
- Illustrate some of the problems that can be encountered in its implementation.
- Present some methods of compensation for linearized filters.
- Describe an efficient numerical technique for MAP trajectory estimation.

10.2 ESTIMATION IN NONLINEAR STOCHASTIC SYSTEMS

10.2.1 The Model

The general state space model for discrete time stochastic systems is of the form

$$\boxed{x(k+1) = f[k, x(k), u(k), v(k)]} \tag{10.2.1-1}$$

where x is the state vector, u the known input, v the process noise. The vector-valued function f is, in general, time varying.

The measurements are described by

$$\boxed{z(k) = h[k, x(k), w(k)]} \tag{10.2.1-2}$$

where w is the measurement noise. The function h, which is also vector-valued, can be time varying.

At least one of the above two functions has to be nonlinear in order for the problem to be nonlinear.

The noise sequences will be assumed to be white with known pdf and mutually independent. The initial state is assumed to have a known pdf and also independent of the noises.

More specific assumptions will be given later when the optimal estimator, which evaluates the pdf of the state conditioned on the observations, is derived.

The Main Result

It will be shown that the **optimal nonlinear state estimator** consists of the computation of the conditional pdf of the state $x(k)$ given all the information available at time k: the prior information about the initial state, the intervening inputs and the measurements through time k.

10.2.2 The Optimal Estimator

The **information set** available at k is

$$I^k = \{Z^k, U^{k-1}\} \tag{10.2.2-1}$$

where Z^k is the cumulative set of observations through time k, which subsumes the initial information Z^0 and U^{k-1} is the set of known inputs prior to time k. Since the information set (10.2.2-1) is growing with k it is desirable to have a *nongrowing information state* that summarizes the past in a manner appropriate for the problem.

For a stochastic system, an **information state** is a function of the available information set that completely summarizes the past of the system in a probabilistic sense: it is a complete substitute for the past data in the pdf of any present and future quantity related to the system.

As shown in the next subsection, the conditional pdf

$$p_k \triangleq p[x(k)|I^k] \tag{10.2.2-2}$$

satisfies this requirement if the two noise sequences (process and measurement) are *white and mutually independent.*

The optimal estimator consists of the *recursive functional relationship* between the information states p_{k+1} and p_k, and is given by

$$\boxed{p_{k+1} = \frac{1}{c}p[z(k+1)|x(k+1)] \int p[x(k+1)|x(k),u(k)]p_k \; dx(k)} \tag{10.2.2-3}$$

where the integration is over the entire domain of $x(k)$ and c is the normalization constant. The derivation of the above is carried out in the next subsection.

Remarks

The following remarks are pertinent to the optimal estimation:

1. The implementation of (10.2.2-3) requires storage of a pdf, which is, in general, equivalent to an *infinite dimensional vector*. Furthermore, one faces the problem of carrying out the integration numerically.

2. In spite of these difficulties, the use of the information state has the following advantages:

    a. It can be approximated by quantization or by a piecewise analytical function;

    b. It yields directly the MMSE estimate of the current state — the conditional mean — and any other desired information pertaining to the current state (e.g., its conditional variance).

3. For linear systems with Gaussian noises and Gaussian initial state, the functional recursion (10.2.2-3) becomes the Kalman Filter. In this case one has a *finite-dimensional sufficient statistic* consisting of the state's conditional mean and covariance that summarize completely the past in a probabilistic sense.

4. If a system is linear but the noises and/or the initial condition are not Gaussian, then, in general, there is no simple sufficient statistic and the recursion (10.2.2-3) has to be used to obtain the optimal MMSE estimator.

A simple example that compares the optimum and the linear MMSE estimators for a linear problem with uniformly distributed random variables is presented in Subsection 10.2.4.

10.2.3 Proof of the Recursion of the Conditional Density of the State

The conditional density of $x(k+1)$ can be written using Bayes' formula

$$p_{k+1} \triangleq p[x(k+1)|I^{k+1}] = p[x(k+1)|I^k, z(k+1), u(k)]$$

$$= \frac{1}{c}p[z(k+1)|x(k+1), I^k, u(k)]p[x(k+1)|I^k, u(k)] \qquad (10.2.3\text{-}1)$$

where c is the normalization constant.

If the measurement noise is white in the following sense: $w(k+1)$ conditioned on $x(k+1)$ has to be independent of $w(j)$, $j \leq k$ (i.e., state-dependent measurement noise is allowed), then

$$p[z(k+1)|x(k+1), I^k, u(k)] = p[z(k+1)|x(k+1)] \qquad (10.2.3\text{-}2)$$

Note that the control (known input) is irrelevant in the above conditioning. The above expression is the likelihood function of $x(k+1)$ given $z(k+1)$.

For an arbitrary value of the control at k one has the **state prediction pdf** (a function, rather than the **point prediction** — the predicted value — as in the linear case)

$$p[x(k+1)|I^k, u(k)] = \int p[x(k+1)|x(k), I^k, u(k)]p[x(k)|I^k, u(k)] \, dx(k) \qquad (10.2.3\text{-}3)$$

The above equation, which is an immediate consequence of the total probability theorem, is known as the **Chapman-Kolmogorov equation**.

It will be assumed that the process noise sequence is white and independent of the measurement noises in the following sense: $v(k)$ conditioned on $x(k)$ has to be independent of $v(j-1)$, $w(j)$, $j \leq k$ (i.e., state-dependent process noise is allowed). This is equivalent to requiring the state vector $x(k)$ to be a Markov process.

Then I^k is irrelevant in the conditioning of the first term on the right hand side of (10.2.3-3), that is,

$$p[x(k+1)|x(k), I^k, u(k)] = p[x(k+1)|x(k), u(k)] \qquad (10.2.3\text{-}4)$$

The above is the **state transition pdf**. Furthermore, since the input $u(k)$ enters the system after the realization of $x(k)$ has occurred, one has

$$p[x(k)|I^k, u(k)] = p[x(k)|I^k] \triangleq p_k \qquad (10.2.3\text{-}5)$$

Inserting (10.2.3-4) and (10.2.3-5) into (10.2.3-3), it follows that the state prediction pdf can be written as

$$p[x(k+1)|I^k, u(k)] = \int p[x(k+1)|x(k), u(k)]p_k \, dx(k) = \phi[k+1, p_k, u(k)] \qquad (10.2.3\text{-}6)$$

where ϕ is a transformation — an **operator** — that maps the function p_k into another function, namely, the state prediction pdf.

Using (10.2.3-2) and (10.2.3-6) in (10.2.3-1), the latter can be rewritten as another transformation ψ that maps p_k into p_{k+1}

$$\boxed{p_{k+1} = \psi[k+1, p_k, z(k+1), u(k)]} \tag{10.2.3-7}$$

The above is the functional recursion for the information state rewritten explicitly in (10.2.2-3).

The Information State

Next it will be shown that p_k is an **information state** according to the definition of the previous subsection, that is, that the pdf of any future state at $j > k$ depends *only* on p_k and the intervening controls. These controls (known inputs) are denoted as

$$U_k^{j-1} \triangleq \{u(i)\}_{i=k}^{j-1} \tag{10.2.3-8}$$

The proof will be done using the *smoothing property of expectations* (1.4.12-4). For $j > k$ one has

$$
\begin{aligned}
p[x(j)|I^k, U_k^{j-1}] &= E[p_j|I^k, U_k^{j-1}] = \int p[x(j)|x(k), I^k, U_k^{j-1}]p[x(k)|I^k]dx(k) \\
&= \int p[x(j)|x(k), U_k^{j-1}]p_k \, dx(k) \triangleq \mu[j, p_k, U_k^{j-1}]
\end{aligned}
\tag{10.2.3-9}
$$

where the whiteness of the process noise sequence and its independence from the measurement noises have been used again. In the above, μ denotes the transformation that maps p_k and the known inputs into the pdf of a future state $x(j)$.

Thus from (10.2.3-7) an (10.2.3-9) it follows that I^k is *summarized* by p_k.

Therefore, the *whiteness and mutual independence of the two noise sequences* is a sufficient condition for p_k to be an information state. It should be emphasized that the whiteness is the crucial assumption.

The above conditions are equivalent to the requirement that $x(k)$ be an **incompletely observed Markov process** — a Markov process observed partially and/or in the presence of noise.

If, for example, the process noise sequence is not white, it is obvious that p_k does not summarize the past data. In this case the vector x is not a state anymore and it has to be augmented.

This discussion points out the reason why the formulation of stochastic estimation and control problems is done with *mutually independent white noise sequences*.

10.2.4 Linear versus Nonlinear Estimation of a Parameter

Consider an unknown parameter x with a prior pdf uniform in the interval $[x_1, x_2]$, that is,

$$p(x) = \mathcal{U}[x; x_1, x_2] \triangleq \frac{1(x - x_1) - 1(x - x_2)}{x_2 - x_1} \qquad (10.2.4\text{-}1)$$

where $1(\cdot)$ denotes the unit step function.

A measurement is made

$$z = x + w \qquad (10.2.4\text{-}2)$$

where w is independent of x and uniformly distributed within the interval $[-a, a]$

$$p(w) = \mathcal{U}[w; -a, a] = \frac{1(w + a) - 1(w - a)}{2a} \qquad (10.2.4\text{-}3)$$

The Optimal MMSE Estimator

First the optimal MMSE estimator (the conditional mean) is obtained. The conditional pdf of x given z is

$$p(x|z) = \frac{p(z|x)p(x)}{p(z)} \qquad (10.2.4\text{-}4)$$

From (10.2.4-3) it follows that the likelihood function of x given z — the pdf of z conditioned on x — is

$$p(z|x) = \mathcal{U}[z; x - a, x + a] = \frac{1[z - (x - a)] - 1[z - (x + a)]}{2a} \qquad (10.2.4\text{-}5)$$

Using (10.2.4-1) and (10.2.4-5) in (10.2.4-4) yields the *conditional pdf* of x given z as

$$\begin{aligned} p(x|z) &= \frac{\{1[x - (z - a)] - 1[x - (z + a)]\}[1(x - x_1) - 1(x - x_2)]}{2a(x_2 - x_1)p(z)} \\ &= \mathcal{U}[x; (z - a) \vee x_1, (z + a) \wedge x_2] \end{aligned} \qquad (10.2.4\text{-}6)$$

where the following notations are used

$$a \vee b \triangleq \max(a, b) \qquad (10.2.4\text{-}7)$$

$$a \wedge b \triangleq \min(a, b) \qquad (10.2.4\text{-}8)$$

The conditional pdf (10.2.4-6) is seen to be uniform in the interval, which is the intersection of the interval $[x_1, x_2]$ from the prior and the interval $[z - a, z + a]$ — the feasibility region of x given z — for obvious reasons.

10.2.4 Linear versus Nonlinear Estimation of a Parameter

The MMSE estimator of x given z (i.e., its *exact conditional mean*) is, from (10.2.4-6),

$$\hat{x}^{\text{MMSE}} = E[x|z] = \frac{(z-a)\vee x_1 + (z+a)\wedge x_2}{2} \qquad (10.2.4\text{-}9)$$

Note that this estimator is a *nonlinear function* of the measurement z.

The *conditional variance* of this estimator is, *for a given z*, from (10.2.4-6) and (1.4.5-7), given by

$$P_{xx|z} = \text{var}(x|z) = \frac{[(z-a)\vee x_1 - (z+a)\wedge x_2]^2}{12} \qquad (10.2.4\text{-}10)$$

Note that the "quality" of the optimal estimate, measured in terms of its variance, depends on the measurement z. This is a major difference between nonlinear estimators, like (10.2.4-9), and linear estimators.

The LMMSE Estimator

The *linear* MMSE estimator of x in terms of z is

$$\hat{x}^{\text{LMMSE}} = \bar{x} + P_{xz}P_{zz}^{-1}(z - \bar{z}) \qquad (10.2.4\text{-}11)$$

where

$$\bar{x} = E[x] = \frac{x_1 + x_2}{2} \qquad (10.2.4\text{-}12)$$

$$\bar{z} = E[z] = \bar{x} \qquad (10.2.4\text{-}13)$$

$$\begin{aligned}
P_{xz} &= E[(x-\bar{x})(z-\bar{z})] = E[(x-\bar{x})(x-\bar{x}+w)] \\
&= E[(x-\bar{x})^2] = P_{xx} = \frac{(x_1-x_2)^2}{12} \qquad (10.2.4\text{-}14)
\end{aligned}$$

$$\begin{aligned}
P_{zz} &= E[(z-\bar{z})^2] = E[(x-\bar{x}+w)^2] \\
&= P_{xx} + P_{ww} = \frac{(x_1-x_2)^2 + 4a^2}{12} \qquad (10.2.4\text{-}15)
\end{aligned}$$

The MSE corresponding to the estimator (10.2.4-11) is

$$\begin{aligned}
P_{xx}^{\text{LMMSE}} &= P_{xx} - \frac{P_{xz}^2}{P_{zz}} = \frac{P_{xx}P_{zz} - P_{xz}^2}{P_{zz}} \\
&= \frac{P_{xx}(P_{xx}+P_{ww}) - P_{xx}^2}{P_{xx}+P_{ww}} \\
&= \frac{P_{xx}P_{ww}}{P_{xx}+P_{ww}} \qquad (10.2.4\text{-}16)
\end{aligned}$$

379

Comparison of the Errors: Optimal versus Best Linear

The comparison between the errors obtained from the *optimal* and *best linear* methods cannot be made between (10.2.4-16) and (10.2.4-10) since the latter is a function of the observations. It has to be made between (10.2.4-16), which is independent of z, and the *expected value* of (10.2.4-10), which is to be *averaged over all the possible measurements z*.

The **average covariance associated with the optimal estimate** is

$$P_{xx}^{\text{OPT}} = E[P_{xx|z}] = \int P_{xx|z} p(z) dz \qquad (10.2.4\text{-}17)$$

where, in this example, $p(z)$ is the convolution of the densities (10.2.4-1) and (10.2.4-3).

Table 10.2.4-1 illustrates the comparison of these two estimators' variances for some numerical values. The interval over which x is uniformly distributed is $x_2 - x_1 = 1$, that is, its "prior" variance was $1/12 = 0.0833$. The measurement ranged from very accurate for $a \ll 1$, to very inaccurate — practically noninformative — for $a \gg 1$.

a	P_{xx}^{LMMSE}	P_{xx}^{OPT}
0.05	8.25E-04	7.64E-04
0.1	3.2E-03	2.81E-03
0.2	11.49E-03	9.5371E-03
0.5	4.16E-02	3.5871E-02
0.8	5.99E-02	5.4693E-02
1	6.66E-02	6.2510E-02
2	7.84E-02	7.6387E-02
5	8.25E-02	8.1249E-02

Table 10.2.4-1: Comparison of best linear and optimal estimators.

Conclusion

As can be seen from Table 10.2.4-1, the benefit from the nonlinear estimation over the linear one is disappointingly modest in this case: it ranges from negligible — 1% for the inaccurate measurement considered, to 6% for the accurate measurement; its maximum, which occurs in the midrange, is about 15% in variance. This latter improvement in variance translates into a 7% improvement in standard deviation (see problem 10-2).

No general conclusions can be drawn here — each problem requires its own evaluation.

10.2.5 Optimal Nonlinear Estimation — Summary

Given a system described by

1. dynamic equation perturbed by white process noise, and
2. measurement equation perturbed by white noise independent of the process noise,

with at least one of these equations nonlinear, the estimation of the system's state consists then of the calculation of its pdf conditioned on the entire available information: the observations, the initial state information and the past inputs.

This conditional pdf has the property of being an *information state*, that is, it summarizes probabilistically the past of the system.

The optimal nonlinear estimator for such a discrete time stochastic dynamic system consists of a recursive functional relationship for the state's conditional pdf. This conditional pdf then yields the MMSE estimator for the state — the conditional mean of the state.

This nonlinear estimator has to be used to obtain the conditional mean of the state unless the system is *linear Gaussian*: both the dynamic and the measurement equations are linear and all the noises and the initial state are Gaussian.

If the system is linear Gaussian, this functional recursion becomes the recursion for the conditional mean and covariance.

Unlike the linear case, in the nonlinear case the accuracy of the estimator is *measurement dependent*.

10.3 THE EXTENDED KALMAN FILTER

10.3.1 Approximation of the Nonlinear Estimation Problem

In view of the very limited feasibility of the implementation of the optimal filter, which consists of a functional recursion, suboptimal[1] algorithms are of interest.

The recursive calculation of the sufficient statistic consisting of the conditional mean and variance in the linear-Gaussian case is the simplest possible state estimation filter. As indicated earlier, in the case of a linear system with nonGaussian random variables the same simple recursion yields an approximate mean and variance.

A framework similar to the one from linear systems is desirable for a nonlinear system. Such an estimator, called the **extended Kalman filter (EKF)**, can be obtained by a series expansion of the nonlinear dynamics and of the measurement equations.

The (first order) EKF is based on

- Linearization — first order series expansion — of the nonlinearities (in the dynamic and/or the measurement equation);
- LMMSE estimation.

The second order EKF relies on a second order expansion, that is, it includes second order correction terms.

---

[1]This is a common euphemism for nonoptimal.

Modeling Assumptions

Consider the system with dynamics

$$\boxed{x(k+1) = f[k, x(k)] + v(k)} \tag{10.3.1-1}$$

where, for simplicity, it is assumed that there is no control, and the noise is assumed additive, zero mean, and white

$$E[v(k)] = 0 \tag{10.3.1-2}$$

$$E[v(k)v(j)'] = Q(k)\delta_{kj} \tag{10.3.1-3}$$

The measurement is

$$\boxed{z(k) = h[k, x(k)] + w(k)} \tag{10.3.1-4}$$

where the measurement noise is additive, zero mean, and white

$$E[w(k)] = 0 \tag{10.3.1-5}$$

$$E[w(k)w(j)'] = R(k)\delta_{kj} \tag{10.3.1-6}$$

and uncorrelated with the process noise.

The initial state, with estimate $\hat{x}(0|0)$ — an approximate conditional mean — and the associated covariance matrix $P(0|0)$, is assumed to be uncorrelated with the two noise sequences.

Similarly to the linear case, it will be assumed that one has the estimate at time k, an *approximate conditional mean*

$$\hat{x}(k|k) \approx E[x(k)|Z^k] \tag{10.3.1-7}$$

and the *associated covariance matrix $P(k|k)$.*

Strictly speaking, $P(k|k)$ is the MSE matrix rather than the covariance matrix in view of the fact that $\hat{x}(k|k)$ is not the exact conditional mean.

Note that (10.3.1-7) implies that the estimation error is *approximately zero mean.* Another assumption that will be made is that the third order moments of the estimation error are approximately zero — this is exact in the case of zero-mean Gaussian random variables.

10.3.2 Derivation of the EKF

State Prediction

To obtain the predicted state $\hat{x}(k+1|k)$, the nonlinear function in (6.3.1-1) is expanded in Taylor series around the latest estimate $\hat{x}(k|k)$ with terms up to first or second order to yield the first or second order EKF, respectively. The second order filter is presented from which the first order one follows as a particular case.

The **vector Taylor series expansion** of (6.3.1-1) up to second order terms is

$$
\begin{aligned}
x(k+1) &= f[k, \hat{x}(k|k)] + f_x(k)[x(k) - \hat{x}(k|k)] \\
&\quad + \tfrac{1}{2}\sum_{i=1}^{n_x} e_i[x(k) - \hat{x}(k|k)]' f_{xx}^i(k)[x(k) - \hat{x}(k|k)] + HOT + v(k)
\end{aligned}
\tag{10.3.2-1}
$$

where e_i is the i-th n_x-dimensional **Cartesian basis vector** (i-th component unity, the rest zero),

$$
f_x(k) \triangleq [\nabla_x f(k, x)']'\big|_{x=\hat{x}(k|k)}
\tag{10.3.2-2}
$$

is the Jacobian of the vector f, evaluated at the latest estimate of the state. Similarly,

$$
f_{xx}^i(k) \triangleq [\nabla_x \nabla_x' f^i(k, x)]\big|_{x=\hat{x}(k|k)}
\tag{10.3.2-3}
$$

is the Hessian of the i-th component of f and HOT represents the **higher order terms**, which we shall not touch — these terms will be neglected.

Note that, given the data Z^k, both the above Jacobian and the Hessian are *deterministic quantities* — only $x(k)$ and $v(k)$ are random variables in (6.3.2-1) in this case. These properties are needed when taking conditional expectations next.

The **predicted state** to time $k+1$ from time k is obtained by taking the expectation of (6.3.2-1) conditioned on Z^k and neglecting the HOT

$$
\boxed{\hat{x}(k+1|k) = f[k, \hat{x}(k|k)] + \frac{1}{2}\sum_{i=1}^{n_x} e_i \, \mathrm{tr}[f_{xx}^i(k) P(k|k)]}
\tag{10.3.2-4}
$$

where use has been made of (1.4.15-1) for the second order term. The first order term in (6.3.2-1) is, in view of (6.3.1-7), (approximately) zero mean and thus vanishes.

The state prediction error is obtained by subtracting (6.3.2-4) from (6.3.2-1) to yield

$$
\tilde{x}(k+1|k) = f_x(k)\tilde{x}(k|k) + \frac{1}{2}\sum_{i=1}^{n_x} e_i\Big\{\tilde{x}(k|k)' f_{xx}^i(k)\tilde{x}(k|k) - \mathrm{tr}[f_{xx}^i(k) P(k|k)]\Big\} + v(k)
\tag{10.3.2-5}
$$

where the HOT have already been dropped.

Multiplying the above with its transpose and taking the expectation conditioned on Z^k, yields the **state prediction covariance** (actually, the **MSE matrix**)

$$P(k+1|k) = f_x(k)P(k|k)f_x(k)' + \frac{1}{2}\sum_{i=1}^{n_x}\sum_{j=1}^{n_x} e_i e_j' \, \text{tr}[f_{xx}^i(k)P(k|k)f_{xx}^j(k)P(k|k)] + Q(k)$$

(10.3.2-6)

where use has been made of identity (1.4.15-8) and of the approximation that the third order moments are zero.

The state prediction (10.3.2-4) includes the second order "correction" term. This is dropped in the first order EKF.

The prediction covariance (10.3.2-6) contains a fourth order term obtained from the squaring of the second order term in (10.3.2-5).

The first order version of (10.3.2-6) is the same as (5.2.3-5) in the linear filter — the Jacobian $f_x(k)$ plays the role of the transition matrix $F(k)$.

Measurement Prediction

Similarly, the **predicted measurement** is, for the second order filter

$$\hat{z}(k+1|k) = h[k+1, \hat{x}(k+1|k)] + \frac{1}{2}\sum_{i=1}^{n_z} e_i \, \text{tr}[h_{xx}^i(k+1)P(k+1|k)]$$

(10.3.2-7)

where e_i denotes here the i-th n_z-dimensional Cartesian basis vector.

The **measurement prediction covariance** or the **innovation covariance** (MSE matrix) is

$$S(k+1) = h_x(k+1)P(k+1|k)h_x(k+1)'$$
$$+ \frac{1}{2}\sum_{i=1}^{n_z}\sum_{j=1}^{n_z} e_i e_j' \text{tr}[h_{xx}^i(k+1)P(k+1|k)h_{xx}^j(k+1)P(k+1|k)] + R(k)$$

(10.3.2-8)

where the Jacobian of h is

$$h_x(k+1) = [\nabla_x h(k+1, x)']' \Big|_{x=\hat{x}(k+1|k)}$$

(10.3.2-9)

and the Hessian of its i-th component is

$$h_{xx}^i(k+1) = [\nabla_x \nabla_x' h^i(k+1, x)] \Big|_{x=\hat{x}(k+1|k)}$$

(10.3.2-10)

Modifications for the First Order EKF

The second order "correction" term in (10.3.2-7) and the corresponding fourth order term in (10.3.2-8) are dropped in the first order EKF. The first order version of (10.3.2-8) is the same as in the linear filter — the Jacobian $h_x(k)$ plays the role of the measurement matrix $H(k)$.

State Update

The expression of the filter gain, the update equation for the state and its covariance are identical to those from the linear filter, given by (5.2.3-11) to (5.2.3-15).

10.3.3 Overview of the EKF

Figure 10.3.3-1 presents the flowchart of the extended Kalman filter (first order version).

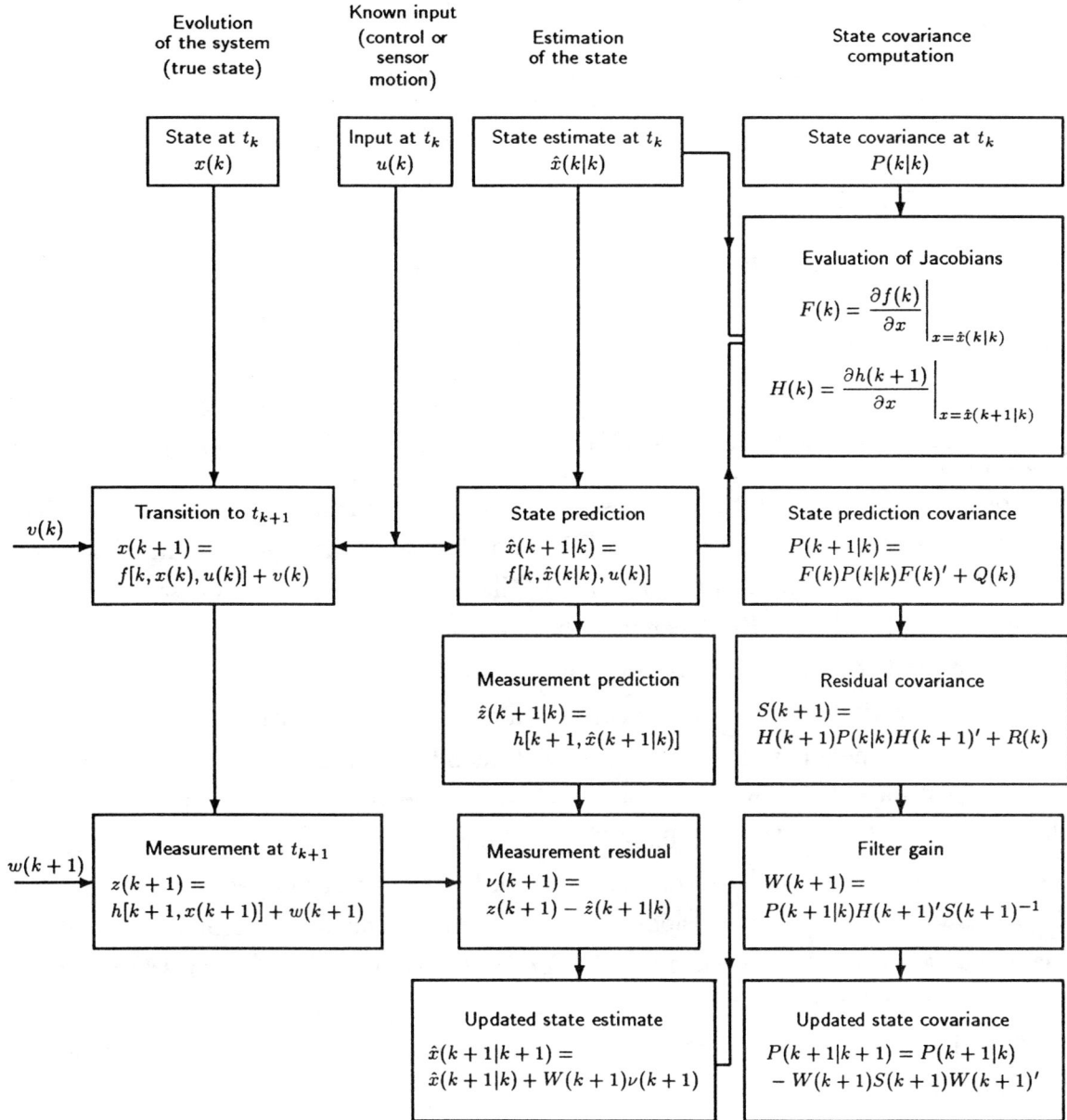

Figure 10.3.3-1: Flowchart of the EKF (one cycle).

The main difference from the Kalman Filter, presented in Figure 5.2.4-1, is the evaluation of the Jacobians of the state transition and the measurement equations. Due to this, the covariance computations are not decoupled anymore from the state estimate calculations and cannot be done, in general, offline.

The linearization (evaluation of the Jacobians) can be done, as indicated here, at the latest state estimate for F and the predicted state for H. Alternatively, it can be done along a **nominal trajectory** — a deterministic precomputed trajectory based on a certain scenario — which allows offline computation of the gain and covariance sequence.

Stability of the EKF

The sufficient conditions for the stability of the KF are not necessarily sufficient for the EKF. The reason is that its inherent approximations can lead to **divergence of the filter** — unbounded estimation errors.

A Cautionary Note about the EKF

It should be pointed out that the use of the series expansion in the state prediction and/or in the measurement prediction has the potential of introducing unmodeled errors that violate some basic assumptions about the prediction errors. These assumptions, which are implicit in the EKF, concern their moments, namely:

1. The prediction errors are zero mean (unbiased); and
2. With covariances equal to the computed ones by the algorithm.

In general, a nonlinear transformation will introduce a bias and the covariance calculation based on a series expansion is not always accurate. This might be the case with the predictions (10.3.2-4) and (10.3.2-7) and the covariances (10.3.2-6) and (10.3.2-8).

There is no guarantee that even the second order terms can compensate for such errors. Also, the fact that these expansions rely on Jacobians (and Hessians in the second order case) that are evaluated at the estimated or predicted state rather than the exact state (which is unavailable) can cause errors.

Thus it is even more important for an EKF to undergo **consistency testing**. The tests are the same as those discussed for the KF. Only if the bias is negligible with respect to the filter-calculated error variances and the mean-square errors are consistent with these variances is the filter consistent, and its results can be trusted.

In practice, if the initial error and the noises are not too large, then the EKF performs well. The actual limits of successful use of the linearization techniques implicit in the EKF can be obtained only via extensive Monte Carlo simulations for consistency verification.

10.3.4 An Example: Tracking with an Angle-Only Sensor

A platform with a sensor moves in a plane according to the discrete time equations

$$x_p(k) = \bar{x}_p(k) + \Delta x_p(k) \qquad k = 0, 1, \ldots, 20 \qquad (10.3.4\text{-}1)$$

$$y_p(k) = \bar{y}_p(k) + \Delta y_p(k) \qquad k = 0, 1, \ldots, 20 \qquad (10.3.4\text{-}2)$$

where $\bar{x}_p(k)$ and $\bar{y}_p(k)$ are the average platform position coordinates and the perturbations (jitter) $\Delta x_p(k)$ and $\Delta y_p(k)$ are assumed to be mutually independent zero-mean Gaussian white noise sequences with variances $r_x = 1$ and $r_y = 1$, respectively.

The average (unperturbed) platform motion is assumed to be horizontal with constant velocity. Its position as a function of the discrete time k is

$$\bar{x}_p(k) = 4k \qquad (10.3.4\text{-}3)$$

$$\bar{y}_p(k) = 20 \qquad (10.3.4\text{-}4)$$

A target moves on the x-axis according to

$$x(k+1) = \begin{bmatrix} 1 & 1 \\ 0 & 1 \end{bmatrix} x(k) + \begin{bmatrix} 1/2 \\ 1 \end{bmatrix} v(k) \qquad (10.3.4\text{-}5)$$

where

$$x(k) = \begin{bmatrix} x_1(k) \\ x_2(k) \end{bmatrix} \qquad (10.3.4\text{-}6)$$

with x_1 denoting the position and x_2 denoting the velocity of the target.

The initial condition is

$$x(0) = \begin{bmatrix} 80 \\ 1 \end{bmatrix} \qquad (10.3.4\text{-}7)$$

and the process noise $v(k)$ is zero mean white with variance $q = 10^{-2}$.

The sensor measurement is

$$z(k) = h[x_p(k), y_p(k), x_1(k)] + w_s(k) \qquad (10.3.4\text{-}8)$$

where

$$h[\cdot] = \tan^{-1} \frac{y_p(k)}{x_1(k) - x_p(k)} \qquad (10.3.4\text{-}9)$$

is the angle between the horizontal and the line of sight from the sensor to the target, and the sensor noise $w_s(k)$ is zero mean white Gaussian with variance $r_s = (3°)^2$. The sensor noise is assumed independent of the sensor platform perturbations.

389

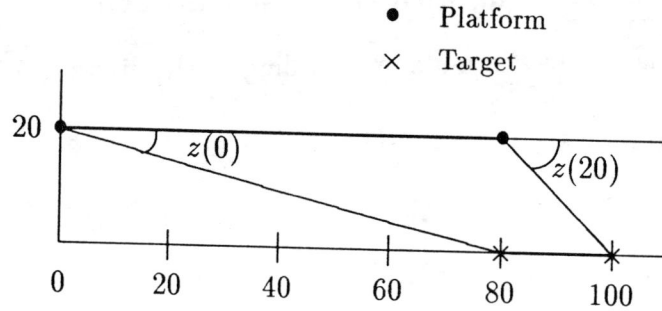

Figure 10.3.4-1: Platform and target motion.

The estimation of the target's state is to be done using the measurements (10.3.4-8) and the knowledge of the unperturbed platform location (10.3.4-3) and (10.3.4-4).

Figure 10.3.4-1 illustrates the average motion of the sensor platform and the target.

The platform location perturbations induce additional errors in the measurements. The effect of these errors is evaluated by expanding the nonlinear measurement function h about the average platform position

$$
\begin{aligned}
z(k) &= h[\bar{x}_p(k), \bar{y}_p(k), x_1(k)] + \frac{\partial h}{\partial x_p}\Delta x_p(k) + \frac{\partial h}{\partial y_p}\Delta y_p(k) + w_s(k) \\
&\triangleq h[\bar{x}_p(k), \bar{y}_p(k), x_1(k)] + w(k)
\end{aligned}
\tag{10.3.4-10}
$$

where the partials of h are evaluated at $\bar{x}_p(k)$, $\bar{y}_p(k)$, and $x_1(k)$, while the HOT have been neglected.

The last three terms in (10.3.4-10) form the *equivalent measurement noise* $w(k)$, with variance

$$
E[w(k)^2] \triangleq r(k) = \frac{\bar{y}_p(k)^2 r_x + [x_1(k) - \bar{x}_p(k)]^2 r_y}{\{[x_1(k) - \bar{x}_p(k)]^2 + \bar{y}_p(k)^2\}^2} + r_s
\tag{10.3.4-11}
$$

The above follows from the explicit evaluation of the partials in (10.3.4-10) and noting that the random variables Δx_p, Δy_p, and w_s are mutually independent. Note that the variance (10.3.4-11) is time varying. This variance (in rad^2) is plotted in Figure 10.3.4-2.

The state x is to be estimated based on the measurement equation (10.3.4-10) with the knowledge of only the average platform motion.

The initialization of the filter is done as follows. Based on Figure 10.3.4-1 one can write the *inverse transformation* from the angle z to the position x_1 as

$$
\hat{x}_1(k) = h^{-1}[z(k), \bar{x}_p(k), \bar{y}_p(k)] = \bar{x}_p(k) + \frac{\bar{y}_p(k)}{\tan z(k)} \qquad k = 0, 1
\tag{10.3.4-12}
$$

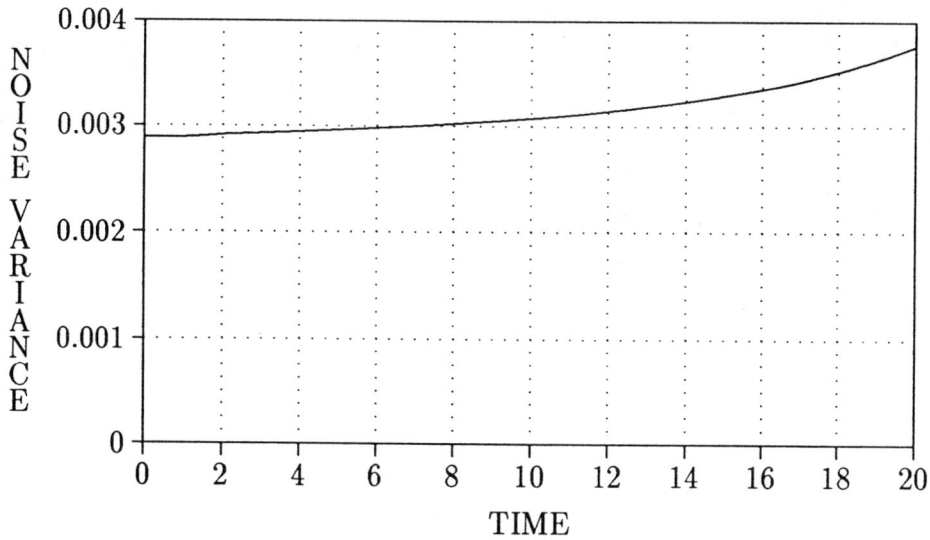

Figure 10.3.4-2: Variance of the equivalent measurement noise.

By differencing one can obtain

$$\hat{x}_2(1) = \frac{\hat{x}_1(1) - \hat{x}_1(0)}{T} \qquad (10.3.4\text{-}13)$$

where $T = 1$ is the sampling period.

The variance associated with the above velocity estimate is

$$P_{22}(1|1) = \frac{P_{11}(1) + P_{11}(0)}{T^2} = P_{11}(1) + P_{11}(0) \qquad (10.3.4\text{-}14)$$

where $P_{11}(k)$, $k = 0, 1$, are the variances associated with (10.3.4-12).

Note, however, that because of the inaccurate angle measurement ($\sqrt{r_s} = \pi/60$ rad), when multiplied by the range (≈ 80), it can be seen that $P_{11}(0) \approx P_{11}(1) > 4^2$. In this situation, assuming (from a priori knowledge of the target) that $|x_2| \le 2$, one is better off with the initial velocity estimate

$$\hat{x}_2(0|0) = 0 \qquad (10.3.4\text{-}15)$$

and associated variance

$$P_{22}(0|0) = 1 \qquad (10.3.4\text{-}16)$$

The off-diagonal term $P_{12}(0|0)$ is taken as zero.

The initial position estimate is then $\hat{x}_1(0|0)$, as given by (10.3.4-12) for $k = 0$ and the variance associated with it is denoted as $P_{11}(0|0)$. This variance is obtained (approximately) by a first order expansion of (10.3.4-12).

The error in the initial position estimate is

$$\tilde{x}_1(0|0) = \frac{\partial h^{-1}}{\partial \bar{x}_p}\Delta x_p + \frac{\partial h^{-1}}{\partial \bar{y}_p}\Delta y_p + \frac{\partial h^{-1}}{\partial z}\Delta z \qquad (10.3.4\text{-}17)$$

where $\Delta z \triangleq w(0)$, denotes the *equivalent measurement noise*.

Evaluating the partials in (10.3.4-17) yields, using also (10.3.4-10), the expression

$$
\begin{aligned}
\tilde{x}_1(0|0) &= \Delta x_p + \frac{1}{\tan z}\Delta y_p - \frac{\bar{y}_p}{\sin^2 z}\Delta z \\
&= \left[1 - \frac{\bar{y}_p}{\sin^2 z}\frac{\partial h}{\partial x_p}\right]\Delta x_p + \left[\frac{1}{\tan z} - \frac{\bar{y}_p}{\sin^2 z}\frac{\partial h}{\partial y_p}\right]\Delta y_p - \frac{\bar{y}_p}{\sin^2 z}w_s \quad (10.3.4\text{-}18)
\end{aligned}
$$

from which

$$P_{11}(0|0) = \left[1 - \frac{\bar{y}_p}{\sin^2 z}\frac{\partial h}{\partial x_p}\right]^2 r_x + \left[\frac{1}{\tan z} - \frac{\bar{y}_p}{\sin^2 z}\frac{\partial h}{\partial y_p}\right]^2 r_y + \frac{\bar{y}_p^2}{\sin^4 z}r_s \qquad (10.3.4\text{-}19)$$

Two Options for Implementation

In the problem considered here, as in many practical problems, the target motion is naturally modeled as linear in Cartesian coordinates, while the measurements are in polar coordinates (angle only in the present case). The nonlinear function h relates the state to the measurements. The following two options are available for implementation of the state estimation filter:

1. By using the inverse transformation $\zeta \triangleq h^{-1}(z)$ one can obtain directly a **converted measurement** of the position, which leads to a purely *linear problem* and a KF can be used. While this is convenient, it should be noted that the inverse transformation of the measurement can lead to errors that are difficult (but not impossible — see [LB93]) to account for.
2. By leaving the measurement in its original form, one has a **mixed coordinate filter**.

The filter was implemented in *mixed coordinates*: the measurement was in the line-of-sight angle coordinate while the target's state was in Cartesian coordinates.

Figures 10.3.4-3 and 10.3.4-4 show, for a single run, the position and velocity estimates compared to the true values. The filter appears to converge, in the sense that the estimation errors tend to decrease, but slowly.

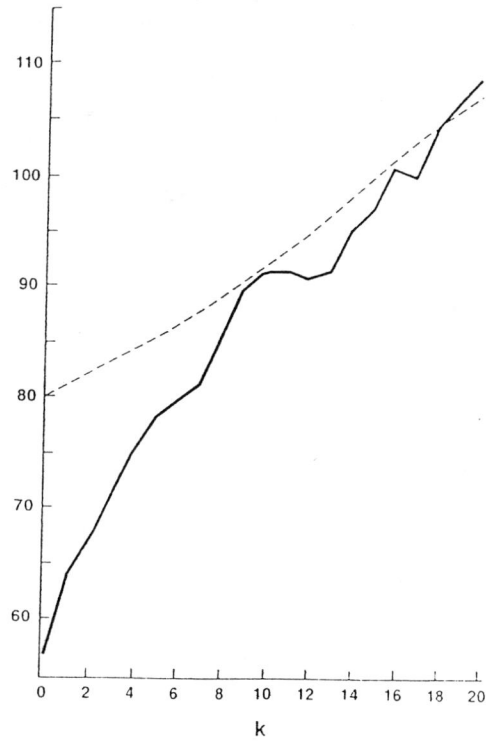

Figure 10.3.4-3: True and estimated position in a single run (- - truth; — estimated).

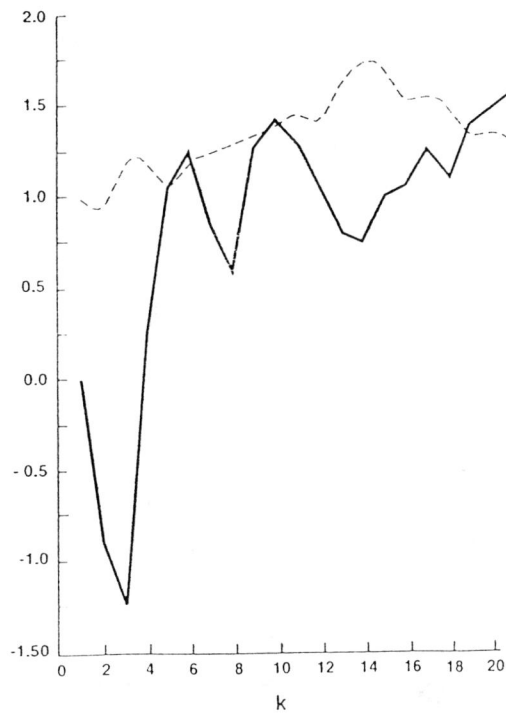

Figure 10.3.4-4: True and estimated velocity in a single run (- - truth; — estimated).

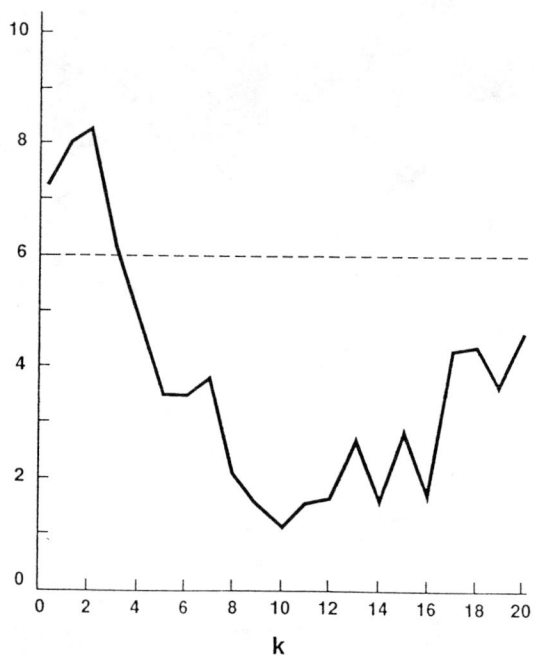

Figure 10.3.4-5: Normalized state estimation error squared (single run) with the one-sided 95% probability region.

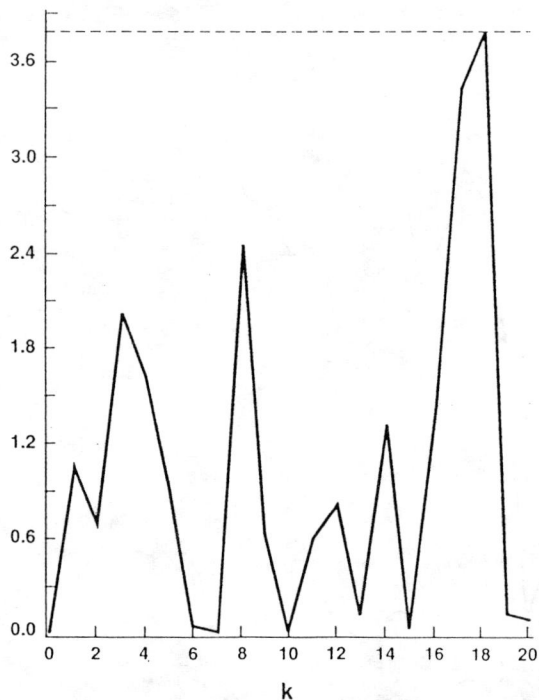

Figure 10.3.4-6: Normalized innovation squared (single run) with the one-sided 95% probability region.

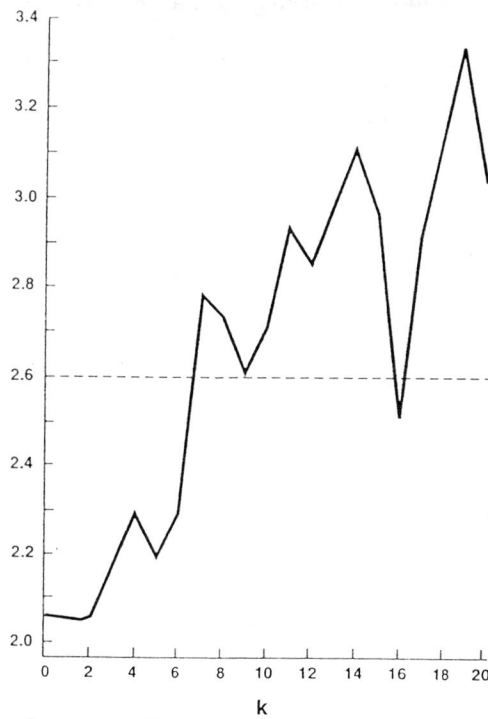

Figure 10.3.4-7: Normalized state estimation error squared (50 runs) with the 97.5% probability level (upper limit of the two-sided 95% region).

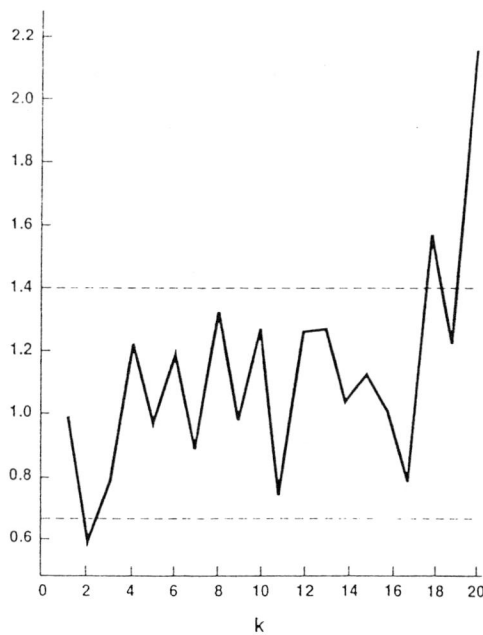

Figure 10.3.4-8: Normalized innovation squared (50 runs) with the two-sided 95% region (2.5% and 97.5% levels).

395

10.3.4 An Example: Tracking with an Angle-Only Sensor

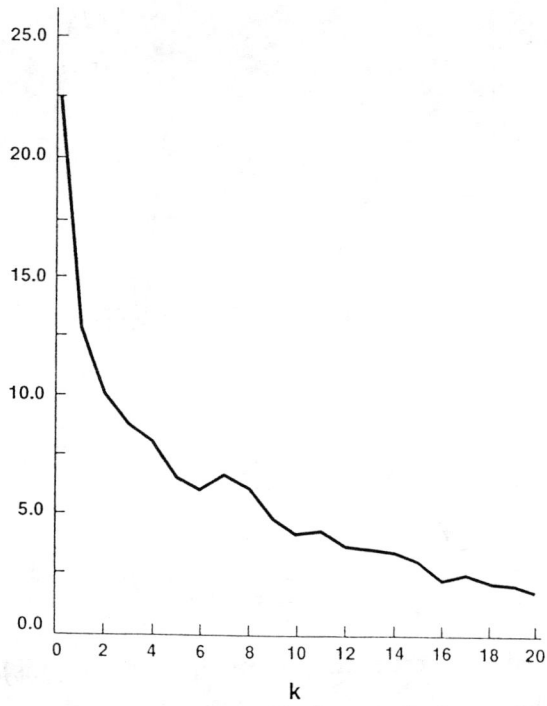

Figure 10.3.4-9: Position estimation RMS error (50 runs).

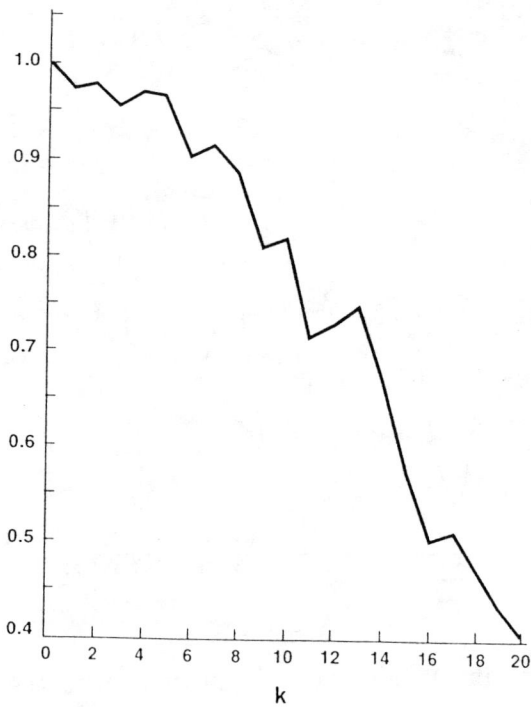

Figure 10.3.4-10: Velocity estimation RMS error (50 runs).

396

10.3.4 An Example: Tracking with an Angle-Only Sensor

The normalized state estimation error squared (NEES) shown in Figure 10.3.4-5 for a single run exceeds the 95% level only for a short time. The normalized innovation squared shown in Figure 10.3.4-6 is below the 95% level. Thus, from this single run, the filter appears to be almost consistent.

The 50-run Monte Carlo averages in Figure 10.3.4-7 show, however, that after a while the errors are practically all the time outside the 95% probability region. This indicates that compared to the filter-calculated covariance the errors are inadmissibly large. The normalized innovation squared shown in Figure 10.3.4-8 exhibits large deviations for the last three sampling times. Thus, this filter is *inconsistent*.

The RMS errors in position and velocity estimates are plotted in Figures 10.3.4-9 and 10.3.4-10.

The conclusion from these simulations is that a *thorough examination* of a nonlinear filter is needed *via Monte Carlo runs* to find out if it is consistent.

Remarks

The inconsistency is due to the fact that the linearization does not work well with large errors — the measurement noise in this case. Even the second order EKF does not work well in this example.

It can be shown that if the measurement is more accurate ($\sqrt{r_s} \leq 1°$), this filter becomes consistent.

10.3.5   The EKF — Summary

Model for the *extended Kalman filter (EKF)*:

- *Initial state* — an estimate (approximate conditional mean) available with an associated covariance.
- *System dynamic equation* — nonlinear with additive zero- mean white process noise with known covariance.
- *Measurement equation* — nonlinear with additive zero-mean white measurement noise, with known covariance.

The initial state error and the two noise sequences are assumed to be mutually independent.

The discrete time recursive state estimation filter is obtained by using a series expansion of

- first order, or
- second order

of the nonlinear dynamic and measurement equations to obtain the state and measurement predictions. These yield, respectively,

- the (standard, or first order) EKF, and
- the second order EKF.

The first order EKF requires evaluation of the Jacobians of the nonlinear functions from the dynamics and the measurement equations. The second order EKF requires also the evaluation of the corresponding Hessians.

The first order discrete time EKF consists of the following:

1. State prediction — the nonlinear function (dynamic equation) of the previous state estimate.

2. Covariance of the state prediction — same covariance propagation equation as in the linear case except for the Jacobian of the dynamic equation (evaluated at the previous state estimate) playing the role of the transition matrix.

3. Measurement prediction — the nonlinear function (measurement equation) of the predicted state.

4. Covariance of the predicted measurement (equal to the innovation covariance) — same equation as in the linear case except for the Jacobian of the measurement equation (evaluated at the predicted state) playing the role of the measurement matrix.

5. Filter gain calculation — same as in the linear case.

6. State update — same as in the linear case.

7. Covariance of the updated state — same as in the linear case.

Note the coupling between the state estimation and covariance calculations: the latest state estimates are used in the *linearization* (Jacobian evaluations) of the nonlinear dynamic and/or measurement equations.

10.4 ERROR COMPENSATION IN LINEARIZED FILTERS

10.4.1 Some Heuristic Methods

The extended Kalman filter presented in the previous section was obtained by a Taylor series expansion up to first order (which amounts to linearization) or up to second order terms. Both these filters will, obviously, introduce errors in the equations where such an expansion is used due to the following:

1. The higher order terms that have been neglected;
2. The evaluations of the Jacobians (and Hessians) being done at estimated or predicted values of the state rather than the exact values, which are not available.

There are several ways of compensating for these errors:

1. Addition of **artificial process noise** or **pseudo-noise** for compensation of the errors in the state prediction (10.3.2-4). This can be done by using in the covariance prediction equation (10.3.2-6) a *larger* **modified process noise covariance**

$$Q_m(k) = Q_p(k) + Q(k) \geq Q(k) \tag{10.4.1-1}$$

where $Q_p \geq 0$ is the positive semidefinite pseudonoise covariance.

2. Multiplication of the state covariance by a scalar $\phi > 1$ at every sampling time. This amounts to letting

$$P_\phi(k+1|k) = \phi P(k+1|k) \tag{10.4.1-2}$$

and then using this "jacked up" prediction covariance matrix $P_\phi(k+1|k)$ in the covariance update equation. The scalar ϕ is called **fudge factor**.[2]

Alternatively, instead of (10.4.1-2), one can use a matrix

$$\Phi = \mathrm{diag}(\sqrt{\phi_i}) \tag{10.4.1-3}$$

to obtain a selectively jacked-up state covariance

$$P^*(k+1|k) = \Phi P(k+1|k)\Phi \tag{10.4.1-4}$$

The multiplication of the state covariance by a scalar greater than unity is equivalent to the filter having a **fading memory**. At every sampling time the past data undergoes a process of **discounting** by attaching to the estimate based on it a higher covariance (lower accuracy).

---

[2]From the Greek ϕudge $\phi\alpha\kappa\tau o\rho$.

Effect of Increasing the Covariance Matrices

The increase of the state covariance will cause the filter gain to be larger, thus giving more weight to the most recent data.

Conversely, increasing the variance of the measurement noise, which will increase the innovation covariance, would lower the filter gain.

These techniques are completely heuristic, and should be used only as a last resort.

If the inconsistency stems from bias due to the nonlinearities in the system, this method will probably not accomplish much.

10.4.2   An Example of Use of the Fudge Factor

The example of Subsection 10.3.4 is considered again. Since the filter did not pass the consistency tests, the covariance increase technique (10.4.1-2) is illustrated for this problem.

The fudge factor was taken as $\phi = 1.03$. The resulting normalized state estimation error, plotted in Figure 10.4.2-1, shows still some points above the 97.5 percentile point. The normalized innovation in Figure 10.4.2-2 is also outside the admissible region at the end of the observation period.

The magnitudes of the RMS errors shown in Figures 10.4.2-3 and 10.4.2-4 are practically the same as in the uncompensated filter, shown earlier in Figures 10.3.4-9 and 10.3.4-10.

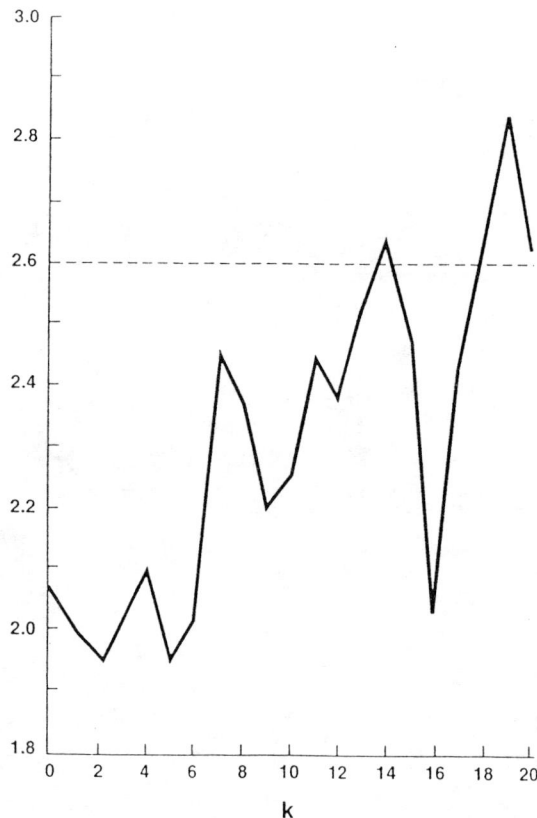

Figure 10.4.2-1: Normalized state estimation error squared ($\phi = 1.03$, 50 runs) with the 97.5% level.

402

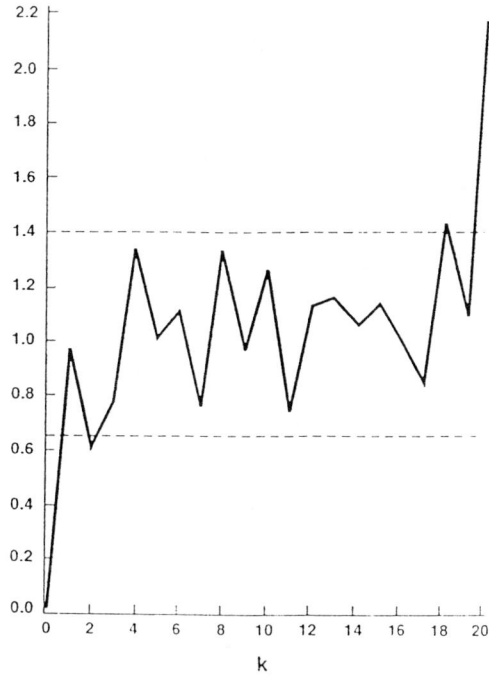

Figure 10.4.2-2: Normalized innovation squared ($\phi = 1.03$, 50 runs) with the two-sided 95% probability region.

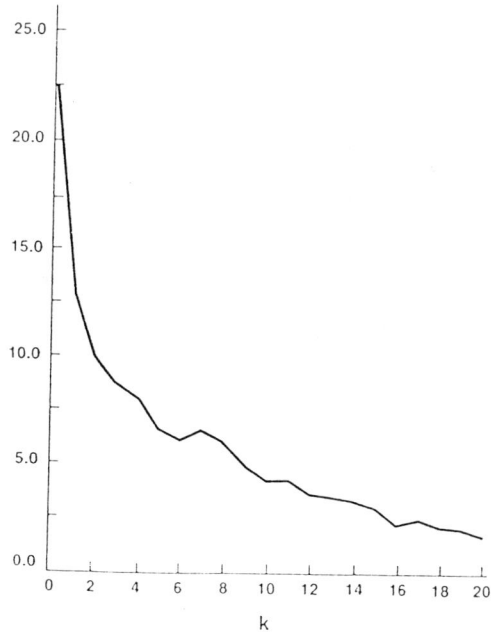

Figure 10.4.2-3: Position estimation RMS error ($\phi = 1.03$, 50 runs).

403

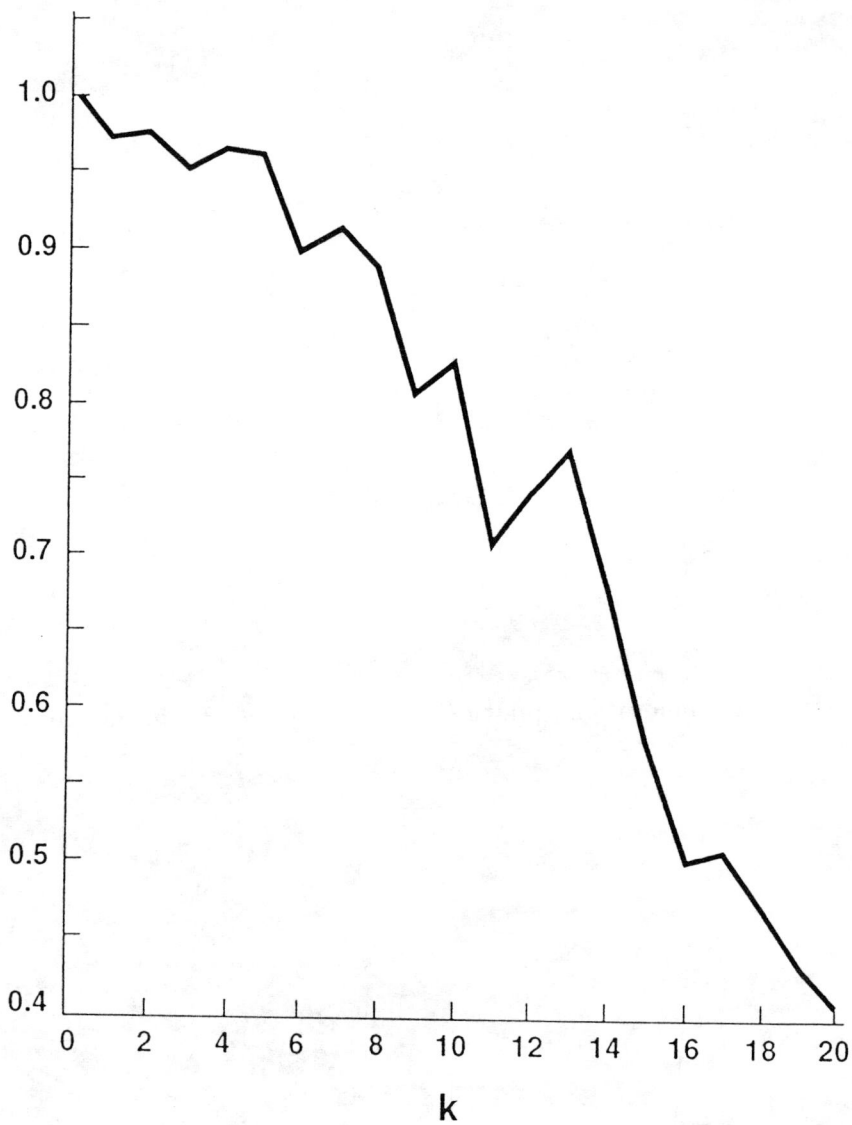

Figure 10.4.2-4: Velocity estimation RMS error ($\phi = 1.03$, 50 runs).

10.4.3   Error Compensation in Linearized Filters — Summary

Linearization of nonlinear systems to obtain linear filters introduces unavoidable errors.
One way to account for these errors is to increase the covariances by

- addition of *pseudonoise covariance*, or
- multiplication of the state covariance by a *fudge factor* (slightly) larger than unity

so that the covariance "covers" the errors (in the sense of filter consistency).
This implicitly assumes that these errors are random.
If the errors are not random, i.e., one has *bias*, then

- bias compensation, e.g., a debiasing technique as in [LB93], or
- bias estimation

is needed.

Bias estimation can be accomplished by adding extra state components. In this case one has to ascertain that the augmented state is completely observable.

10.5 SOME ERROR REDUCTION METHODS

10.5.1 Improved State Prediction

The extended Kalman filter carries out the state prediction and measurement prediction according to (10.3.2-4) and (10.3.2-7), respectively. Both these equations involve nonlinear transformations.

Applying a nonlinear transformation on an estimate introduces errors that are not completely accounted for in the prediction covariance equation. The second order state prediction equation (10.3.2-4) does capture some of the nonlinearity effects. However, since it can be expensive to implement and there is no guarantee that this will solve the problem, a different approach is adopted in some cases.

For **continuous time nonlinear systems** whose state evolves according to nonlinear differential equations and is observed at discrete times, one can obtain the predicted state by discretization and linearization, which amounts to a first order numerical integration.

If this leads to unsatisfactory accuracy, then one can use a more accurate **numerical integration** on the continuous time differential equation of the state from t_k to t_{k+1} starting at the initial condition $\hat{x}(k|k)$ with the process noise replaced by its mean (zero).

The calculation of the state prediction covariance can be done as in (10.3.2-6), that is, relying upon linearization or series expansion if the process noise is specified in discrete time, or by numerical integration of (9.3.1-8) with the system matrix replaced by the Jacobian of the nonlinear dynamic equation (see problem 10-4).

10.5.2   The Iterated Extended Kalman Filter

A modified state updating approach can be obtained by an iterative procedure as follows.

The measurement prediction based on expansion up to first order is

$$\hat{z}(k+1|k) = h[k, \hat{x}(k+1|k)] \qquad (10.5.2\text{-}1)$$

that is, the predicted state $\hat{x}(k+1|k)$ is used for $x(k+1)$. Since the state prediction errors might be already significant, when compounded with additional errors due to the measurement nonlinearity this can lead to undesirable errors.

One approach to alleviate this situation is to compute the updated state not as an approximate conditional mean, that is, a linear combination of the prediction and the innovation (5.2.3-12), but as a maximum a posteriori (MAP) estimate. An *approximate MAP estimate* can be obtained by an iteration that amounts to **relinearization of the measurement equation**.

The conditional pdf of $x(k+1)$ given Z^{k+1} can be written, assuming all the pertinent random variables to be Gaussian, as

$$
\begin{aligned}
p[x(k+1)|Z^{k+1}] &= p[x(k+1)|z(k+1), Z^k] \\
&= \frac{1}{c} p[z(k+1)|x(k+1)] p[x(k+1)|Z^k] \\
&= \frac{1}{c} \mathcal{N}\Big[z(k+1); h[k+1, x(k+1)], R(k+1)\Big] \\
&\quad \cdot \mathcal{N}[x(k+1); \hat{x}(k+1|k), P(k+1|k)]
\end{aligned}
\qquad (10.5.2\text{-}2)
$$

Maximizing the above with respect to $x(k+1)$ is equivalent to minimizing

$$
\begin{aligned}
J[x(k+1)] &= \frac{1}{2}\Big\{z(k+1) - h[k+1, x(k+1)]\Big\}' R(k+1)^{-1} \\
&\quad \cdot \Big\{z(k+1) - h[k+1, x(k+1)]\Big\} \\
&\quad + \frac{1}{2}[x(k+1) - \hat{x}(k+1|k)]' P(k+1|k)^{-1} \\
&\quad \cdot [x(k+1) - \hat{x}(k+1|k)]
\end{aligned}
\qquad (10.5.2\text{-}3)
$$

The iterative minimization of (10.5.2-3), say, using a Newton-Raphson algorithm, will yield an *approximate MAP estimate* of $x(k+1)$. This is done by expanding J in a Taylor series up to second order about the i-th iterated value of the estimate of $x(k+1)$, denoted (without time argument) as x^i,

$$J = J^i + J_x^{i\prime}(x - x^i) + \frac{1}{2}(x - x^i)' J_{xx}^i (x - x^i) \qquad (10.5.2\text{-}4)$$

where, using abbreviated notation,

$$J^i = J \Big|_{x=x^i} \qquad (10.5.2\text{-}5)$$

and

$$J_x^i = \nabla_x J \Big|_{x=x^i} \qquad (10.5.2\text{-}6)$$

$$J_{xx}^i = \nabla_x \nabla_x' J \Big|_{x=x^i} \qquad (10.5.2\text{-}7)$$

are the gradient and Hessian of J with respect to $x(k+1)$.

Setting the gradient of (10.5.2-4) with respect to x to zero yields the next value of x in the iteration to minimize (10.5.2-3) as

$$x^{i+1} = x^i - (J_{xx}^i)^{-1} J_x^i \qquad (10.5.2\text{-}8)$$

The gradient of J is, using now the full notation,

$$\begin{aligned}
J_x^i = \ & -h_x^i[k+1, \hat{x}^i(k+1|k+1)]' R(k+1)^{-1} \\
& \cdot \left\{ z(k+1) - h[k+1, \hat{x}^i(k+1|k+1)] \right\} \\
& + P(k+1|k)^{-1} [\hat{x}^i(k+1|k+1) - \hat{x}(k+1|k)] \qquad (10.5.2\text{-}9)
\end{aligned}$$

The Hessian of J, retaining only up to the first derivative of h, is

$$\begin{aligned}
J_{xx}^i &= h_x^i[k+1, \hat{x}^i(k+1|k+1)]' R(k+1)^{-1} h_x[k+1, \hat{x}^i(k+1|k+1)] + P(k+1|k)^{-1} \\
&= H^i(k+1)' R(k+1)^{-1} H^i(k+1) + P(k+1|k)^{-1} \qquad (10.5.2\text{-}10)
\end{aligned}$$

where

$$\boxed{H^i(k+1) \triangleq h_x[k+1, \hat{x}^i(k+1|k+1)]} \qquad (10.5.2\text{-}11)$$

is the Jacobian of the *relinearized measurement equation*.

408

Using the matrix inversion lemma (1.3.3-11), one has

$$
\begin{aligned}
(J^i_{xx})^{-1} &= P(k+1|k) - P(k+1|k)H^i(k+1)' \\
&\quad \cdot [H^i(k+1)P(k+1|k)H^i(k+1)' + R(k+1)]^{-1}H^i(k+1)P(k+1|k) \\
&\triangleq P^i(k+1|k+1)
\end{aligned}
\tag{10.5.2-12}
$$

Substituting (10.5.2-9) and (10.5.2-12) into (10.5.2-8) yields

$$
\boxed{
\begin{aligned}
\hat{x}^{i+1}(k+1|k+1) &= \hat{x}^i(k+1|k+1) + P^i(k+1|k+1)H^i(k+1)'R(k+1)^{-1} \\
&\quad \cdot \left\{ z(k+1) - h[k+1, \hat{x}^i(k+1|k+1)] \right\} \\
&\quad - P^i(k+1|k+1)P(k+1|k)^{-1}[\hat{x}^i(k+1|k+1) - \hat{x}(k+1|k)]
\end{aligned}
}
\tag{10.5.2-13}
$$

which is the **iterated extended Kalman filter**.

Starting the iteration for $i = 0$ with

$$
\hat{x}^0(k+1|k+1) \triangleq \hat{x}(k+1|k)
\tag{10.5.2-14}
$$

causes the last term in (10.5.2-13) to be zero and yields after the first iteration $\hat{x}^1(k+1|k+1)$, that is, the same as the first order (noniterated) EKF.

The covariance associated with $\hat{x}^i(k+1|k+1)$ is, from (10.5.2-12), given by

$$
\boxed{
\begin{aligned}
P^i(k+1|k+1) &= P(k+1|k) - P(k+1|k)H^i(k+1)' \\
&\quad \cdot [H^i(k+1)P(k+1|k)H^i(k+1)' + R(k+1)]^{-1} \\
&\quad \cdot H^i(k+1)P(k+1|k)
\end{aligned}
}
\tag{10.5.2-15}
$$

Overview of the Iteration Sequence

For $i = 0, N-1$

    (10.5.2-11)

    (10.5.2-15)

    (10.5.2-13)

For $i = N$

    (10.5.2-15)

with N decided either a priori or based on a convergence criterion.

10.5.3   Some Error Reduction Methods — Summary

In systems with continuous time nonlinear dynamic equations the state prediction can be done by *numerical integration* of the dynamic equation.

This is more accurate than discretizing the continuous time equation and using a truncated series expansion.

Another way to reduce the effects of linearization errors in the EKF is to *relinearize the measurement equation* around the *updated state* rather than relying only on the *predicted state*.

This is equivalent to a MAP estimation of the state under the (obviously approximate) Gaussian assumption carried out via a Newton-Raphson search.

It results in a set of coupled iterations for

- The updated state;
- The updated state covariance; and
- The Jacobian of the measurement equation.

10.6 MAXIMUM A POSTERIORI TRAJECTORY ESTIMATION VIA DYNAMIC PROGRAMMING

10.6.1 The Approach

This approach finds the sequence of MAP estimates

$$
\begin{aligned}
\hat{x}^{k|k} &\triangleq \{\hat{x}(0|k), \ldots, \hat{x}(k|k)\} \\
&= \arg \max_{x(0),\ldots,x(k)} p[x(0), \ldots, x(k)|Z^k]
\end{aligned}
\tag{10.6.1-1}
$$

These estimates, which are actually smoothed values, constitute the **modal trajectory**. This is the maximum — the **mode** — of the joint pdf of the state trajectory, that is, the most probable sequence of state values up to the current time.

Equation (10.6.1-1) can be rewritten, equivalently as

$$
p[\hat{x}^{k|k}|Z^k] = \max_{X^k} p[X^k|Z^k]
\tag{10.6.1-2}
$$

where

$$
X^k \triangleq \{x(j)\}_{j=0}^k
\tag{10.6.1-3}
$$

The technique of **dynamic programming** is used. This optimization technique replaces the simultaneous maximization (10.6.1-2) by the *sequential maximization*:

$$
p[\hat{x}^{k|k}|Z^k] = \max_{x(k)} \max_{X^{k-1}} p[X^k|Z^k]
\tag{10.6.1-4}
$$

This technique can be used if the function to be maximized can be decomposed (separated) into a sum or product of two functions, the first of which depends only on the first variable. Then one can obtain a sequential maximization.

The function to be maximized recursively is

$$
I[x(k), k] = \max_{X^{k-1}} p[X^k|Z^k]
\tag{10.6.1-5}
$$

from which (10.6.1-4) follows by one more maximization.

10.6.2   The Dynamic Programming for Trajectory Estimation

Using Bayes' formula and eliminating irrelevant variables in the conditioning, (10.6.1-5) can be rewritten as

$$
\begin{aligned}
I[x(j+1), j+1] &= \max_{X^j} p[X^{j+1} | Z^{j+1}] = \max_{X^j} p[x(j+1), X^j | z(j+1), Z^j] \\
&= \max_{X^j} \frac{p[z(j+1)|x(j+1), X^j, Z^j] p[x(j+1), X^j | Z^j]}{p[z(j+1)|Z^j]} \\
&= \frac{1}{c_j} \max_{X^j} \Big\{ p[z(j+1)|x(j+1)] p[x(j+1)|X^j, Z^j] p[X^j | Z^j] \Big\} \\
&= \frac{1}{c_j} \max_{x(j)} \Big\{ \max_{X^{j-1}} p[z(j+1)|x(j+1)] p[x(j+1)|x(j)] p[X^j | Z^j] \Big\} \\
&= \frac{1}{c_j} \max_{x(j)} \Big\{ p[z(j+1)|x(j+1)] p[x(j+1)|x(j)] \max_{X^{j-1}} p[X^j | Z^j] \Big\}
\end{aligned}
$$

$$(10.6.2\text{-}1)$$

where in the last line advantage has been taken of the "separability" property mentioned earlier. The term c_j is the normalization constant; it does not affect the maximization and can be dropped.

Combining (10.6.1-5) and (10.6.2-1) yields the (forward) functional recursion

$$
\boxed{ I^*[x(j+1), j+1] = \max_{x(j)} \Big\{ p[z(j+1)|x(j+1)] p[x(j+1)|x(j)] I^*[x(j), j] \Big\} }
\tag{10.6.2-2}
$$

where $I^*(j+1)$ is proportional to $I(j+1)$, with the proportionality factor consisting of normalization constants that are irrelevant to the estimation. This is the desired iterative relationship that yields (10.6.1-5).

The resulting argument from the maximization (10.6.2-2) is

$$
\hat{x}(j|j+1) = x[j, x(j+1), Z^j] \qquad j = 0, \ldots, k-1
\tag{10.6.2-3}
$$

that is, it is not yet the modal trajectory.

To obtain the modal trajectory, use is made of (10.6.1-4) with the final result of (10.6.2-2), which yields:

$$
\boxed{ \hat{x}(k|k) = \arg\max_{x(k)} I^*[x(k), k] }
\tag{10.6.2-4}
$$

This is then used in (10.6.2-3), which is a backward recursion, as follows

$$
\boxed{ \hat{x}(j|k) = x[j, \hat{x}(j+1|k), Z^j] \qquad j = k-1, \ldots, 0 }
\tag{10.6.2-5}
$$

Note that the above is equivalent to a **MAP smoothing technique**.

In the communication theory literature a similar approach is known as the **Viterbi algorithm**.

Remarks

This approach is very general, in the sense that it does not require any particular form of the state or measurement equations. The state transition equation is summarized by the conditional pdf

$$p[x(j+1)|x(j)] \tag{10.6.2-6}$$

while the measurement equation is summarized by

$$p[z(j+1)|x(j+1)] \tag{10.6.2-7}$$

This technique has been applied in Chapter 4 of [Bar90] to a problem where (10.6.2-6) describes the motion of a target and (10.6.2-7) models the observed intensities (from a target or background) in each resolution cell of an electro-optical sensor. The resulting algorithm, which can detect target trajectories in a very low SNR environment and estimate their state, carries out what amounts to a **multiscan signal processing**.

The practical implementation of this algorithm requires discretization of the state space as well as the measurement space.

10.7 NOTES AND PROBLEMS

10.7.1 Bibliographical Notes

The discrete time nonlinear estimation for Markov processes is a straightforward consequence of Bayes' formula. More on this topic can be found in, for example, [SM71]. The concept of information state has been discussed in [Str65]. The application of numerical techniques for optimal estimation with update of the pdf is illustrated for some problems in [KS87, JD91].

Historical notes on the extended Kalman filter can be found in [Sor85]. The second order EKF derivation presented in Section 10.3 is based on [AWB68].

The issues of error compensation discussed in Section 10.4 are treated in numerous places, see, for example [Jaz70]. Bias estimation and compensation has been discussed in [Fri69]. The debiasing for the polar-to-Cartesian transformation is presented in [LB93].

The iterated EKF described in Section 10.5 has been used for many years, see, for example, [WLA70]. The application of this to ballistic missile tracking has been reported in [CDY84]. Other techniques for improving the performance of the EKF in the presence of significant nonlinearities were presented in [THS77, WM80, SS85]. Issues related to bearings-only tracking have been discussed in [NA81, AH83]. The use of two angular observations to track a target moving in a 3-dimensional space has been treated in [Sta87]. A comparison of Cartesian and polar coordinates in terms of the conditioning number of the EKF-calculated covariance can be found in [COD91].

The use of dynamic programming for maximum a posteriori trajectory estimation, discussed in Section 10.6, was proposed in [LP66]. This technique was used in [Bar85] and Chapter 4 of [Bar90] in a "track-before-detect" approach for extracting trajectories from a low SNR optical sensor. Direct implementation of the functional recursion for the update of the state's conditional pdf was investigated in [KS87, BR87].

10.7.2 Problems

10-1 **Signal interpolation for measurement generation.** Given three noisy signal intensity observations

$$z_i = y(x_i) + w_i \qquad i = 1, 2, 3$$

of the unknown function $y(x)$ at $x_1 < x_2 < x_3$. We want to find, using these observations, the maximum of $y(x)$ and its accuracy. Assuming that $z_1 < z_2 > z_3$, the unknown function $y(x)$ can be approximated locally by a parabola.

1. Using a parabolic interpolation find

$$\hat{x}_m = f(x_1, x_2, x_3, z_1, z_2, z_3)$$

the estimate of the location of the maximum of $y(x)$.

2. Denoting the vector of noises as $w \triangleq [w_1 \quad w_2 \quad w_3]'$ and assuming

$$E[w] = 0 \qquad\qquad E[ww'] = P$$

find, using a first order series expansion, the variance associated with the estimate $\hat{x}_m$.

(iii) Calculate the location of the maximum and its variance for

$$x_1 = n - 1 \qquad\qquad x_2 = n \qquad\qquad x_3 = n + 1$$

$$z_1 = 0.6 \qquad z_2 = 0.8 \qquad z_3 = 0.6 \qquad E[w_i w_j] = 0.01\delta_{ij}$$

10-2 Standard deviation improvement versus variance improvement. Assume that the variance of the estimate of a variable in a certain baseline method is equal to 1. A more sophisticated method yields a smaller variance $1 - \alpha$ (i.e., its improvement is $(100\alpha)\%$). Find the improvement in % in the corresponding standard deviation, assuming $\alpha < 0.2$.

10-3 Coordinate transformation from polar to Cartesian. Let

$$\rho = \sqrt{x_1^2 + x_2^2} \qquad\qquad \theta = \tan^{-1}\frac{x_2}{x_1}$$

Given the measurements

$$z = \begin{bmatrix} z_1 \\ z_2 \end{bmatrix} = \begin{bmatrix} \rho + w_1 \\ \theta + w_2 \end{bmatrix} = h(x_1, x_2) + w$$

with $w \sim \mathcal{N}(0, R), R = \text{diag}(\sigma_\rho^2, \sigma_\theta^2)$.

1. Find the transformation into Cartesian coordinates

$$y = h^{-1}(z) = [x_1 x_2]' + w_C$$

2. Find the covariance matrix R_C of the Cartesian converted measurement noise w_C using linearization at $\rho_0 = 10^5$, $\theta_0 = 45°$.

3. Simulate $N = 100$ realizations of w with $R = \text{diag}[100^2, (0.5°)^2]$. Calculate the corresponding polar measurements z at $\rho = \rho_0$, $\theta = \theta_0$ and obtain the resulting Cartesian positions with the solution of (1). Find the average of y.

4. Analyze the results whether the errors w_C are zero mean.

5. Analyze whether the covariance matrix calculated using as in (2) can be accepted as correct. Hint: use the average of $w_C' R_C^{-1} w_C$ as the statistic for the covariance test.

10-4 Continuous time EKF. Given the continuous time nonlinear dynamic system

$$\dot{x}(t) = \phi[t, x(t)] + \tilde{v}(t)$$

with measurement

$$z(t) = \eta[t, x(t)] + \tilde{w}(t)$$

where the noises are mutually uncorrelated with autocorrelations

$$E[\tilde{v}(t)\tilde{v}(\tau)'] = \tilde{q}(t)\delta(t - \tau) \qquad\qquad E[\tilde{w}(t)\tilde{w}(\tau)'] = \tilde{R}(t)\delta(t - \tau)$$

and uncorrelated from the initial state, whose estimate and covariance are $\hat{x}(0)$ and $P(0)$, respectively. Find the continuous time extended Kalman-Bucy filter using linearization for the above system.

10-5 **Angular rate estimation.** Given the estimate $\hat{x}(t)$ of the state (4.2.2-16) of an object at time t. Assume that in the interval $(t, t+T]$ it moves with a constant speed and constant angular rate Ω.

1. Obtain the extrapolated state $\hat{x}(t+T) = e^{AT}\hat{x}(t)$ where the transition matrix for this motion is given in (4.2.2-19).

2. Given the covariance matrix $P(t)$ associated with $\hat{x}(t)$ find the covariance associated with $\hat{x}(t+T)$.

3. A position measurement

$$z(t+T) = \begin{bmatrix} 1 & 0 & 0 & 0 \\ 0 & 0 & 1 & 0 \end{bmatrix} x(t+T) + w(t+T)$$

with $w \sim \mathcal{N}(0, \sigma^2 I)$ is made. Write the likelihood function

$$p[z(t+T)|\hat{x}(t), \Omega]$$

assuming $\tilde{x}(t) \sim \mathcal{N}[0, P(t)]$. Indicate the explicit expressions of $\hat{z}(t+T)$ and its associated covariance $S(t+T)$.

4. How can one obtain $\hat{\Omega}^{\text{ML}}$ from the above?

5. Assume, for simplicity, that $\Omega T \ll 1$ and $S(t+T) = cI$ and obtain the LS estimate $\hat{\Omega}$ and its variance.

Chapter 11

ADAPTIVE ESTIMATION AND MANEUVERING TARGETS

11.1 INTRODUCTION

11.1.1 Adaptive Estimation — Outline

In the models considered previously, the only uncertainties consisted of additive white noises (process and measurement) with known statistical properties. In other words, the system model, consisting of the state transition matrix, the input gain (and input, if any), the measurement matrix, and noise covariances, were all assumed to be known. In the case of "colored" noise with known autocorrelation, it has been shown that it can be reduced to the standard situation in Chapter 8.

In many practical situations the above listed "parameters of the problem" are partially unknown and possibly time varying. There are numerous techniques of **system identification** (e.g., [Lju87]) that deal with the identification (estimation) of the *structural parameters* of a system.

The purpose of this chapter is to present state estimation techniques that can "adapt" themselves to certain types of uncertainties beyond those treated earlier — *adaptive estimation algorithms*.

One type of uncertainty to be considered is the case of unknown inputs into the system, which typifies maneuvering targets. The other type will be a combination of system parameter uncertainties with unknown inputs where the system parameters take values in a discrete set.

Maneuvering targets are characterized by an equation of the same form as (4.3.1-9), namely

$$x(k+1) = F(k)x(k) + G(k)u(k) + v(k) \qquad (11.1.1-1)$$

but the input $u(k)$, which enters the system in addition to the process noise $v(k)$, is *unknown*.

For simplicity, linear models are considered — in the case of nonlinear models the same techniques that are discussed in the sequel can be used with linearization.

The approaches that can be used in such a situation fall into two broad categories:

1. The unknown input (the maneuver command) is modeled as a random process.
2. The unknown input is estimated in real time.

The random process type models can be classified according to the statistical properties of the process modeling the maneuver as

- White noise, or
- Autocorrelated noise.

Note that this **unknown input as a random process** approach amounts to treating the unknown input as (an additional) process noise.

The use of fixed-level white noise to model maneuvers has been discussed in the context of kinematic models. However, maneuvers, by their nature, are of different magnitudes at different times. This can be accommodated by adjusting the level of the process noise. One option is a **continuous noise level adjustment** technique, discussed in Section 11.2. Alternatively, several discrete levels of noise can be assumed in the filter, with a **noise level switching** according to a certain rule, also discussed in Section 11.2.

The use of autocorrelated (colored) noise to model maneuvers has been presented in Section 8.2 — the approach is to "prewhiten" the noise and augment the state with the prewhitening subsystem. The noise level adjustment techniques can be used with the augmented system, which is driven by white noise.

All these methods relying on *modeling maneuvers as random processes* are approximations because maneuvers are, in general, not stochastic processes. Nevertheless, such approaches are simple and can be quite effective.

The second type of approach, **input estimation**, is implemented assuming the input to be constant over a certain period of time. The estimation can be done based on the least squares criterion and the result can be used in the following ways:

- The state estimate is corrected with the estimated input, or
- The state is augmented and the input becomes an extra state component that is reestimated sequentially within the augmented state.

The input estimation with state estimate correction technique is presented in Section 11.3. The technique of estimating the input and, when "statistically significant," augmenting the state with it (which leads to **variable state dimension**), is the topic of Section 11.4. These two algorithms and the noise level switching technique of Section 11.2 are compared in Section 11.5.

The so-called **multiple model (MM)** algorithms are the topic of Section 11.6. These algorithms assume that the system behaves according to one of a finite number of models — it is in one of several **modes**. The models can differ in *noise levels* or their *structure* — different state dimensions and unknown inputs can be accommodated as well. Such systems are called **hybrid systems** — they have both discrete (structure/parameters) and continuous uncertainties (additive noises).

First the MM algorithm for fixed (nonswitching) models is considered. Then the optimal algorithm for switching models (according to a Markov chain) is presented. Since the optimal algorithm is not practical for implementation, two suboptimal approaches, one called **generalized pseudo-Bayesian (GPB)** and the other the **interacting multiple model (IMM)**, are also presented. The design of an IMM estimator for **air traffic control (ATC)** is discussed in detail.

The use of the extended Kalman filter for state and system parameter estimation is briefly presented in Section 11.7.

11.1.2   Adaptive Estimation — Summary of Objectives

Show modeling of maneuvers as

- A random process (white or autocorrelated process noise)
  - continuously variable level;
  - several discrete levels with switching between them.

- A fixed input, estimated as an unknown constant, with
  - state estimate compensation;
  - variable state dimension.

Compare some of these algorithms.

Illustrate the optimal method of comparison of algorithms.

Multiple-model (hybrid system) approach:

   Several models are used in parallel with a probabilistic weighting:

   - Optimal
   - GPB1
   - GPB2
   - IMM

Present the design of an IMM estimator for air traffic control.

Show the use of the EKF for state and system parameter estimation.

11.2 ADJUSTABLE LEEL PROCESS NOISE

11.2.1 Continuous Noise Level Adjustment

In this **continuous noise level adjustment** approach, the target is tracked with a filter in which a certain low level of process noise is assumed. The level of the noise is determined by its variance (or covariance matrix).

A maneuver manifests itself as a "large" innovation. A simple detection procedure for such an occurrence is based on the **normalized innovation squared**

$$\epsilon_\nu(k) = \nu(k)'S(k)^{-1}\nu(k) \qquad (11.2.1\text{-}1)$$

A threshold is set up such that based on the target model (for the nonmaneuvering situation)

$$P\{\epsilon_\nu(k) < \epsilon_{max}\} = 1 - \mu \qquad (11.2.1\text{-}2)$$

with, say, tail probability $\mu = 0.01$.

If the threshold is exceeded then the process noise variance $Q(k-1)$ is scaled up until ϵ_ν is reduced to the threshold ϵ_{max}.

Using a **scaling factor** $\phi(k)$, (i.e., a **fudge factor**) the covariance of the innovations becomes

$$S(k) = H(k)[F(k-1)P(k-1|k-1)F(k-1)' + \phi(k)Q(k-1)]H(k)' + R(k) \quad (11.2.1\text{-}3)$$

In place of the single-time test statistic (11.2.1-1) one can also use a time average over a "sliding window." This is illustrated in the next subsection.

A similar technique can be used to lower the process noise level after the maneuver.

11.2.2 Process Noise with Several Discrete Levels

Another approach is to assume two or more levels of process noise and use a rule for **noise level switching**.

Under normal conditions the filter is operating with the low level noise Q_1.

The normalized innovation squared

$$\epsilon_\nu(k) = \nu(k)'S(k)^{-1}\nu(k) \tag{11.2.2-1}$$

is monitored and if it *exceeds* a certain threshold the filter switches to a prechosen higher level of process noise, Q_2.

Under the linear-Gaussian assumptions, the pdf of ϵ_ν is chi-square distributed with n_z, the dimension of the measurement, degrees of freedom

$$\epsilon_\nu \sim \chi^2_{n_z} \tag{11.2.2-2}$$

The switching threshold is chosen based on (11.2.2-2) such that the probability of being exceeded under normal conditions is small.

The decision statistic (11.2.2-1), which is based on a single sampling time, can be replaced by a **moving average** (or **moving sum**) of the normalized innovations squared over a **sliding window** of s sampling times

$$\boxed{\epsilon_\nu^s(k) = \sum_{j=k-s+1}^{k} \epsilon_\nu(j)} \tag{11.2.2-3}$$

The above is chi-square distributed with sn_z degrees of freedom

$$\epsilon_\nu^s \sim \chi^2_{sn_z} \tag{11.2.2-4}$$

since (11.2.2-3) is the sum of s independent terms with distribution (11.2.2-2).

Alternatively, a **fading memory average** (also called **exponentially discounted average**)

$$\boxed{\epsilon_\nu^\alpha(k) = \alpha\epsilon_\nu^\alpha(k-1) + \epsilon_\nu(k)} \tag{11.2.2-5}$$

where

$$0 < \alpha < 1 \tag{11.2.2-6}$$

can be used.

The expected value of (11.2.2-5) in steady state is

$$E[\epsilon_\nu^\alpha(k)] = \frac{n_z}{1-\alpha} \tag{11.2.2-7}$$

The **effective window length** of (11.2.2-5) can be considered as

$$s_\alpha = \frac{1}{1 - \alpha} \tag{11.2.2-8}$$

For example, for $\alpha = 0.8$ one has $s_\alpha = 5$.

Based on (11.2.2-7), one can assume, as a first approximation (matching the first moment only), that $\epsilon_\nu^\alpha(k)$ is chi-square distributed with number of degrees of freedom given by (11.2.2-7).

Using the first *and* second moment-matching approximation described in Subsection 1.4.18, one obtains

$$\epsilon_\nu^\alpha \sim \frac{1}{1 + \alpha} \chi^2_{n_\alpha} \tag{11.2.2-9}$$

where the number of degrees of freedom is

$$n_\alpha = n_z \frac{1 + \alpha}{1 - \alpha} \tag{11.2.2-10}$$

For example, for $\alpha = 0.8$ one obtains $n_\alpha = 18$. Using the first approximation for the same α with $n_z = 2$ leads to 10 degrees of freedom. In other words, the moment-matching approximation (with the first two moments) indicates a (relatively) narrower pdf. The "width" of a chi-square pdf can be taken as the ratio of its standard deviation to its mean. For χ^2_n, this is (see Subsection 1.4.17) $\sqrt{2n}/n$, which decreases as n increases.

Estimator Operation

Thus, the filter switches from the lower process noise covariance Q_1 to the higher Q_2 if the average (11.2.2-3) or (11.2.2-5) exceeds an **upcrossing threshold**, determined based on (11.2.2-4) or (11.2.2-9), respectively, with a small tail probability.

After the filter starts running with the higher level process noise the innovations are monitored again to see whether there is reason to switch back.

The change to the model with lower level process noise is done when the normalized innovation, or a certain average of it, falls *below* another threshold — the **downcrossing threshold**.

There is no exact way to choose these thresholds — even if the distributions were known exactly, the tail probabilities are still subjective. They can be chosen, to begin with, based on tail probabilities and then set following experimentation (Monte Carlo runs) and subjective evaluation of the results.

Remarks

This technique of noise level switching can be extended to more than two levels of process noise. It can be also used with white process noise or autocorrelated process noise — the latter has to be prewhitened, as discussed in Subsection 8.2.1.

In general, n_α, given in (11.2.2-10), is not an integer. In this case one has a gamma rather than chi-square distribution; for the sake of simplicity, however, one can use the chi-square tables with an interpolation.

11.2.3 Adjustable Level Process Noise — Summary

A maneuver can be detected by monitoring the *normalized innovation* — if it exceeds a threshold then it can be assumed (in the absence of other factors) that the target has deviated from its previous "pattern."

A *fudge factor* can be used to scale up (continuously) the process noise (at the previous time) such that the modified prediction covariance is sufficiently large for the normalized innovation to be below the set threshold — the increased innovation covariance should "cover" the observed deviation.

Alternatively, a number of levels of process noise can be assumed. In this approach, at any given time a *single filter* operates with the *"current" noise level*.

A simple rule of switching from the current noise level to the one above or below is followed.

The innovations are monitored via

- a moving average over a sliding window, or
- a fading-memory average

of their normalized squared value.

This average is compared to a threshold based on the chi-square density for a small upper tail probability. If this *upcrossing threshold* is exceeded then the filter switches to a higher level of process noise.

The downswitch is done using a *downcrossing threshold*.

11.3 INPUT ESTIMATION

11.3.1 The Model

Consider the system with state equation

$$x(k+1) = Fx(k) + Gu(k) + v(k) \qquad (11.3.1\text{-}1)$$

where u is an **unknown input** modeling the target maneuvers and v is the process noise, zero mean white with known covariance Q.

The observations are

$$z(k) = Hx(k) + w(k) \qquad (11.3.1\text{-}2)$$

with the observation noise w zero mean, white, with covariance R, and independent of the process noise.

The estimation of the state is done using the model without input (nonmaneuvering)

$$x(k+1) = Fx(k) + v(k) \qquad (11.3.1\text{-}3)$$

Two Kalman filters are considered:

1. The *actual one* based on the nonmaneuvering model (11.3.1-3).
2. A *hypothetical one* based on the maneuvering model (11.3.1-1) with known u.

From the innovations of the **nonmaneuvering filter** based on (11.3.1-3), the input u is to be

- detected;
- estimated; and
- used to correct the state estimate.

This will be done using a sliding window of the latest s measurements and during this window period the input will be assumed *constant*.

11.3.2 The Innovations as a Linear Measurement of the Unknown Input

Denote the present time by k and assume that the target starts maneuvering at time $k-s$, that is, the **maneuver onset time** is $k-s$. The unknown inputs during the interval $[k-s, \ldots, k]$ are $u(i)$, $i = k-s, \ldots, k-1$.

The state estimates from the *mismatched nonmaneuvering filter* based on (11.3.1-3) will be denoted by an asterisk. The recursion for these estimates is

$$
\begin{aligned}
\hat{x}^*(i+1|i) &= F[I - W(i)H]\hat{x}^*(i|i-1) + FW(i)z(i) \\
&\triangleq \Phi(i)\hat{x}^*(i|i-1) + FW(i)z(i) \qquad i = k-s, \ldots, k-1 \quad (11.3.2\text{-}1)
\end{aligned}
$$

with the initial condition

$$
\hat{x}^*(k-s|k-s-1) = \hat{x}(k-s|k-s-1) \qquad\qquad (11.3.2\text{-}2)
$$

being the correct estimate (one-step prediction) before the maneuver started. This follows from the assumption that the maneuver onset time is $k-s$.

Similarly to (4.3.3-1), recursion (11.3.2-1) yields, in terms of (11.3.2-2)

$$
\hat{x}^*(i+1|i) = \left[\prod_{j=0}^{i-k+s} \Phi(i-j)\right]\hat{x}(k-s|k-s-1) + \sum_{j=k-s}^{i}\left[\prod_{m=0}^{i-j-1}\Phi(i-m)\right]FW(j)z(j)
$$
$$
i = k-s, \ldots, k-1 \qquad\qquad (11.3.2\text{-}3)
$$

If the inputs were known, the *hypothetical correct filter* based on (11.3.1-1) would yield estimates according to the recursion

$$
\begin{aligned}
\hat{x}(i+1|i) &= \Phi(i)\hat{x}(i|i-1) + FW(i)z(i) + Gu(i) \\
&= \left[\sum_{j=0}^{i-k+s}\Phi(i-j)\right]\hat{x}(k-s|k-s-1) \\
&\quad + \sum_{j=k-s}^{i}\left[\prod_{m=0}^{i-j-1}\Phi(i-m)\right][FW(j)z(j) + Gu(j)] \\
&\qquad\qquad i = k-s, \ldots, k-1 \qquad (11.3.2\text{-}4)
\end{aligned}
$$

which is the same as (11.3.2-3) except for the last term containing the inputs.

The innovations

$$
\nu(i+1) = z(i+1) - H\hat{x}(i+1|i) \qquad\qquad (11.3.2\text{-}5)
$$

corresponding to the correct (but *hypothetical*) filter (11.3.2-4) are a zero-mean white sequence with covariance $S(i+1)$ given as in (5.2.3-9).

11.3.2   The Innovations as a Linear Measurement of the Unknown Input

The innovations corresponding to the *nonmaneuvering filter* (11.3.2-3) are

$$\nu^*(i+1) = z(i+1) - H\hat{x}^*(i+1|i) \tag{11.3.2-6}$$

From (11.3.2-3) and (11.3.2-4) it follows that the innovations (11.3.2-6) of the nonmaneuvering filter — the *only real filter* — are the same as the "white noise sequence" (11.3.2-5) plus a "bias term" related to the inputs

$$\nu^*(i+1) = \nu(i+1) + H \sum_{j=k-s}^{i} \left[\prod_{m=0}^{i-j-1} \Phi(i-m) \right] Gu(j) \tag{11.3.2-7}$$

Assuming the input to be constant over the interval $[k-s, \ldots, k-1]$, that is,

$$u(j) = u \qquad j = k-s, \ldots, k-1 \tag{11.3.2-8}$$

yields

$$\boxed{\nu^*(i+1) = \Psi(i+1)u + \nu(i+1) \qquad i = k-s, \ldots, k-1} \tag{11.3.2-9}$$

where

$$\Psi(i+1) \triangleq H \sum_{j=k-s}^{i} \left[\prod_{m=0}^{i-j-1} \Phi(i-m) \right] G \tag{11.3.2-10}$$

Equation (11.3.2-9) shows that

the *nonmaneuvering filter innovation* ν^*

is

a *linear measurement of the input u*

in the presence of the additive "white noise" ν given by (11.3.2-5) above.

428

11.3.3 Estimation of the Unknown Input

Based on (11.3.2-9), the input can be estimated via LS from

$$y = \Psi u + \epsilon \qquad (11.3.3\text{-}1)$$

where

$$y = \begin{bmatrix} \nu^*(k) \\ \vdots \\ \nu^*(k-s+1) \end{bmatrix} \qquad (11.3.3\text{-}2)$$

is the stacked "measurement" vector,

$$\Psi = \begin{bmatrix} \Psi(k) \\ \vdots \\ \Psi(k-s+1) \end{bmatrix} \qquad (11.3.3\text{-}3)$$

is the measurement matrix, and the "noise" ϵ, whose components are the innovations (11.3.2-5), is zero mean with block-diagonal covariance matrix

$$S = \mathrm{diag}[S(i)] \qquad (11.3.3\text{-}4)$$

The **input estimate** in batch form is

$$\boxed{\hat{u} = (\Psi' S^{-1} \Psi)^{-1} \Psi' S^{-1} y} \qquad (11.3.3\text{-}5)$$

with the resulting covariance matrix

$$L = (\Psi' S^{-1} \Psi)^{-1} \qquad (11.3.3\text{-}6)$$

An estimate is accepted, that is, a **maneuver detection** is declared if and only if (11.3.3-5) is *statistically significant* (see Subsection 1.5.2). The significance test for the estimate $\hat{u}$, which is an n_u-dimensional vector, is

$$\epsilon_{\hat{u}} = \hat{u}' L^{-1} \hat{u} \geq c \qquad (11.3.3\text{-}7)$$

where c is a threshold chosen as follows. If the input is zero, then

$$\hat{u} \sim \mathcal{N}(0, L) \qquad (11.3.3\text{-}8)$$

The statistic $\epsilon_{\hat{u}}$ from (11.3.3-7) is then chi-square distributed with n_u degrees of freedom and c is such that the probability of false alarm is

$$P\{\epsilon_{\hat{u}} \geq c\} = \alpha \qquad (11.3.3\text{-}9)$$

with, e.g., $\alpha \leq 0.01$.

11.3.4 Correction of the State Estimate

If a maneuver is detected, then the state estimate has to be corrected. This is done, as before, assuming that the maneuver onset time was $k - s$. The term reflecting the effect of the input in (11.3.2-4) is used to correct the predicted state as follows

$$\boxed{\hat{x}^u(k|k-1) = \hat{x}^*(k|k-1) + M(k)\hat{u}} \qquad (11.3.4\text{-}1)$$

where

$$M(k) \triangleq \sum_{j=k-s}^{k-1} \left[\prod_{m=0}^{k-j-2} \Phi(k-1-m) \right] G \qquad (11.3.4\text{-}2)$$

The covariance associated with the prediction (11.3.4-1) is

$$P^u(k|k-1) = P(k|k-1) + M(k)LM(k)' \qquad (11.3.4\text{-}3)$$

where L is given in (11.3.3-6). Note that the covariance increases — the last term above is positive semidefinite or positive definite — to account for the uncertainty in the correction term used in (11.3.4-1).

The assumption about the maneuver onset time being $k - s$ *at every* k makes the algorithm very elegant, but not optimal.

Maneuver Onset Time Estimation

It is possible to estimate the onset time using a maximum likelihood criterion — this requires running in parallel a number of such algorithms. Each algorithm evaluates the likelihood function for its assumed maneuver onset time and the most likely s is chosen.

It is clear that this becomes quite expensive.

11.3.5 Input Estimation — Summary

The input estimation method is based on the following:

Given a linear system driven by a constant input, a filter based on the state model of the system *without input (input equal to zero)* estimates the state.

Then the innovations of this *mismatched filter* are

- a linear measurement of the input
- with additive zero-mean white noise with covariance equal to the filter's innovation covariance.

This allows the estimation of the input, assuming that

- it started at the beginning of the assumed sliding window; and
- it is approximately constant over this window in time.

The estimation is done via least squares.

The estimated input (e.g., acceleration) is tested for *statistical significance* (i.e., if it is "large enough" compared to its standard deviation).

If the estimated input is statistically significant then the state estimate is corrected with the effect of the input from the time it was assumed to have started (at the beginning of the window).

The covariance of the state is adjusted (increased) because the correction was done with an estimated input, rather than the exact input, which is not available.

11.4 THE ARIABLE STATE DIMENSION APPROACH

11.4.1 The Approach

The target maneuver is considered here an inherent part of its dynamics, rather than noise. In the absence of maneuvers the filter operates using a "quiescent state model." Once a maneuver is detected, new state components are added — thus the name **variable state dimension (VSD)** approach.

The extent of the maneuver as detected is then used to yield an estimate for the extra state components, and corrections are made on the other state components. The tracking is then done with the augmented state model until it will be reverted to the quiescent model by another decision.

The rationale for using a lower order quiescent model and a higher order maneuvering model is the following: this will allow good tracking performance in both situations rather than a compromise. For example, if the target does not have acceleration, using a third order model increases the estimation errors for both position and velocity.

The two models used here, described in Section 6.3, are a (nearly) constant velocity model for the quiescent situation and a (nearly) constant acceleration model for the maneuvering situation.

The state vector, for a planar motion, is in the **quiescent model**

$$x = \begin{bmatrix} \xi & \dot{\xi} & \eta & \dot{\eta} \end{bmatrix}' \tag{11.4.1-1}$$

In the **maneuvering model**, the state is

$$x^m = \begin{bmatrix} \xi & \dot{\xi} & \eta & \dot{\eta} & \ddot{\xi} & \ddot{\eta} \end{bmatrix}' \tag{11.4.1-2}$$

The method is not limited to these models. For example, the maneuvering situation could be modeled using autocorrelated acceleration, as in Section 8.2.

11.4.2   The Maneuver Detection and Model Switching

The target tracking filter is initialized under the constant velocity model assumption as in Section 5.5, using a two-point differencing procedure.

The maneuver detection is done as follows:

A **fading memory average** of the innovations from the estimator based on the quiescent model is computed

$$\epsilon_\nu^\alpha(k) = \alpha\epsilon_\nu^\alpha(k-1) + \epsilon_\nu(k) \tag{11.4.2-1}$$

with

$$\epsilon_\nu(k) \stackrel{\Delta}{=} \nu(k)'S(k)^{-1}\nu(k) \tag{11.4.2-2}$$

where $0 < \alpha < 1$, $\nu(k)$ is the innovation and $S(k)$ its covariance.

Since $\epsilon_\nu(k)$ is, under the Gaussian assumptions, chi-square distributed with n_z (dimension of the measurement) degrees of freedom, one has

$$\lim_{k\to\infty} E[\epsilon_\nu^\alpha(k)] = \frac{n_z}{1-\alpha} \tag{11.4.2-3}$$

As in Section 11.2, one can look at

$$s_\alpha = \frac{1}{1-\alpha} \tag{11.4.2-4}$$

as the **effective window length** over which the presence of a maneuver is tested.

The hypothesis that a maneuver is taking place is accepted if $\epsilon_\nu^\alpha(k)$ exceeds a certain threshold that can be determined based on the chi-square distribution as in Section 11.2. Then the estimator switches from the quiescent model to the maneuvering model.

The scheme for reverting to the quiescent model is as follows: The estimated accelerations are compared to their variances, and if they are not statistically significant the maneuver hypothesis is rejected.

Maneuver Termination versus Maneuver Onset Detection

It is more difficult to detect **maneuver termination** than **maneuver onset** because a maneuvering model has a larger innovation covariance than a nonmaneuvering model. This is due to the fact that the latter has a larger state vector and assumes more motion uncertainty than the former.

Significance Test for the Acceleration Estimate

The test statistic for significance of the acceleration estimates is

$$\epsilon_{\hat{a}}(k) = \hat{a}(k|k)' P_a^m(k|k)^{-1} \hat{a}(k|k) \qquad (11.4.2\text{-}5)$$

where $\hat{a}$ is the estimate of the acceleration components and P_a^m is the corresponding block from the covariance matrix of the maneuvering model.

When the **moving average** over a window of length p

$$\epsilon_{\hat{a}}^p(k) = \sum_{j=k-p+1}^{k} \epsilon_{\hat{a}}(j) \qquad (11.4.2\text{-}6)$$

falls below a threshold, the acceleration is deemed *insignificant*. A fading memory average can also be used.

The situation where the acceleration drops suddenly to zero can lead to large innovations for the maneuvering model. This can be taken care of by allowing a switch to the lower order model also when the maneuvering model's innovation exceeds, say, the 99% confidence region.

11.4.3 Initialization of the Augmented Model

When a maneuver is detected at time k, the filter assumes that the target had a constant acceleration starting at $k - s - 1$ where s is the effective window length. The state estimates for time $k - s$ are then modified as follows. It is assumed that only position measurements are available.

First, the estimates at $k - s$ for the acceleration are

$$\hat{x}_{4+i}^m(k - s|k - s) = \frac{2}{T^2}[z_i(k - s) - \hat{z}_i(k - s|k - s - 1)] \qquad i = 1, 2 \qquad (11.4.3\text{-}1)$$

The position components of the estimate at $k - s$ are taken as the corresponding measurements, that is,

$$\hat{x}_{2i-1}^m(k - s|k - s) = z_i(k - s) \qquad i = 1, 2 \qquad (11.4.3\text{-}2)$$

while the velocity components are corrected with the acceleration estimates as follows:

$$\hat{x}_{2i}^m(k - s|k - s) = \hat{x}_{2i}(k - s - 1|k - s - 1) + T\hat{x}_{4+i}^m(k - s|k - s) \qquad i = 1, 2 \quad (11.4.3\text{-}3)$$

The covariance matrix associated with the modified state estimate as given in (11.4.3-1) to (11.4.3-3) is $P^m(k - s|k - s)$ and its elements are presented in [BB82] together with their derivation. These equations specify in full the initialization of the filter based on the maneuvering model for time $k - s$.

Then, a recursive estimation algorithm (Kalman filter) based on this model reprocesses the measurements sequentially up to time k. Following this, the measurements are processed sequentially as they arrive. This is depicted in Figure 11.4.3-1.

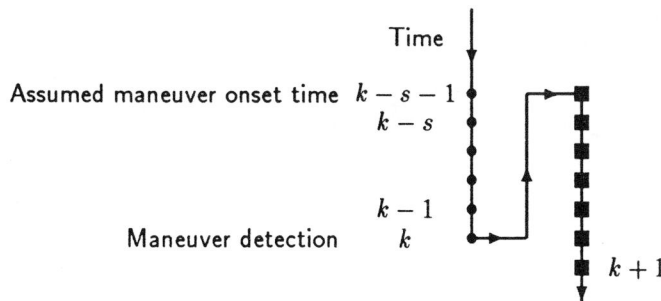

• Sequential estimation with quiescent model;   ■ with maneuvering model

Figure 11.4.3-1: Switching from quiescent to maneuvering model.

435

11.4.4 VSD Approach — Summary

In the absence of maneuvers, the filter is based on a *quiescent* (low order) model of the state.

If a maneuver is detected, the estimator switches to a higher dimension (augmented) *maneuvering* model that incorporates additional state components (acceleration).

The use of the lower order model yields maximum estimation accuracy in the case where the target undergoes no accelerations — no information is "wasted" on estimating state components that are zero.

A maneuver is declared detected when a *fading memory average* of the normalized innovations exceeds a threshold. The fading memory has an *effective window length* and the onset of the maneuver is then taken as the beginning of this sliding window.

The augmented filter is initialized at the beginning of the maneuver detection window.

A maneuver is declared terminated when the extra state component's (acceleration) estimates become statistically insignificant.

11.5 A COMPARISON OF SEVERAL ADAPTIVE ESTIMATION METHODS FOR MANEUVERING TARGETS

11.5.1 The Problem

In this section the following methods of maneuver detection are illustrated and compared:

1. White process noise with two levels (Section 11.2).
2. Input estimation (Section 11.3).
3. Variable dimension filtering (Section 11.4).

The example considers a target whose position is sampled every $T = 10s$. The target is moving in a plane with constant course and speed until $k = 40$ when it starts a slow 90° turn that is completed in 20 sampling periods. A second fast turn of 90° starts at $k = 61$ and is completed in 5 sampling times.

The initial condition of the target, with state

$$x = \begin{bmatrix} \xi & \dot{\xi} & \eta & \dot{\eta} \end{bmatrix}' \tag{11.5.1-1}$$

is, with position and velocity units m and m/s, respectively

$$x(0) = [2000 \ 0 \ 10000 \ -15]' \tag{11.5.1-2}$$

The slow turn is the result of acceleration inputs

$$u_\xi = u_\eta = 0.075 m/s^2 \qquad 400s \le t \le 600s \tag{11.5.1-3}$$

and the fast turn has accelerations

$$u^\xi = u^\eta = -0.3 m/s^2 \qquad 610s \le t \le 660s \tag{11.5.1-4}$$

Note that the changes in velocity are 0.75 and $3m/s$ per sampling period for the slow and fast turn, respectively.

The measurements are position only according to the equation

$$z(k) = [1 \ 0] x(k) + w(k) \tag{11.5.1-5}$$

with $w(k)$ zero mean, white, independent of the process noise, and with variance

$$E[w(k)^2] = r = 10^4 m^2 \tag{11.5.1-6}$$

11.5.2 The White Noise Model with Two Levels

The adaptive algorithm based on white process noise with switching between two levels for maneuver modeling used the following state representation per coordinate (same for each coordinate and decoupled between them)

$$x(k+1) = \begin{bmatrix} 1 & T \\ 0 & 1 \end{bmatrix} x(k) + \begin{bmatrix} T/2 \\ 1 \end{bmatrix} v(k) \qquad (11.5.2\text{-}1)$$

with $v(k)$ zero mean white with variance

$$E[v(k)^2] = q \qquad (11.5.2\text{-}2)$$

Note that here $v(k)$ represents the **velocity increment over a sampling period**.

The two levels of process noise were q_0 for the quiescent period and q_1 after a maneuver was detected.

A maneuver was declared detected if the normalized innovation squared (11.2.2-1) exceeded a threshold ϵ_m. Then q_0 was replaced by q_1 in the filter. A maneuver was considered terminated when the normalized innovation squared, obtained from the filter with the higher q_1, fell below the same threshold ϵ_m — for simplicity, the upcrossing and downcrossing thresholds were taken the same.

The value for the quiescent process noise was $q_0 = 0$. Several values for the variance q_1 of the process noise modeling the maneuver and the threshold ϵ_m were tried. The best combination was determined based on the MSE in the position estimate of the target from 50 Monte Carlo runs. The resulting values were $q_1 = 10$ and $\epsilon_m = 3$.

The RMS position error (coordinate ξ) corresponding to these values is shown in Figure 11.5.2-1. Note that the value $\sqrt{q_1} \approx 3$ matches approximately the *maximum change in velocity per sampling period* during the fast maneuver, indicated in Subsection 11.5.1.

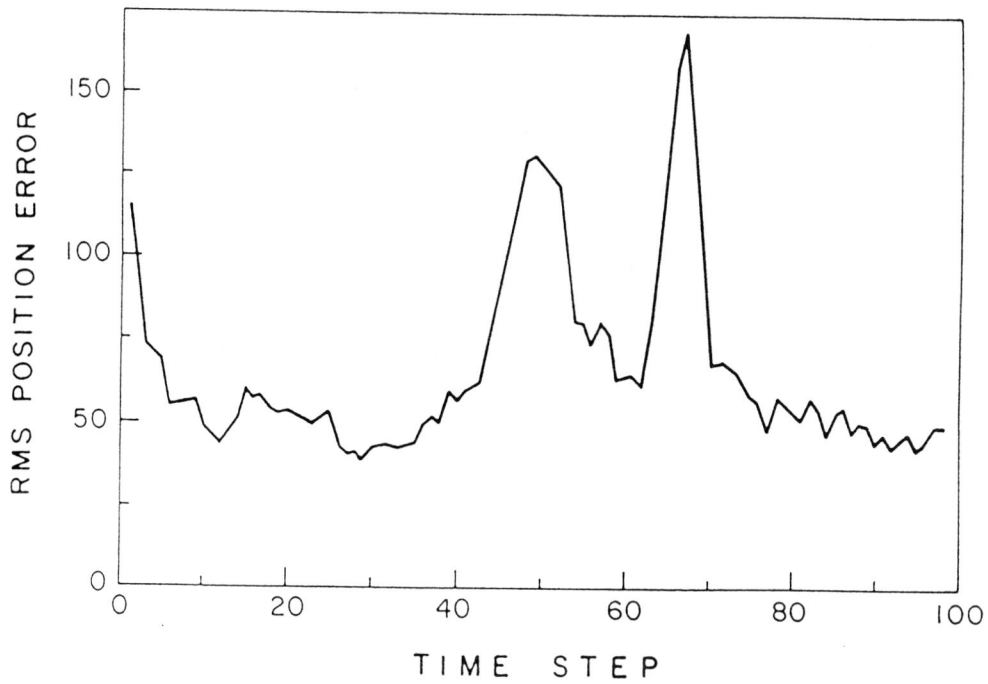

Figure 11.5.2-1: Position estimation error in coordinate ξ for the adaptive filter with two process noise levels (from 50 runs).

11.5.3 The IE and VSD Methods

The same problem was simulated using the ***input estimation (IE)*** and ***variable state dimension (VSD)*** methods.

The IE method was run with a window of $s = 5$ samples to estimate the input. The state equation was the constant velocity model (6.3.2-1) for each coordinate and no process noise.

The VSD method was based on a quiescent model with state (11.4.1-1) and a maneuvering model with state (11.4.1-2). The quiescent model was, for each coordinate, the constant velocity state equation (6.3.2-1) with no process noise. The maneuvering model was a nearly constant acceleration state equation, given by (6.3.3-1) with process noise standard deviation σ_v in (6.3.3-4) taken equal to 5% of the estimated acceleration.

The following maneuver detection parameters were used in the VSD method:

1. The fading memory parameter from (11.4.2-1) was $\alpha = 0.8$, which corresponds to an effective window length (11.4.2-4) of $s = 5$.

2. The threshold for the fading memory average (11.4.2-1) was $\chi_{10}^2(95\%) = 18.3$. The maneuver detection was done together in the two coordinates based on both measurements (i.e., $n_z = 2$), which, with (11.4.2-3), yields that ϵ_ν^α is approximately chi-square distributed with 10 degrees of freedom.

3. The window for the calculation of the average normalized acceleration (11.4.2-6) for the significance test was $p = 2$.

4. The threshold for the acceleration significance test was $\chi_4^2(95\%) = 9.49$, based on a 4 degrees of freedom chi-square random variable (the dimension of the acceleration, 2, multiplied by $p = 2$).

Figure 11.5.3-1a presents the position RMS errors (in coordinate ξ) for the IE and VSD filters based on a 50 run Monte Carlo average. These two filters were run with the same random numbers for further comparison.

The RMS errors are smaller in the VSD filter compared to the IE filter. Comparing with Figure 11.5.2-1, it is seen that the simple approach of two-level white process noise modeling of the maneuver is quite effective — it is slightly worse than the VSD filter but better than the IE filter.

Figure 11.5.3-1b presents the velocity RMS errors for the IE and VSD filters based on the same 50 Monte Carlo runs. For velocity, the VSD filter is only slightly superior compared to the IE algorithm.

440

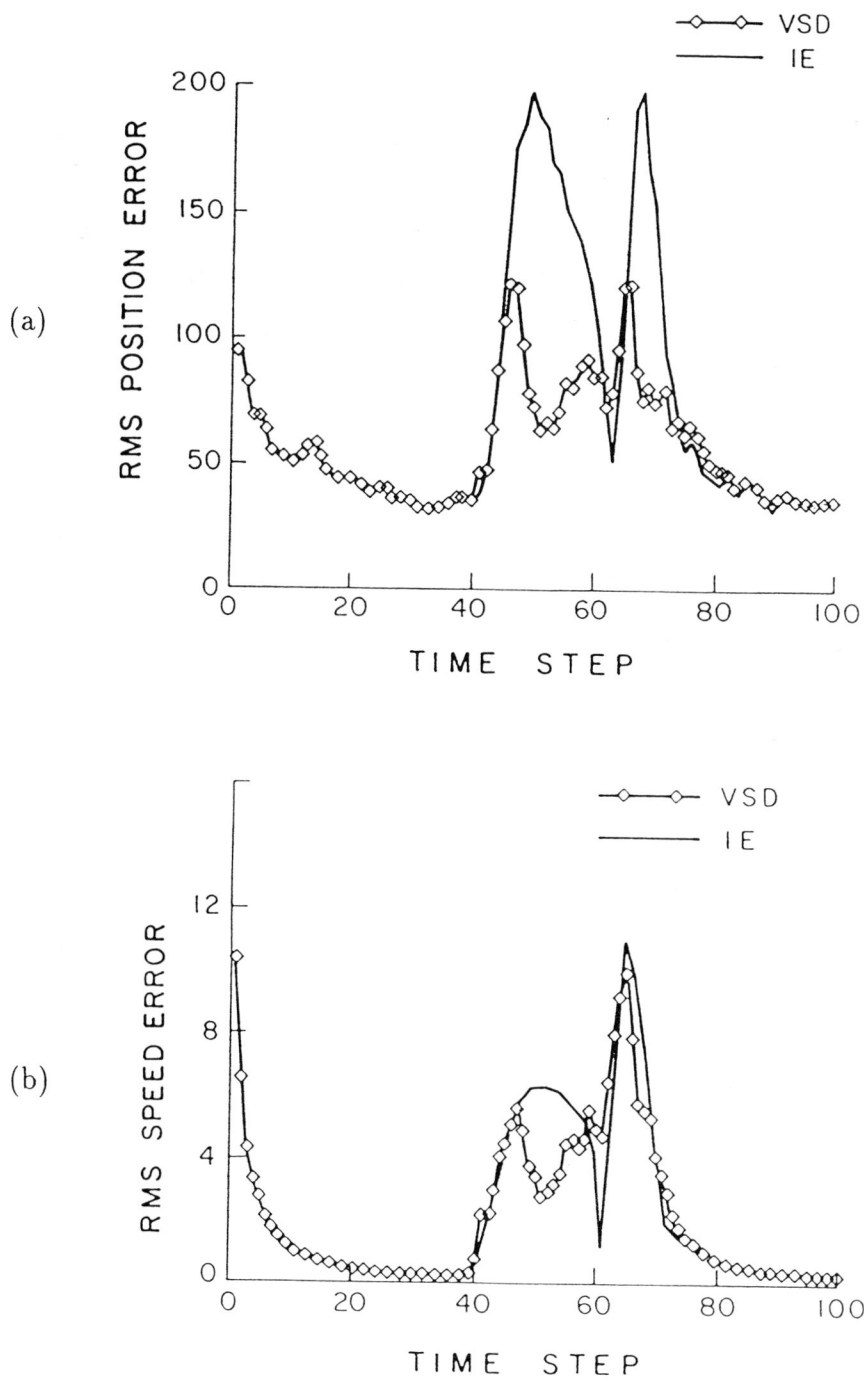

Figure 11.5.3-1: Estimation errors in coordinate ξ for the VSD and IE filters (from 50 runs).

441

Remark

It should be noted that none of these three adaptive algorithms considered here manage to keep the RMS position error in the estimate below the position measurement noise RMS value (which is, in one coordinate, $100m$) *during the entire maneuver period.* In other words, these algorithms have difficulty providing **noise reduction** during the critical time of a maneuver.

This is a general problem with adaptive estimation algorithms: the adaptation might not be rapid enough. In the present example, the errors reach their peak shortly after the start of the maneuver.

This issue will be considered again in a later example.

11.5.4 Statistical Test for Comparison of the IE and VSD Methods

Next, we shall examine more closely the results of the IE and VSD filters and carry out the systematic comparison of algorithms presented in Subsection 1.5.3. The question is whether, in view of the limited sample size, one can state that the VSD algorithm is superior to the IE algorithm.

The performance of interest will be the mean square error in the estimate of one state component, not indicated explicitly, for simplicity

$$\epsilon^{\text{VSD}}(k) \triangleq E\left[[\tilde{x}^{\text{VSD}}(k|k)]^2\right] \tag{11.5.4-1}$$

$$\epsilon^{\text{IE}}(k) \triangleq E\left[[\tilde{x}^{\text{IE}}(k|k)]^2\right] \tag{11.5.4-2}$$

The **sample mean** of the performance from N **Monte Carlo runs** is

$$\bar{\epsilon}^{\text{VSD}}(k) = \frac{1}{N}\sum_{i=1}^{N}[\tilde{x}^{\text{VSD}i}(k|k)]^2 \tag{11.5.4-3}$$

$$\bar{\epsilon}^{\text{IE}}(k) = \frac{1}{N}\sum_{i=1}^{N}[\tilde{x}^{\text{IE}i}(k|k)]^2 \tag{11.5.4-4}$$

where $\tilde{x}^{\text{VSD}i}(k|k)$ is the estimation error of the state component under consideration at time k in run i with algorithm VSD, and similarly for IE. These are the quantities shown in Figures 11.5.3-1a and 11.5.3-1b.

We want to test if one can accept the hypothesis

$$H_1 : \Delta^{\text{IV}}(k) \triangleq \epsilon^{\text{IE}}(k) - \epsilon^{\text{VSD}}(k) > 0 \tag{11.5.4-5}$$

that is, that algorithm VSD is superior to algorithm IE. Note that "superiority" is based on the *true mean square errors* defined in (11.5.4-1) and (11.5.4-2). A comparison of the *sample mean square errors* (11.5.4-3) and (11.5.4-4) corresponding to the two algorithms is not sufficient (actually is rather naive) since the inaccuracy in these sample means is not considered at all.

The correct statistical test is, as discussed in Subsection 1.5.3, based on the **sample performance differences**

$$\Delta^{\text{IV}i}(k) \triangleq [\tilde{x}^{\text{IE}i}(k|k)]^2 - [\tilde{x}^{\text{VSD}i}(k|k)]^2 \tag{11.5.4-6}$$

The test uses the sample mean $\bar{\Delta}$ of the above differences and its standard error $\sigma_{\bar{\Delta}}$ given in (1.5.3-9) and (1.5.3-10), respectively.

443

If the ratio $\bar{\Delta}/\sigma_{\bar{\Delta}}$ exceeds a threshold, then the *difference of the true means* in (11.5.4-5) is accepted as positive (H_1 is accepted) and the null hypothesis

$$H_0 : \Delta^{\mathrm{IV}} \leq 0 \qquad (11.5.4\text{-}7)$$

is rejected because it has a "low level of significance." This is equivalent to saying that the estimated mean $\bar{\Delta}$ is positive and statistically significant (i.e., H_1 can be accepted because it is "significant").

Hypothesis H_1 is usually accepted only if the significance level of H_0 is less than 5%, in which case the test is

$$\frac{\bar{\Delta}}{\sigma_{\bar{\Delta}}} > \mathcal{G}(95\%) = 1.65 \qquad (11.5.4\text{-}8)$$

where $\mathcal{G}(1-\alpha)$ represents the point on the standard Gaussian distribution corresponding to upper tail probability of α.

The above statistical test was applied for comparing the performance over intervals $[k, l]$, rather than single points in time, by replacing (11.5.4-6) with

$$\Delta^{\mathrm{IV}i}(k,l) = \frac{1}{l-k+1} \sum_{m=k}^{l} \left\{ [\tilde{x}^{\mathrm{IE}i}(m|m)]^2 - [\tilde{x}^{\mathrm{VSD}i}(m|m)]^2 \right\} \qquad (11.5.4\text{-}9)$$

Table 11.5.4-1 shows the test for the difference of the MSE between the two algorithms over several time intervals. Note that the slow maneuver took place during the time interval $[40, 60]$ and the fast one during the interval $[61, 66]$. Statistically significant improvements of the VSD algorithm over the IE algorithm are observed in position over all the intervals considered and in velocity over two intervals. In the remaining two intervals, while there was improvement, it was not sufficient to reject H_0 at the 5% level. At the 8% level, however, H_0 would have been rejected in all cases.

Time interval	Velocity			Position		
	$\bar{\Delta}$	$\sigma_{\bar{\Delta}}$	Test statistic	$\bar{\Delta}$	$\sigma_{\bar{\Delta}}$	Test statistic
40-60	8.06	1.19	6.77	13,560	735	18.45
60-70	4.99	3.28	1.52	10,336	1,013	10.20
40-70	7.72	1.43	5.40	12,677	619	20.48
60-80	2.45	1.75	1.40	6,091	626	9.73

Table 11.5.4-1: Test of means for comparison of VSD and IE algorithms from 50 runs.

11.5.5 Comparison of Several Algorithms — Summary

The optimal statistical method of comparing two algorithms based on Monte Carlo runs has been illustrated in the evaluation of the variable state dimension and input estimation filters.

The VSD filter turned out to be superior (based on statistical significance analysis) and less demanding computationally than the IE filter.

The much simpler two-level white noise filter appears to have performance close to the VSD filter.

A multiple-level white or autocorrelated noise model seems to be the best compromise from the point of view of performance versus cost of implementation.

The computational requirements of the IE filter, VSD filter and the two-level process noise filter were, approximately, 8 : 2 : 1.

Neither of these three filters considered here managed to keep the *peak position estimation error* (during the maneuver) below the *raw position measurement RMS error* — the measurement noise standard deviation. This is a major shortcoming of most adaptive estimation schemes.

As it will be shown in the next section, the multiple model approach can accomplish this (almost [1]), while yielding good noise reduction during the constant velocity portions of the trajectory.

---

[1] In the spirit of "Our algorithms can almost do almost everything."

11.6 THE MULTIPLE MODEL APPROACH

11.6.1 Formulation of the Approach

In the **multiple model approach (MM)** it is assumed that the system obeys one of a finite number of models. Such systems are called **hybrid**: they have both *continuous* (noise) uncertainties as well as *discrete* uncertainties — **model** or **mode** uncertainties.

A *Bayesian framework* is used: starting with prior probabilities of each model being correct (i.e., the system is in a particular mode), the corresponding posterior probabilities are obtained.

First it will be assumed that the model the system obeys is *fixed*, that is, no switching from one mode to another occurs during the estimation process (time-invariant mode).

The model, assumed to be in effect throughout the process, is one of r possible models (the system is in one of r modes)

$$M \in \{M_j\}_{j=1}^r \qquad (11.6.1\text{-}1)$$

The prior probability that M_j is correct (the system is in mode j) is

$$P\{M_j|Z^0\} = \mu_j(0) \qquad j = 1, \ldots, r \qquad (11.6.1\text{-}2)$$

where Z^0 is the prior information and

$$\sum_{j=1}^r \mu_j(0) = 1 \qquad (11.6.1\text{-}3)$$

since the correct model is among the assumed r possible models.

It will be assumed that all models are linear-Gaussian. This approach can be used for nonlinear systems as well via linearization.

Subsequently, the situation of **switching models** or **mode jumping** is considered. In the latter case, the system undergoes transitions from one mode to another.

11.6.2 The Multiple Model Approach for Fixed Models

Using Bayes' formula, the posterior probability of model j being correct, given the measurement data up to k, is obtained recursively as

$$\mu_j(k) \triangleq P\{M_j|Z^k\} = P\{M_j|z(k), Z^{k-1}\} = \frac{p[z(k)|Z^{k-1}, M_j]P\{M_j|Z^{k-1}\}}{p[z(k)|Z^{k-1}]}$$

$$= \frac{p[z(k)|Z^{k-1}, M_j]P\{M_j|Z^{k-1}\}}{\sum_{i=1}^{r} p[z(k)|Z^{k-1}, M_i]P\{M_i|Z^{k-1}\}} \qquad (11.6.2\text{-}1)$$

or

$$\boxed{\mu_j(k) = \frac{p[z(k)|Z^{k-1}, M_j]\mu_j(k-1)}{\sum_{i=1}^{r} p[z(k)|Z^{k-1}, M_i]\mu_i(k-1)} \qquad j = 1,\ldots,r} \qquad (11.6.2\text{-}2)$$

starting with the given prior probabilities (11.6.1-2).

The first term on the right hand side above is the **likelihood function of mode j** at time k, which, under the linear-Gaussian assumptions, is given by the expression

$$\Lambda_j(k) \triangleq p[z(k)|Z^{k-1}, M_j] = p[\nu_j(k)] = \mathcal{N}[\nu_j(k); 0, S_j(k)] \qquad (11.6.2\text{-}3)$$

where ν_j and S_j are the innovation and its covariance from the **mode-matched filter** corresponding to mode j.

Thus a filter matched to each mode is set up yielding **mode-conditioned state estimates** and the associated **mode-conditioned covariances**. The probability of each mode being correct — the **mode estimates** — is obtained according to (11.6.2-2) based on its likelihood function (11.6.2-3) relative to the other filters' likelihood functions.

This **bank of filters** is illustrated in Figure 11.6.2-1.

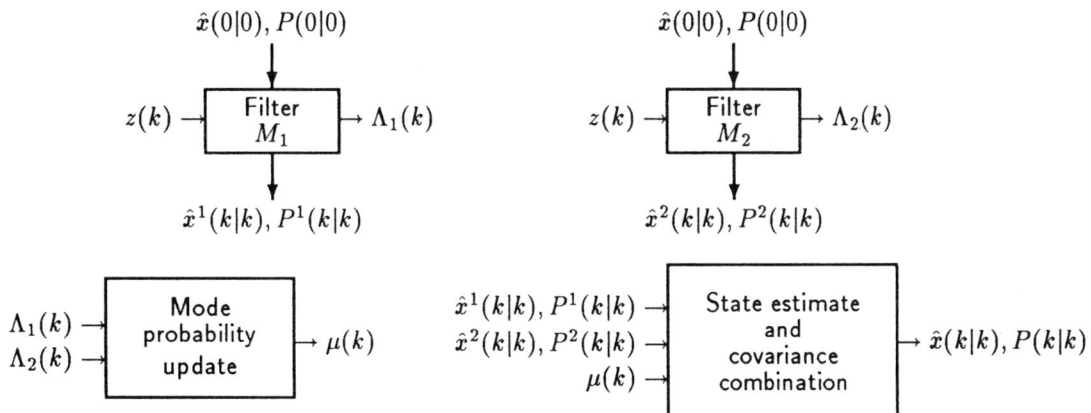

Figure 11.6.2-1: The multiple model approach for $r = 2$ fixed models.

The output of each mode-matched filter is the **mode-conditioned state estimate** $\hat{x}^j$, the associated covariance P^j and the **mode likelihood function** Λ_j.

After the filters are initialized, they run recursively *on their own estimates*. Their likelihood functions are used to update the mode probabilities. The latest mode probabilities are used to combine the mode-conditioned estimates and covariances.

Under the above assumptions the pdf of the state of the system is a Gaussian mixture with r terms

$$p[x(k)|Z^k] = \sum_{j=1}^{r} \mu_j(k)\mathcal{N}[x(k); \hat{x}^j(k|k), P^j(k|k)] \qquad (11.6.2\text{-}4)$$

The combination of the mode-conditioned estimates is done therefore as follows

$$\boxed{\hat{x}(k|k) = \sum_{j=1}^{r} \mu_j(k)\hat{x}^j(k|k)} \qquad (11.6.2\text{-}5)$$

and the covariance of the combined estimate is (see Subsection 1.4.16)

$$\boxed{P(k|k) = \sum_{j=1}^{r} \mu_j(k)\{P^j(k|k) + [\hat{x}^j(k|k) - \hat{x}(k|k)][\hat{x}^j(k|k) - \hat{x}(k|k)]'\}} \qquad (11.6.2\text{-}6)$$

where the last term above is the **spread of the means** term.

The above is exact under the following assumptions:

1. The correct model is among the set of models considered,
2. The same model has been in effect from the initial time.

Assumption (1) can be considered a reasonable approximation; however, (2) is obviously not true if a maneuver has started at some time within the interval $[1, k]$, in which case a model change — **mode jump** — occurred.

Convergence of the Mode Estimates

If the mode set includes the correct one and no mode jump occurs, then the probability of the true mode will converge to unity, that is, this approach yields consistent estimates of the system parameters. Otherwise the probability of the model "nearest" to the correct one will converge to unity (this is discussed in detail in [BS78]).

Ad hoc Modifications for the Case of Switching Models

The following ad hoc modifications can be made to the above algorithm for estimating the state in the case of switching models:

1. An artificial lower bound is imposed on the model probabilities (with a suitable renormalization of the remaining probabilities).
2. A finite memory (sliding window) or exponentially discounted likelihood function is used in the calculation of the model probabilities.

A shortcoming of the static MM algorithm when used with switching models is that, in spite of these ad hoc modifications, the mismatched filter's errors can grow to unacceptable levels. Thus, reinitialization of the filters that are mismatched is in general needed.

11.6.3   The Multiple Model Approach for Switching Models

The system is modeled by the equations

$$x(k) = F[M(k)]x(k-1) + v[k-1, M(k)] \qquad (11.6.3\text{-}1)$$

$$z(k) = H[M(k)]x(k) + w[k, M(k)] \qquad (11.6.3\text{-}2)$$

where $M(k)$ denotes the model "at time k" — in effect *during the sampling period ending at k*. Such systems are also called **jump-linear systems**. The mode jump process is assumed **left-continuous** (i.e., the impact of the new model starts at t_k^+).

The mode at time k is assumed to be among the possible r modes

$$M(k) \in \{M_j\}_{j=1}^r \qquad (11.6.3\text{-}3)$$

The continuous-valued vector $x(k)$ is sometimes referred to as the **base state**, and the discrete variable $M(k)$ as the **modal state**.

For example,

$$F[M_j] = F_j \qquad (11.6.3\text{-}4)$$

$$v(k-1, M_j) \sim \mathcal{N}(u_j, Q_j) \qquad (11.6.3\text{-}5)$$

that is, the structure of the system and/or the statistics of the noises might be different from model to model. The mean u_j of the noise can model a maneuver as a deterministic input.

The l-th **mode history** — or **sequence of models** — through time k is denoted as

$$M^{k,l} = \{M_{i_{1,l}}, \ldots, M_{i_{k,l}}\} \qquad l = 1, \ldots, r^k \qquad (11.6.3\text{-}6)$$

where $i_{\kappa,l}$ is the model index at time κ from history l and

$$1 \leq i_{\kappa,l} \leq r \qquad \kappa = 1, \ldots, k \qquad (11.6.3\text{-}7)$$

Note that the number of histories increases *exponentially with time*.

For example, with $r = 2$ one has at time $k = 2$ the following $r^k = 4$ possible sequences (histories)

l	$i_{1,l}$	$i_{2,l}$
1	1	1
2	1	2
3	2	1
4	2	2

450

It will be assumed that the **mode (model) jump process** is a Markov process (Markov chain) with known mode transition probabilities

$$p_{ij} \triangleq P\{M(k) = M_j | M(k-1) = M_i\}$$

(11.6.3-8)

These **mode transition probabilities** will be assumed time invariant and independent of the base state. In other words, this is a **homogeneous Markov chain**.

The system (11.6.3-1), (11.6.3-2), and (11.6.3-8) is a generalized version of a **hidden Markov model**.

The event that model j is in effect at time k is denoted as

$$M_j(k) \triangleq \{M(k) = M_j\}$$

(11.6.3-9)

The conditional probability of the l-th history

$$\mu^{k,l} \triangleq P\{M^{k,l} | Z^k\}$$

(11.6.3-10)

will be evaluated next.

The l-th sequence of models through time k can be written as

$$M^{k,l} = \{M^{k-1,s}, M_j(k)\}$$

(11.6.3-11)

where sequence s through $k-1$ is its *parent sequence* and M_j its last element.

Then, in view of the Markov property,

$$P\{M_j(k) | M^{k-1,s}\} = P\{M_j(k) | M_i(k-1)\} \triangleq p_{ij}$$

(11.6.3-12)

where $i = s_{k-1}$, the index of the last model in the parent sequence s through $k-1$.

The conditional pdf of the state at k is obtained using the total probability theorem with respect to the mutually exclusive and exhaustive set of events (11.6.3-6), as a **Gaussian mixture** with an *exponentially increasing number of terms*

$$p[x(k) | Z^k] = \sum_{l=1}^{r^k} p[x(k) | M^{k,l}, Z^k] P\{M^{k,l} | Z^k\}$$

(11.6.3-13)

Since *to each mode sequence one has to match a filter*, it can be seen that an exponentially increasing number of filters are needed to estimate the (base) state, which makes the optimal approach impractical.

The probability of a mode history is obtained using Bayes' formula as

$$
\begin{aligned}
\mu^{k,l} &= P\{M^{k,l}|Z^k\} \\
&= P\{M^{k,l}|z(k), Z^{k-1}\} \\
&= \frac{1}{c}\, p[z(k)|M^{k,l}, Z^{k-1}] P\{M^{k,l}|Z^{k-1}\} \\
&= \frac{1}{c}\, p[z(k)|M^{k,l}, Z^{k-1}] P\{M_j(k), M^{k-1,s}|Z^{k-1}\} \\
&= \frac{1}{c}\, p[z(k)|M^{k,l}, Z^{k-1}] P\{M_j(k)|M^{k-1,s}, Z^{k-1}\}\mu^{k-1,s} \\
&= \frac{1}{c}\, p[z(k)|M^{k,l}, Z^{k-1}] P\{M_j(k)|M^{k-1,s}\}\mu^{k-1,s}
\end{aligned}
\tag{11.6.3-14}
$$

where c is the normalization constant.

If the current mode depends only on the previous one (i.e., it is a Markov chain), then

$$
\mu^{k,l} = \frac{1}{c}\, p[z(k)|M^{k,l}, Z^{k-1}] P\{M_j(k)|M_i(k-1)\}\mu^{k-1,s}
\tag{11.6.3-15}
$$

or

$$
\boxed{\mu^{k,l} = \frac{1}{c}\, p[z(k)|M^{k,l}, Z^{k-1}] p_{ij}\mu^{k-1,s}}
\tag{11.6.3-16}
$$

where $i = s_{k-1}$ is the last model of the parent sequence s.

The above equation shows that *conditioning on the entire past history* is needed even if the random parameters are Markov.

Practical Approaches

The only way to avoid the exponentially increasing number of histories, which have to be accounted for, is by going to suboptimal techniques.

A simple-minded suboptimal technique is to keep, say, the N histories with the largest probabilities, discard the rest, and renormalize the probabilities such that they sum up to unity.

The **generalized pseudo-Bayesian (GPB)** approaches combine histories of models that differ in "older" models. The first order GPB, denoted as GPB1, considers only the possible models in the last sampling period. The second order version, GPB2, considers all the possible models in the last two sampling periods. These algorithms require r and r^2 filters to operate in parallel, respectively.

Finally, the **interacting multiple model (IMM)** algorithm will be presented. This algorithm is conceptually similar to GPB2, but requires only r filters to operate in parallel.

The Mode Transition Probabilities

The **mode transition probabilities** (11.6.3-8), indicated as assumed to be known, are actually **estimator design parameters** to be selected in the design process of the algorithm. This will be discussed in detail in Subsections 11.6.7 and 11.6.8.

Note

The GPB1 and IMM algorithms have approximately the same computational requirements as the static (fixed model) MM algorithm, but do not require ad hoc modifications as the latter, which is actually obsolete for switching models.

11.6.4 The GPB1 Multiple Model Approach for Switching Models

In the **generalized pseudo-Bayesian approach of first order (GPB1)**, at time k the state estimate is computed under each possible current model — a total of r possibilities (hypotheses) are considered. All histories that differ in "older" models are combined together.

The total probability theorem is thus used as follows:

$$
\begin{aligned}
p[x(k)|Z^k] &= \sum_{j=1}^{r} p[x(k)|M_j(k), Z^k] P\{M_j(k)|Z^k\} \\
&= \sum_{j=1}^{r} p[x(k)|M_j(k), z(k), Z^{k-1}] \mu_j(k) \\
&\approx \sum_{j=1}^{r} p[x(k)|M_j(k), z(k), \hat{x}(k-1|k-1), P(k-1|k-1)] \mu_j(k)
\end{aligned}
$$

(11.6.4-1)

Thus at time $k-1$ there is a single **lumped estimate** $\hat{x}(k-1|k-1)$ that summarizes (approximately) the past Z^{k-1}, and the associated covariance. From this, one carries out the prediction to time k and the update at time k under r hypotheses, namely

$$
\hat{x}^j(k|k) = \hat{x}[k|k; M_j(k), \hat{x}(k-1|k-1), P(k-1|k-1)] \qquad j = 1, \ldots, r \quad (11.6.4\text{-}2)
$$

$$
P^j(k|k) = P[k|k; M_j(k), P(k-1|k-1)] \qquad j = 1, \ldots, r \quad (11.6.4\text{-}3)
$$

After the update, the estimates are combined with the weightings $\mu_j(k)$ (detailed later), resulting in the new combined estimate $\hat{x}(k|k)$. In other words, *the r hypotheses are merged into a single hypothesis at the end of each cycle.*

Finally, the mode (or model) probabilities are updated.

Figure 11.6.4-1 describes this algorithm, which requires r filters in parallel.

The output of each model-matched filter is the **mode-conditioned state estimate** $\hat{x}^j$, the associated covariance P^j and the **mode likelihood function** Λ_j.

After the filters are initialized, they run recursively using *the previous combined estimate*. Their likelihood functions are used to update the mode probabilities. The latest mode probabilities are used to combine the model-conditional estimates and covariances.

The structure of this algorithm is

$$
(N_e; N_f) = (1; r)
$$

(11.6.4-4)

where N_e is the *number of estimates* at the start of the cycle of the algorithm and N_f is the *number of filters* in the algorithm.

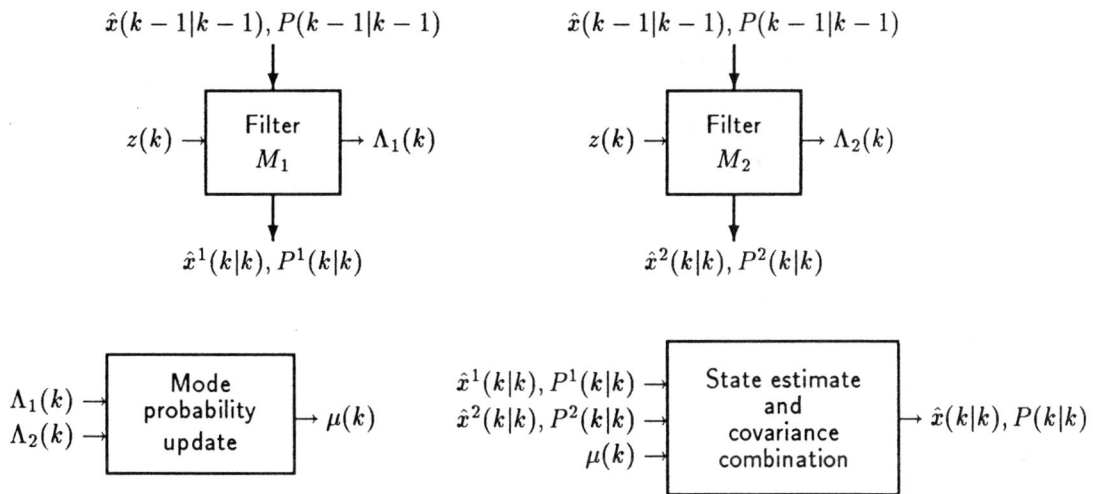

Figure 11.6.4-1: The GPB1 MM approach for $r = 2$ switching models (one cycle).

The Algorithm

One cycle of the algorithm consists of the following:

1. **Mode-matched filtering** $(j = 1, \ldots, r)$. Starting with $\hat{x}(k-1|k-1)$ one computes $\hat{x}^j(k|k)$ and the associated covariance $P^j(k|k)$ through a filter matched to $M_j(k)$. The likelihood functions

$$\Lambda_j(k) = p[z(k)|M_j(k), Z^{k-1}] \tag{11.6.4-5}$$

corresponding to these r filters are evaluated as

$$\boxed{\Lambda_j(k) = p[z(k)|M_j(k), \hat{x}(k-1|k-1), P(k-1|k-1)]} \tag{11.6.4-6}$$

2. **Mode probability update** $(j = 1, \ldots, r)$. This is done as follows:

$$
\begin{aligned}
\mu_j(k) &\triangleq P\{M_j(k)|Z^k\} \\
&= P\{M_j(k)|z(k), Z^{k-1} \\
&= \frac{1}{c} p[z(k)|M_j(k), Z^{k-1}]P\{M_j(k)|Z^{k-1}\} \\
&= \frac{1}{c} \Lambda_j(k) \sum_{i=1}^{r} P\{M_j(k)|M_i(k-1), Z^{k-1}\}P\{M_i(k-1)|Z^{k-1}\}
\end{aligned} \tag{11.6.4-7}
$$

which yields with p_{ij} the known *mode transition probabilities*,

$$\boxed{\mu_j(k) = \frac{1}{c} \Lambda_j(k) \sum_{i=1}^{r} p_{ij}\mu_i(k-1)} \tag{11.6.4-8}$$

where c is the normalization constant

$$c = \sum_{j=1}^{r} \Lambda_j(k) \sum_{i=1}^{r} p_{ij}\mu_i(k-1) \tag{11.6.4-9}$$

3. **State estimate and covariance combination**. The latest combined state estimate and covariance are obtained according to the summation in (11.6.4-1) as

$$\boxed{\hat{x}(k|k) = \sum_{j=1}^{r} \hat{x}^j(k|k)\mu_j(k)} \tag{11.6.4-10}$$

$$\boxed{P(k|k) = \sum_{j=1}^{r} \mu_j(k)\{P^j(k|k) + [\hat{x}^j(k|k) - \hat{x}(k|k)][\hat{x}^j(k|k) - \hat{x}(k|k)]'\}} \tag{11.6.4-11}$$

11.6.5   The GPB2 Multiple Model Approach for Switching Models

In the **generalized pseudo-Bayesian approach of second order (GPB2)**, at time k the state estimate is computed under *each possible current and previous model — a* total of r^2 hypotheses (histories) are considered. All histories that differ only in "older" models are merged.

The total probability theorem is thus used as follows:

$$
\begin{aligned}
p[x(k)|Z^k] &= \sum_{j=1}^{r}\sum_{i=1}^{r} p[x(k)|M_j(k), M_i(k-1), Z^k] P\{M_i(k-1)|M_j(k), Z^k\} P\{M_j(k)|Z^k\} \\
&= \sum_{j=1}^{r}\sum_{i=1}^{r} p[x(k)|M_j(k), z(k), M_i(k-1), Z^{k-1}]\mu_{i|j}(k-1|k)\mu_j(k) \\
&\approx \sum_{j=1}^{r}\sum_{i=1}^{r} p[x(k)|M_j(k), z(k), \hat{x}^i(k-1|k-1), P^i(k-1|k-1)] \\
&\quad \cdot \mu_{i|j}(k-1|k)\mu_j(k)
\end{aligned}
\tag{11.6.5-1}
$$

that is, the past $\{M_i(k-1), Z^{k-1}\}$ is approximated by the **mode-conditioned estimate** $\hat{x}^i(k-1|k-1)$ and associated covariance.

Thus at time $k-1$ there are r estimates and covariances, each to be predicted to time k and updated at time k under r hypotheses, namely

$$
\hat{x}^{ij}(k|k) \triangleq \hat{x}[k|k; M_j(k), \hat{x}^i(k-1|k-1), P^i(k-1|k-1)] \qquad i,j = 1,\ldots,r \tag{11.6.5-2}
$$

$$
P^{ij}(k|k) \triangleq P[k|k; M_j(k), P^i(k-1|k-1)] \qquad i,j = 1,\ldots,r \tag{11.6.5-3}
$$

After the update, the estimates corresponding to the same latest model hypothesis are combined with the weightings $\mu_{i|j}(k-1|k)$, detailed later, resulting in r estimates $\hat{x}^j(k|k)$. In other words, the r^2 hypotheses are *merged into r at the end of each estimation cycle.*

Figure 11.6.5-1 describes this algorithm, which requires r^2 parallel filters.

The structure of the GPB2 algorithm is

$$
(N_e; N_f) = (r; r^2) \tag{11.6.5-4}
$$

where N_e is the *number of estimates* at the start of the cycle of the algorithm and N_f is the *number of filters* in the algorithm.

$$\hat{x}^1(k-1|k-1) \qquad \hat{x}^1(k-1|k-1) \qquad \hat{x}^2(k-1|k-1) \qquad \hat{x}^2(k-1|k-1)$$
$$P^1(k-1|k-1) \qquad P^1(k-1|k-1) \qquad P^2(k-1|k-1) \qquad P^2(k-1|k-1)$$

$z(k) \rightarrow$ | Filter M_1 | $z(k) \rightarrow$ | Filter M_2 | $z(k) \rightarrow$ | Filter M_1 | $z(k) \rightarrow$ | Filter M_2 |

$$\hat{x}^{11}(k-1|k-1) \qquad \hat{x}^{12}(k-1|k-1) \qquad \hat{x}^{21}(k-1|k-1) \qquad \hat{x}^{22}(k-1|k-1)$$
$$P^{11}(k-1|k-1) \qquad P^{12}(k-1|k-1) \qquad P^{21}(k-1|k-1) \qquad P^{22}(k-1|k-1)$$
$$\Lambda_{11}(k) \qquad\qquad \Lambda_{12}(k) \qquad\qquad \Lambda_{21}(k) \qquad\qquad \Lambda_{22}(k)$$

$$\hat{x}^{11}(k-1|k-1), P^{11}(k-1|k-1) \qquad\qquad \hat{x}^{12}(k-1|k-1), P^{12}(k-1|k-1)$$
$$\hat{x}^{21}(k-1|k-1), P^{21}(k-1|k-1) \qquad\qquad \hat{x}^{22}(k-1|k-1), P^{22}(k-1|k-1)$$

| Merging | | Merging |

$$\hat{x}^1(k|k), P^1(k|k) \qquad\qquad\qquad \hat{x}^2(k|k), P^2(k|k)$$

$\Lambda_1(k) \rightarrow$ | Mode probability update and merging probability calculation | $\rightarrow \mu(k)$ $\rightarrow \mu(k-1|k)$
$\Lambda_2(k) \rightarrow$

$\hat{x}^1(k|k), P^1(k|k) \rightarrow$ | State estimate and covariance combination | $\hat{x}(k|k)$ $\rightarrow P(k|k)$
$\hat{x}^2(k|k), P^2(k|k) \rightarrow$
$\mu(k) \rightarrow$

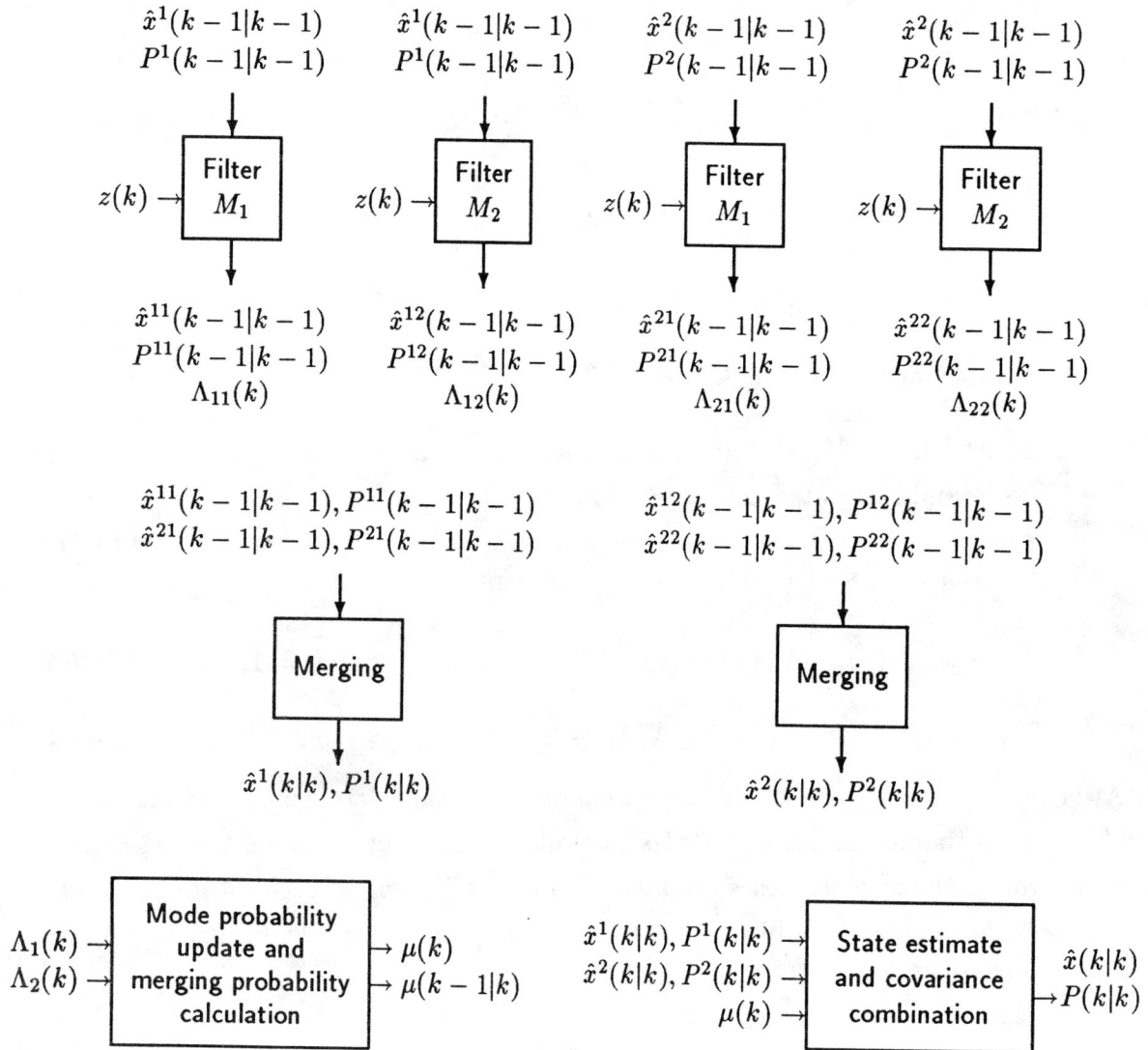

Figure 11.6.5-1: The GPB2 MM algorithm for $r = 2$ models (one cycle).

458

The Algorithm

One cycle of the algorithm consists of the following:

1. **Mode-matched filtering** $(i, j = 1, \ldots, r)$. Starting with $\hat{x}^i(k-1|k-1)$ one computes $\hat{x}^{ij}(k|k)$ and the associated covariance $P^{ij}(k|k)$ through a filter matched to $M_j(k)$. The likelihood functions corresponding to these r^2 filters

$$\Lambda_{ij}(k) = p[z(k)|M_j(k), M_i(k-1), Z^{k-1}] \qquad (11.6.5\text{-}5)$$

are evaluated as

$$\boxed{\Lambda_{ij}(k) = p[z(k)|M_j(k), \hat{x}^i(k-1|k-1), P^i(k-1|k-1)] \qquad i,j = 1, \ldots, r} \quad (11.6.5\text{-}6)$$

2. **Calculation of the merging probabilities** $(i, j = 1, \ldots, r)$. The probability that mode i was in effect at $k-1$ if mode j is in effect at k is, conditioned on Z^k

$$
\begin{aligned}
\mu_{i|j}(k-1|k) \;&\triangleq\; P\{M_i(k-1)|M_j(k), Z^k\} \\
&= P\{M_i(k-1)|z(k), M_j(k), Z^{k-1}\} \\
&= \frac{1}{c_j} P[z(k), M_j(k)|M_i(k-1), Z^{k-1}] P\{M_i(k-1)|Z^{k-1}\} \\
&= \frac{1}{c_j} p[z(k)|M_j(k), M_i(k-1), Z^{k-1}] \\
&\quad \cdot P\{M_j(k)|M_i(k-1), Z^{k-1}\} P\{M_i(k-1)|Z^{k-1}\} \qquad (11.6.5\text{-}7)
\end{aligned}
$$

Thus the **merging probabilities** are

$$\boxed{\mu_{i|j}(k-1|k) = \frac{1}{c_j} \Lambda_{ij}(k) p_{ij} \mu_i(k-1) \qquad i,j = 1, \ldots, r} \quad (11.6.5\text{-}8)$$

where $P[\cdot]$ denotes a mixed pdf-probability and

$$c_j = \sum_{i=1}^{r} \Lambda_{ij}(k) p_{ij} \mu_i(k-1) \qquad (11.6.5\text{-}9)$$

The **mode transition probabilities** p_{ij} are assumed to be known — their selection is part of the algorithm design process.

3. **Merging** $(j = 1, \ldots, r)$. The state estimate corresponding to $M_j(k)$ is obtained by combining the estimates (11.6.5-2) according to the inner summation in (11.6.5-1) as follows

$$\boxed{\hat{x}^j(k|k) = \sum_{i=1}^{r} \hat{x}^{ij}(k|k)\mu_{i|j}(k-1|k) \qquad j = 1, \ldots, r} \qquad (11.6.5\text{-}10)$$

The covariance corresponding to the above is

$$\boxed{P^j(k|k) = \sum_{i=1}^{r} \mu_{i|j}(k-1|k)\Big\{P^{ij}(k|k) + [\hat{x}^{ij}(k|k) - \hat{x}^j(k|k)][\hat{x}^{ij}(k|k) - \hat{x}^j(k|k)]'\Big\}}$$
$$(11.6.5\text{-}11)$$

4. **Mode probability updating** $(j = 1, \ldots, r)$. This is done as follows

$$
\begin{aligned}
\mu_j(k) \;&\triangleq\; P\{M_j(k)|z(k), Z^{k-1}\} \\
&= \frac{1}{c}\, P[z(k), M_j(k)|Z^{k-1}] \\
&= \frac{1}{c}\sum_{i=1}^{r} P[z(k), M_j(k)|M_i(k-1), Z^{k-1}]P\{M_i(k-1)|Z^{k-1}\} \\
&= \frac{1}{c}\sum_{i=1}^{r} p(z(k)|M_j(k), M_i(k-1), Z^{k-1})P\{M_j(k)|M_i(k-1), Z^{k-1}\}\mu_i(k-1)
\end{aligned}
$$
$$(11.6.5\text{-}12)$$

Thus the updated **mode probabilities** are

$$\boxed{\mu_j(k) = \frac{1}{c}\sum_{i=1}^{r} \Lambda_{ij}(k)p_{ij}\mu_i(k-1) = \frac{c_j}{c} \qquad j = 1, \ldots, r} \qquad (11.6.5\text{-}13)$$

where c_j is the expression from (11.6.5-10) and c is the normalization constant

$$c = \sum_{j=1}^{r} c_j \qquad (11.6.5\text{-}14)$$

5. **State estimate and covariance combination**. The latest state estimate and covariance *for output only* are

$$\boxed{\hat{x}(k|k) = \sum_{j=1}^{r} \hat{x}^j(k|k)\mu_j(k)} \qquad (11.6.5\text{-}15)$$

$$\boxed{P(k|k) = \sum_{j=1}^{r} \mu_j(k)\Big\{P^j(k|k) + [\hat{x}^j(k|k) - \hat{x}(k|k)][\hat{x}^j(k|k) - \hat{x}(k|k)]'\Big\}} \qquad (11.6.5\text{-}16)$$

11.6.6   The Interacting Multiple Model Algorithm

In the **interacting multiple model (IMM)** approach, at time k the state estimate is computed under *each possible current model* using r filters, with each filter using a different combination of the previous model-conditioned estimates — *mixed initial condition.*

The total probability theorem is used as follows to yield r filters running in parallel:

$$p[x(k)|Z^k] = \sum_{j=1}^r p[x(k)|M_j(k), Z^k]P\{M_j(k)|Z^k\} = \sum_{j=1}^r p[x(k)|M_j(k), z(k), Z^{k-1}]\mu_j(k)$$

$$(11.6.6\text{-}1)$$

The model-conditioned posterior pdf of the state, given by

$$p[x(k)|M_j(k), z(k), Z^{k-1}] = \frac{p[z(k)|M_j(k), x(k)]}{p[z(k)|M_j(k), Z^{k-1}]} p[x(k)|M_j(k), Z^{k-1}] \qquad (11.6.6\text{-}2)$$

reflects one cycle of the state estimation filter matched to model $M_j(k)$ starting with the prior, which is the last term above.

The total probability theorem is now applied to the last term above (the prior), yielding

$$p[x(k)|M_j(k), Z^{k-1}] = \sum_{i=1}^r p[x(k)|M_j(k), M_i(k-1), Z^{k-1}]P\{M_i(k-1)|M_j(k), Z^{k-1}\}$$

$$\approx \sum_{i=1}^r p\Big[x(k)|M_j(k), M_i(k-1), \{\hat{x}^l(k-1|k-1), P^l(k-1|k-1)\}_{l=1}^r\Big]\mu_{i|j}(k-1|k-1)$$

$$= \sum_{i=1}^r p[x(k)|M_j(k), M_i(k-1), \hat{x}^i(k-1|k-1), P^i(k-1|k-1)]\mu_{i|j}(k-1|k-1)$$

$$(11.6.6\text{-}3)$$

The second line above reflects the approximation that the past through $k-1$ is summarized by r model-conditioned estimates and covariances. The last line of (11.6.6-3) is a mixture with weightings, denoted as $\mu_{i|j}(k-1|k-1)$, different for each current model $M_j(k)$. This mixture is assumed to be a mixture of Gaussian pdfs (a Gaussian sum) and then approximated via moment matching by a single Gaussian (details given later):

$$p[x(k)|M_j(k), Z^{k-1}] = \sum_{i=1}^r \mathcal{N}\Big[x(k); E[x(k)|M_j(k), \hat{x}^i(k-1|k-1)], \text{cov}[\cdot]\Big]\mu_{i|j}(k-1|k-1)$$

$$\approx \mathcal{N}\Big[x(k); \sum_{i=1}^r E\Big[x(k)|M_j(k), \hat{x}^i(k-1|k-1)\Big]\mu_{i|j}(k-1|k-1), \text{cov}[\cdot]\Big]$$

$$= \mathcal{N}\Big[x(k); E\Big[x(k)|M_j(k), \sum_{i=1}^r \hat{x}^i(k-1|k-1)\mu_{i|j}(k-1|k-1)\Big], \text{cov}[\cdot]\Big]$$

$$(11.6.6\text{-}4)$$

The last line above follows from the linearity of the Kalman filter and amounts to the following:

The input to the filter matched to model j is obtained from an **interaction** of the r filters, which consists of the **mixing** of the estimates $\hat{x}^i(k-1|k-1)$ with the weightings (probabilities) $\mu_{i|j}(k-1|k-1)$, called the **mixing probabilities**.

The above is equivalent to hypothesis merging *at the beginning* of each estimation cycle [BB88]. More specifically, the r hypotheses, instead of "fanning out" into r^2 hypotheses (as in the GPB2 — see Figure 11.6.5-1), are "mixed" into a new set of r hypotheses as shown in Figure 11.6.6-1. This is the key feature that yields r hypotheses with r filters, rather than with r^2 filters as in the GPB2 algorithm.

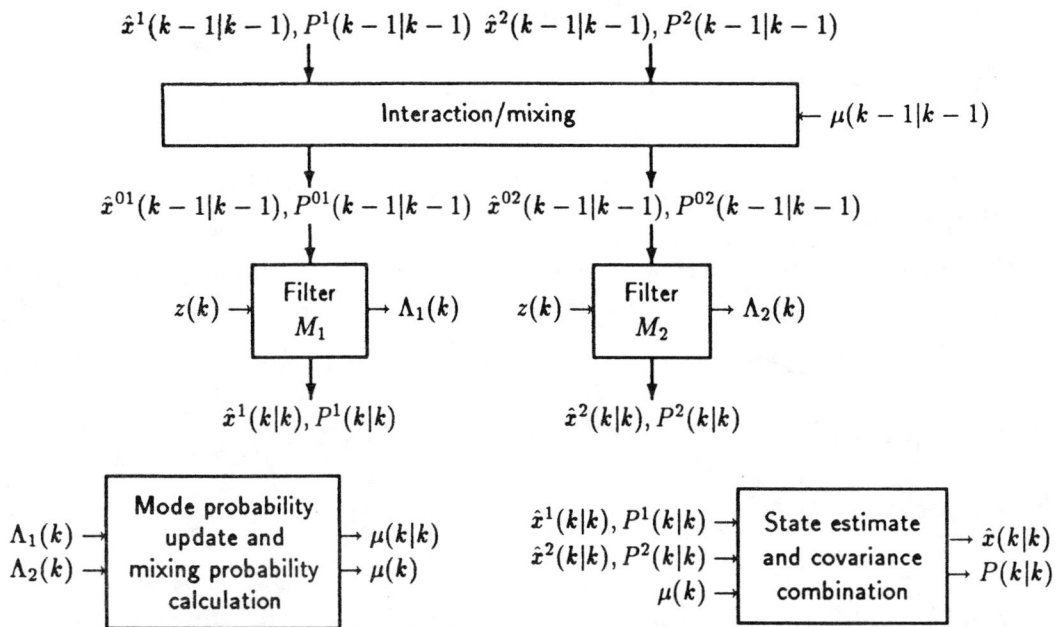

Figure 11.6.6-1: The IMM algorithm (one cycle).

Figure 11.6.6-1 describes this algorithm, which consists of r interacting filters operating in parallel. The mixing is done at the input of the filters with the probabilities, detailed later in (11.6.6-6), conditioned on Z^{k-1}. In contrast to this, the GPB2 algorithm has r^2 filters and a somewhat similar mixing is done, but at their outputs, with the probabilities (11.6.5-7), conditioned on Z^k.

The structure of the IMM algorithm is

$$(N_e; N_f) = (r; r) \qquad (11.6.6\text{-}5)$$

where N_e is the *number of estimates* at the start of the cycle of the algorithm and N_f is the *number of filters* in the algorithm.

The Algorithm

One cycle of the algorithm consists of the following:

1. ***Calculation of the mixing probabilities*** $(i, j = 1, \ldots, r)$. The probability that mode M_i was in effect at $k - 1$ given that M_j is in effect at k conditioned on Z^{k-1} is

$$
\begin{aligned}
\mu_{i|j}(k-1|k-1) &\triangleq P\{M_i(k-1)|M_j(k), Z^{k-1}\} \\
&= \frac{1}{\bar{c}_j} P\{M_j(k)|M_i(k-1), Z^{k-1}\} P\{M_i(k-1)|Z^{k-1}\}
\end{aligned}
$$

$$(11.6.6\text{-}6)$$

The above are the **mixing probabilities**, which can be written as

$$
\boxed{\mu_{i|j}(k-1|k-1) = \frac{1}{\bar{c}_j} p_{ij}\mu_i(k-1) \qquad i, j = 1, \ldots, r}
$$

$$(11.6.6\text{-}7)$$

where the normalizing constants are

$$
\bar{c}_j = \sum_{i=1}^{r} p_{ij}\mu_i(k-1) \qquad j = 1, \ldots, r
$$

$$(11.6.6\text{-}8)$$

Note the difference between (11.6.6-6), where the conditioning is Z^{k-1}, and (11.6.5-7), where the conditioning is Z^k. This is what makes it possible to carry out the mixing at the *beginning* of the cycle, rather than the standard merging at the *end* of the cycle.

2. **Mixing** $(j = 1, \ldots, r)$. Starting with $\hat{x}^i(k-1|k-1)$ one computes the mixed initial condition for the filter matched to $M_j(k)$ according to (11.6.6-4) as

$$
\boxed{\hat{x}^{0j}(k-1|k-1) = \sum_{i=1}^{r} \hat{x}^i(k-1|k-1)\mu_{i|j}(k-1|k-1) \qquad j = 1, \ldots, r}
$$

$$(11.6.6\text{-}9)$$

The covariance corresponding to the above is

$$
\boxed{
\begin{aligned}
P^{0j}(k-1|k-1) = \sum_{i=1}^{r} \mu_{i|j}(k-1|k-1) & \Big\{ P^i(k-1|k-1) \\
& + \big[\hat{x}^i(k-1|k-1) - \hat{x}^{0j}(k-1|k-1)\big] \\
& \cdot \big[\hat{x}^i(k-1|k-1) - \hat{x}^{0j}(k-1|k-1)\big]' \Big\} \\
& j = 1, \ldots, r
\end{aligned}
}
$$

$$(11.6.6\text{-}10)$$

3. **Mode-matched filtering** $(j = 1, \ldots, r)$. The estimate (11.6.6-9) and covariance (11.6.6-10) are used as input to the filter matched to $M_j(k)$, which uses $z(k)$ to yield $\hat{x}^j(k|k)$ and $P^j(k|k)$.

The likelihood functions corresponding to the r filters

$$\Lambda_j(k) = p[z(k)|M_j(k), Z^{k-1}] \tag{11.6.6-11}$$

are computed using the mixed initial condition (11.6.6-9) and the associated covariance (11.6.6-10) as

$$\Lambda_j(k) = p[z(k)|M_j(k), \hat{x}^{0j}(k-1|k-1), P^{0j}(k-1|k-1)]$$
$$j = 1, \ldots, r \tag{11.6.6-12}$$

that is,

$$\boxed{\Lambda_j(k) = \mathcal{N}\Big[x(k); \hat{z}^j[k|k-1; \hat{x}^{0j}(k-1|k-1)], S^j[k; P^{0j}(k-1|k-1)]\Big] \qquad j = 1, \ldots, r}$$
$$\tag{11.6.6-13}$$

4. **Mode probability update** $(j = 1, \ldots, r)$. This is done as follows:

$$\begin{aligned}
\mu_j(k) &\triangleq P\{M_j(k)|Z^k\} \\
&= \frac{1}{c}\, p[z(k)|M_j(k), Z^{k-1}]P\{M_j(k)|Z^{k-1}\} \\
&= \frac{1}{c}\, \Lambda_j(k) \sum_{i=1}^{r} P\{M_j(k)|M_i(k-1), Z^{k-1}\}P\{M_i(k-1)|Z^{k-1}\} \\
&= \frac{1}{c}\, \Lambda_j(k) \sum_{i=1}^{r} p_{ij}\mu_i(k-1) \qquad j = 1, \ldots, r
\end{aligned} \tag{11.6.6-14}$$

or

$$\boxed{\mu_j(k) = \frac{1}{c}\, \Lambda_j(k)\bar{c}_j \qquad j = 1, \ldots, r} \tag{11.6.6-15}$$

where $\bar{c}_j$ is the expression from (11.6.6-8) and

$$c = \sum_{j=1}^{r} \Lambda_j(k)\bar{c}_j \tag{11.6.6-16}$$

is the normalization constant for (11.6.6-15).

5. ***Estimate and covariance combination.*** Combination of the model-conditioned estimates and covariances is done according to the mixture equations

$$\hat{x}(k|k) = \sum_{j=1}^{r} \hat{x}^j(k|k)\mu_j(k)$$

(11.6.6-17)

$$P(k|k) = \sum_{j=1}^{r} \mu_j(k)\left\{P^j(k|k) + [\hat{x}^j(k|k) - \hat{x}(k|k)][\hat{x}^j(k|k) - \hat{x}(k|k)]'\right\}$$

(11.6.6-18)

Note that this combination is only for output purposes — it is not part of the algorithm recursions.

11.6.7 An Example with the IMM

The use of the IMM is illustrated on the example simulated in Section 11.5 where several of the earlier techniques were compared. The results presented in the sequel deal with the turn of 90° over 20 sampling periods.

A two-model IMM, designated as IMM2, was first used. This algorithm consisted of

1. A constant velocity model (second order, with no process noise); and
2. A Wiener process acceleration model (third order model) with process noise (acceleration increment over a sampling period) $q = 10^{-3} = 0.316^2$.

Note that the acceleration in this case ($0.075m/s^2$) corresponds to about $2.4\sqrt{q}$.

The Markov chain transition matrix between these models was taken as

$$[p_{ij}] = \begin{bmatrix} 0.95 & 0.05 \\ 0.05 & 0.95 \end{bmatrix} \tag{11.6.7-1}$$

The final results were not very sensitive to these values (e.g., p_{11} can be between 0.8 and 0.98). The lower (higher) value will yield less (more) peak error during maneuver but higher (lower) RMS error during the quiescent period, that is, it has a higher (lower) bandwidth.

A three-model IMM, designated as IMM3, was also used. This consisted of the above two models plus another one:

3. A constant acceleration (third order) model without process noise.

The Markov chain transition matrix was

$$[p_{ij}] = \begin{bmatrix} 0.95 & 0.05 & 0 \\ 0.33 & 0.34 & 0.33 \\ 0 & 0.05 & 0.95 \end{bmatrix} \tag{11.6.7-2}$$

Figures 11.6.7-1 and 11.6.7-2 show the position (coordinate ξ) RMS error from 50 Monte Carlo runs for the IMM2 and IMM3, respectively, with the estimator design parameters as indicated above. Comparing with Figure 11.5.3-1a, it can be seen that the peak errors are approximately equal to the measurement noise standard deviation (which is $100m$), that is, substantially smaller than with the IE or VSD. During the nonaccelerating period the errors are somewhat larger, but still there is an *RMS **noise reduction factor*** of 2 — this corresponds to a *variance **noise reduction factor*** of 4.

Similar results for the velocity (in coordinate ξ) can be observed by comparing Figures 11.6.7-3 and 11.6.7-4 with Figure 11.5.3-1b.

466

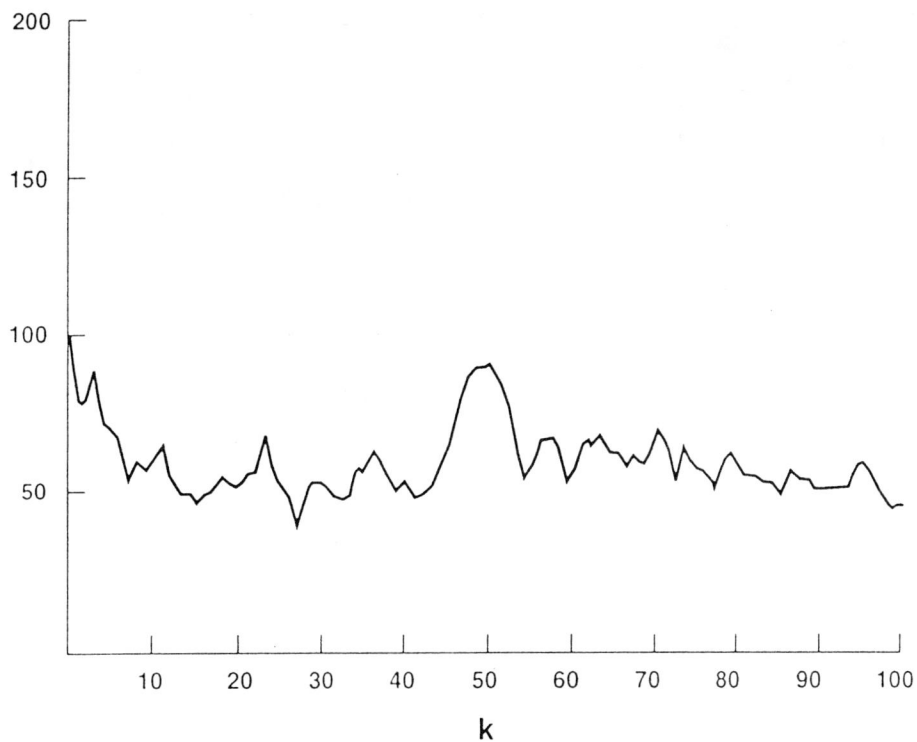

Figure 11.6.7-1: Position RMS error for IMM2.

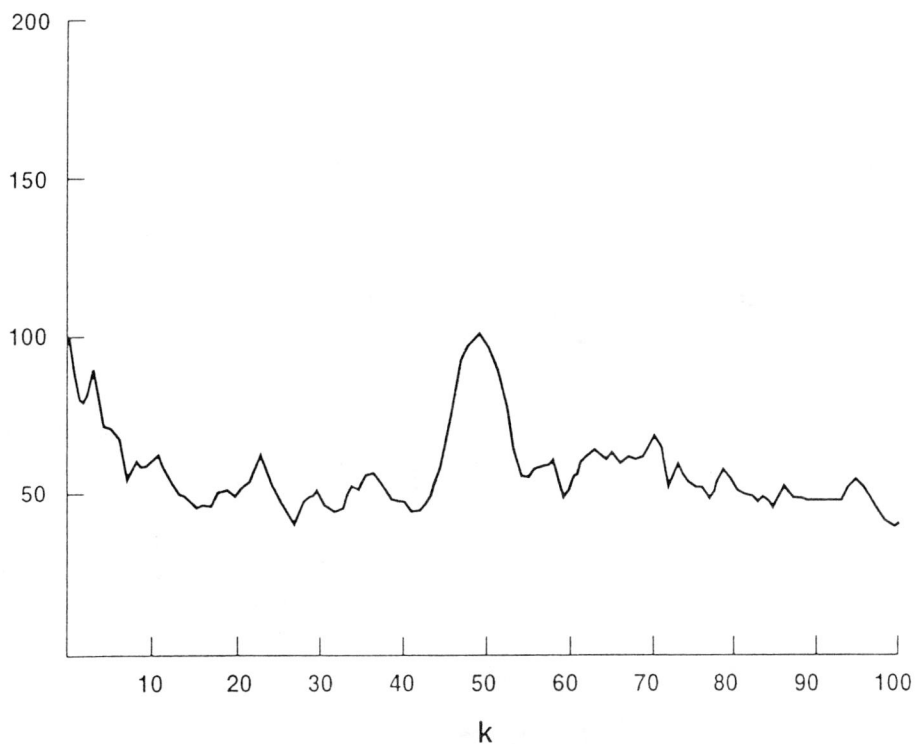

Figure 11.6.7-2: Position RMS error for IMM3.

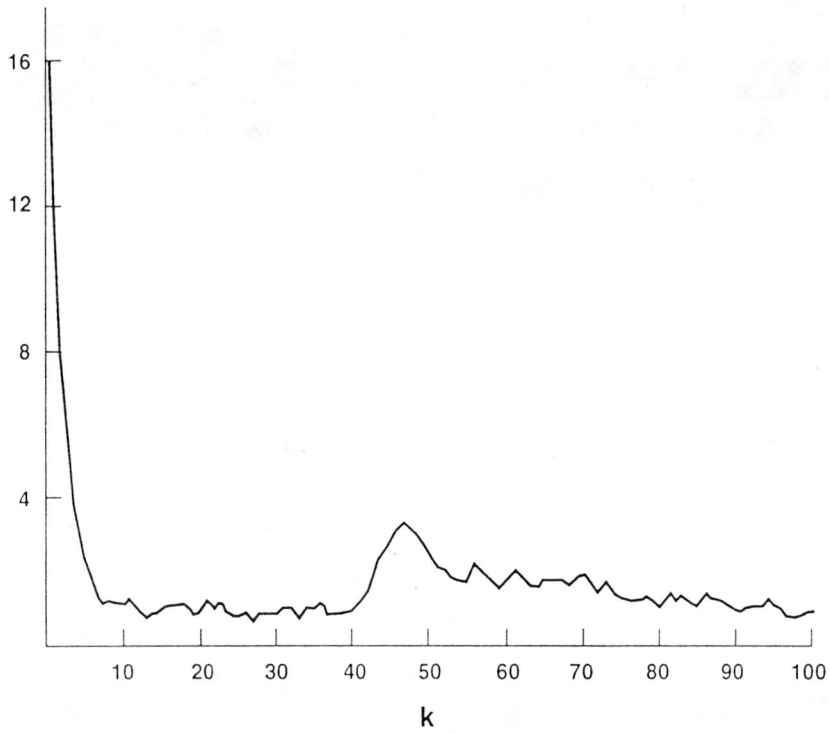

Figure 11.6.7-3: Velocity RMS error for IMM2.

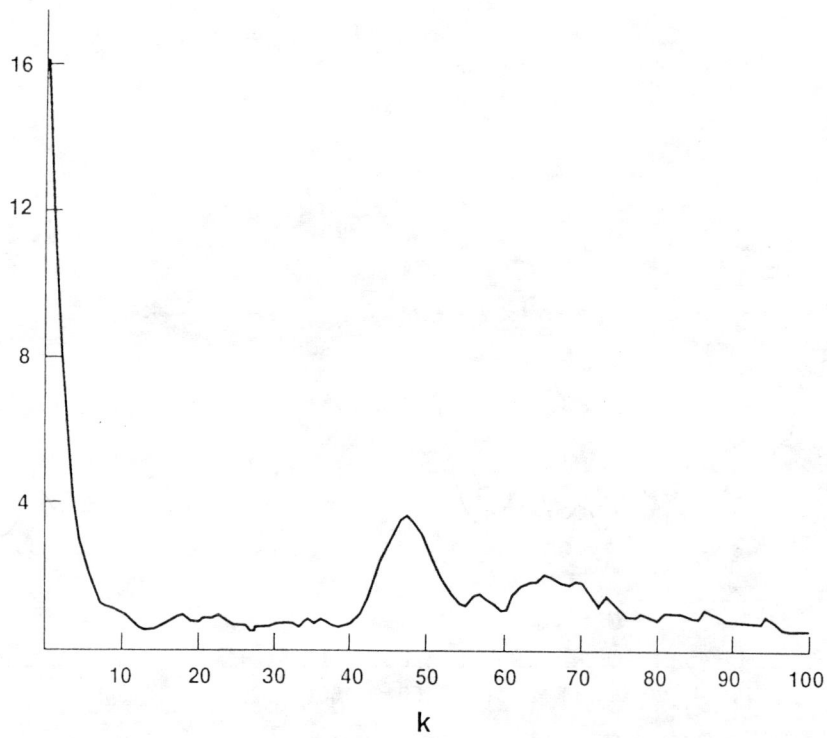

Figure 11.6.7-4: Velocity RMS error for IMM3.

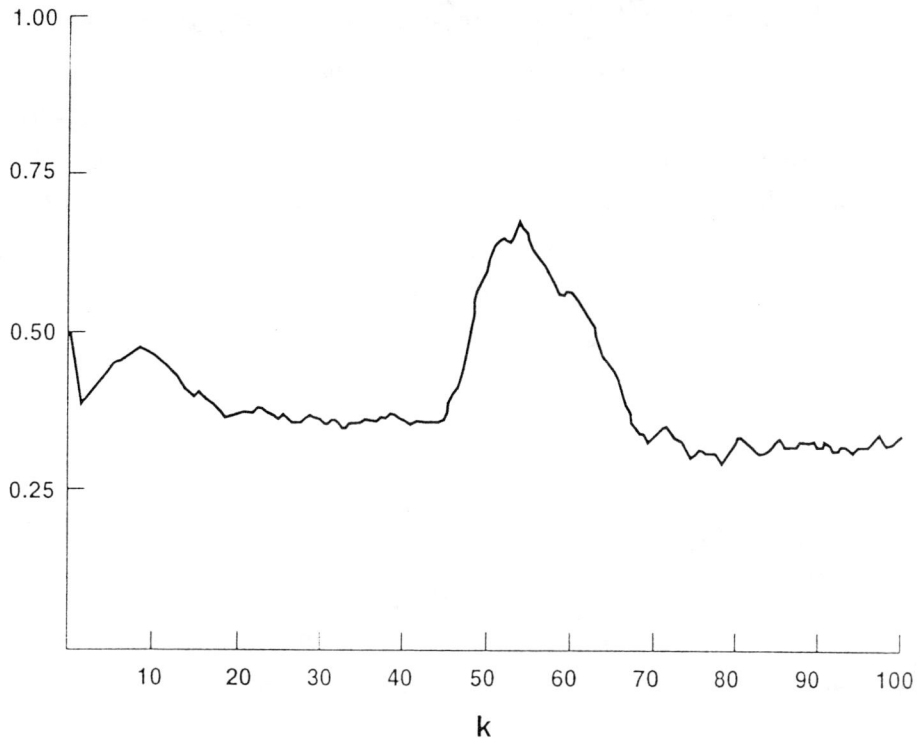

Figure 11.6.7-5: Maneuvering mode probability in IMM2.

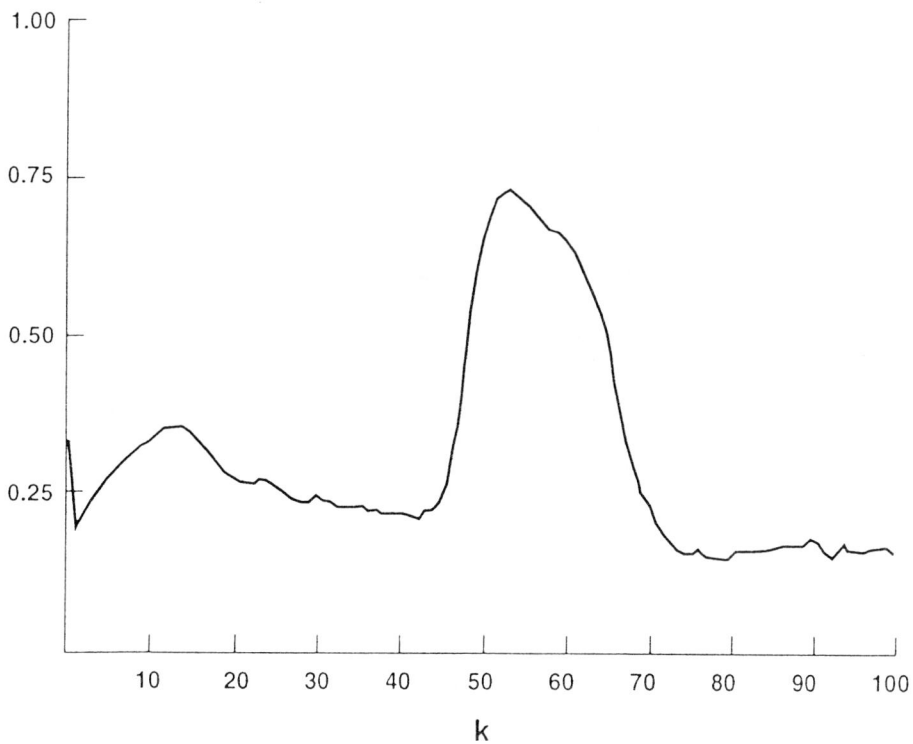

Figure 11.6.7-6: Maneuvering mode probability in IMM3.

Overall, the IMM2 and the IMM3 perform similarly in the position estimation; the IMM3 has a somewhat better velocity estimation capability.

Figures 11.6.7-5 and 11.6.7-6 show the evolution in time of the maneuvering mode probabilities in IMM2 and IMM3, respectively. These figures illustrate the ***soft switching*** that takes place when a maneuver is initiated: it takes a few samples in this case to "detect" the maneuver. The "detection" of a maneuver manifests itself here as a sharp increase in the probability of the maneuvering mode — mode 2 in IMM2 and mode 3 in IMM3.

Remark

These results, from [BCB89], are not only superior to the IE technique of Section 11.3 (illustrated in Subsection 11.5.3), but also superior to the augmented version of the IE algorithm that estimates the maneuver onset time [Bog87].

This shows that the IMM, at the cost of about 3 Kalman filters is preferable to the augmented IE technique, which has a much larger implementation complexity — of 30 to 100 Kalman filters.

11.6.8   Example — Design of an IMM Estimator for ATC Tracking

Stochastic Hybrid Systems and Motion Models

In **air traffic control (ATC)**, civilian aircraft have two basic modes of flight:

- **Uniform motion (UM)** — the straight and level flight with a constant speed and heading, and
- **Maneuver** — turning or climbing/descending.

A class of **stochastic hybrid systems** with additive noise can be described as

$$x(k) = f[k-1, x(k-1), M(k)] + g\Big[k-1, x(k-1), v[k-1, M(k)], M(k)\Big] \quad (11.6.8\text{-}1)$$

with noisy measurements

$$z(k) = h[k, x(k), M(k)] + w[k, M(k)] \quad (11.6.8\text{-}2)$$

and the "system mode" $M(k)$ is a homogeneous Markov chain with probabilities of transition given by

$$P\{M_j(k+1)|M_i(k)\} = p_{ij} \qquad \forall M_i, M_j \in \mathbf{M} \quad (11.6.8\text{-}3)$$

where

- $x(k) \in R^{n_x}$ is the continuous-valued **base state** vector at k, such as the position, velocity, and turn rate of the aircraft;
- $z(k) \in R^{n_z}$ is the vector-valued noisy measurements at k, such as the radar reports;
- $M(k)$ is the scalar-valued **modal state** (system mode index) at k;
- $\mathbf{M}$ is the set of modal states (i.e., models);
- $v[k-1, M(k)] \in R^{n_v}$ is the (mode-dependent) process noise sequence with mean $\bar{v}[k-1, M(k)]$ and covariance $Q[k-1, M(k)]$;
- $w[k, M(k)] \in R^{n_z}$ is the (mode-dependent) measurement noise sequence with mean $\bar{w}[k, M(k)]$ and covariance $R[k, M(k)]$;
- f, g, and h are (nonlinear) vector-valued functions for the problem considered.

In ATC tracking, $M(k)$ stands for the mode of flight at time k. Equation (11.6.8-2) implies that the base state observations are noisy and mode dependent. Therefore, the mode information is imbedded in the measurement sequence. In other words, the modal state is an **indirectly observed Markov chain**. For radar measurements, in fact, the measurement noise are also base-state dependent.

471

The flight modes in the *horizontal plane* can be modeled by

- A second order kinematic (nearly constant velocity) model for the uniform motion,
- A *nearly "coordinated turn"* model for the maneuver.

The nearly constant velocity model is given by

$$x(k) = \begin{bmatrix} 1 & T & 0 & 0 \\ 0 & 1 & 0 & 0 \\ 0 & 0 & 1 & T \\ 0 & 0 & 0 & 1 \end{bmatrix} x(k-1) + \begin{bmatrix} \frac{1}{2}T^2 & 0 \\ T & 0 \\ 0 & \frac{1}{2}T^2 \\ 0 & T \end{bmatrix} v(k-1) \tag{11.6.8-4}$$

where T is the sampling interval; x is the state of the aircraft, defined as

$$x = [\xi \ \dot{\xi} \ \eta \ \dot{\eta}]' \tag{11.6.8-5}$$

with ξ and η denoting the Cartesian coordinates of the horizontal plane; and v is a zero-mean Gaussian white noise used to model ("cover") small accelerations, the turbulence, wind change, and so on with an appropriate covariance Q, which is a design parameter.

The turning of a civilian aircraft usually follows a pattern known as **"coordinated turn"** (Subsection 4.2.2) — a turn with a constant turn rate and a constant speed. Although the actual turning of a civilian aircraft is not exactly "coordinated" since the ground speed is the airspeed plus the wind speed, it can be suitably described by the "coordinated turn" model plus a fairly small noise representing the modeling error, resulting in the nearly coordinated turn model. Such a model is necessarily a nonlinear one if the turn rate is not a known constant. Augmenting the state vector (11.6.8-5) by one more component — the turn rate ω, that is,

$$x = [\xi \ \dot{\xi} \ \eta \ \dot{\eta} \ \omega]' \tag{11.6.8-6}$$

the **nearly coordinated turn model** is then given by

$$x(k) = \begin{bmatrix} 1 & \dfrac{\sin \omega T}{\omega} & 0 & -\dfrac{1 - \cos \omega T}{\omega} & 0 \\ 0 & \cos \omega T & 0 & -\sin \omega T & 0 \\ 0 & \dfrac{1 - \cos \omega T}{\omega} & 1 & \dfrac{\sin \omega T}{\omega} & 0 \\ 0 & \sin \omega T & 0 & \cos \omega T & 0 \\ 0 & 0 & 0 & 0 & 1 \end{bmatrix} x(k-1) + \begin{bmatrix} \frac{1}{2}T^2 & 0 \\ T & 0 \\ 0 & \frac{1}{2}T^2 \\ 0 & T \\ 0 & 0 \end{bmatrix} v(k-1) + \begin{bmatrix} 0 \\ 0 \\ 0 \\ 0 \\ 1 \end{bmatrix} u(k) \tag{11.6.8-7}$$

where u is the control input for the turn rate.

Note that the process noise v in (11.6.8-4) and (11.6.8-7) has in general different statistics to reflect such factors as different modeling errors. That is exactly why the process noise v in (11.6.8-1) should be mode dependent in general.

Assuming only position measurements are available to reflect the radar technology currently in use for ATC, this yields the following observation equation

$$z(k) = \begin{bmatrix} 1 & 0 & 0 & 0 & 0 \\ 0 & 0 & 1 & 0 & 0 \end{bmatrix} x(k) + w(k) \tag{11.6.8-8}$$

Although the original radar measurements are in polar coordinates, they can be converted to the Cartesian coordinates by

$$z^1 = r \cos \theta \tag{11.6.8-9}$$

$$z^2 = r \sin \theta \tag{11.6.8-10}$$

where $z = [z^1 \ z^2]'$; r and θ are the range and azimuth measurements, respectively. The measurement errors in polar coordinates can be converted accordingly. The first order approximation of the covariance of these converted measurement errors can be found in [AM79, Bla86]. An equivalent but more revealing form is the following (see problem 10-3)

$$R = \text{cov}[w] \simeq \frac{\sigma_r^2 - r^2 \sigma_\theta^2}{2} \begin{bmatrix} b + \cos 2\theta & \sin 2\theta \\ \sin 2\theta & b - \cos 2\theta \end{bmatrix} \tag{11.6.8-11}$$

in which σ_r and σ_θ are the standard deviations of the range measurement and the azimuth measurement, respectively, and

$$b \triangleq \frac{\sigma_r^2 + r^2 \sigma_\theta^2}{\sigma_r^2 - r^2 \sigma_\theta^2} \tag{11.6.8-12}$$

The approximation (11.6.8-11) is good when $r\sigma_\theta^2/\sigma_r < 0.4$ [LB93a]. Otherwise, the debiased consistent transformation technique of [LB93a] can be used instead. This technique guarantees the accuracy of the conversion for tracking purpose for all practical sensors.

Clearly, (11.6.8-4) and (11.6.8-7) are both special forms of (11.6.8-1), and (11.6.8-8) is a special form of (11.6.8-2). Also, it can be assumed that the transition of the mode of flight has the Markovian property, governed by (11.6.8-3). Consequently, the kinematic behavior of civilian aircraft can be suitably described in the framework of the stochastic hybrid systems and therefore the IMM algorithm can be used.

Selection of Models and Parameters

To obtain the best possible results, the IMM algorithm has to be properly designed to meet the following requirements of the ATC tracking simultaneously:

- Reduce as much as possible the estimation errors during the uniform motion;
- Maintain the peak estimation error during the maneuver lower than that of the unfiltered raw measurements;
- Provide correct and timely indication of the flight mode, especially rapid detection of the maneuver.

These requirements are fulfilled by means of

- Design of aircraft motion models for all modes of flight;
- Selection of the model parameters, such as the noise levels;
- Determination of the parameters of the underlying Markov chain, that is, the transition probabilities, defined in (11.6.8-3).

Model selection should consider both the quality and complexity of the model. Typically, the models used in the IMM configuration for ATC tracking will include one (a nearly constant velocity model) for the uniform motion and one or more for the maneuver — the nearly coordinated turn model of (11.6.8-7) is a typical one.

Selection of noise levels for each model is an important part of the estimator design. Although the uniform motion is better modeled in (11.6.8-4) with a small process noise v to model the air turbulence, winds aloft changes, and so forth, it may be legitimate to use a somewhat larger process noise for the uniform motion to cover slow turns as well as small linear accelerations so as to ease the burden of modeling a broad range of maneuvers. The right choice of the noise level of the nearly coordinated turn model depends on what turn rate range is expected and how many models are to be used for the maneuvers.

The performance of the IMM algorithm is not very sensitive to the *choice of the transition probabilities*. However, this choice provides to a certain degree the trade-off between the peak estimation errors at the onset of the maneuver and the maximum reduction of the estimation errors during the uniform motion. The guideline for a proper choice is to match roughly the transition probabilities with the actual mean sojourn time of each mode (see problem 11-6).

11.6.8 Example — Design of an IMM Estimator for ATC Tracking

As a generic ATC tracking problem, the following scenario is considered. The radar provides position only measurements (after the polar-to-Cartesian conversion) with RMS errors of $100m$ in each of the two Cartesian coordinates. The intervals between the samples are $T = 5$ seconds. In such a scenario the aircraft, with an assumed speed of $120m/s$ (240 knots) and a standard turn rate of $3°/s$ ($0.6g$), executes a $90°$ turn in 6 sampling periods. This scenario leads to a maneuvering index (see Chapter 6) that is quite high (almost 1.5) and thus very little noise reduction can be achieved by a single model based state estimator.

Expedite turn rates of $5°/s$ and $10°/s$, corresponding to $1g$ and $2g$ accelerations, respectively, are also considered. It should be noted that turns with such large accelerations occur only in an emergency situation for a civilian aircraft.

The following IMM configurations were considered first:

1. Two second order linear kinematic models with two noise levels. The one with the lower noise level is used to model the uniform motion and the other one for the maneuvers.

2. One second order kinematic model (a nearly constant velocity model) for the uniform motion and a nearly coordinated turn model with the turn rate being estimated using the least squares (LS) method.

The first order extended Kalman filter is used for the nearly coordinated turn model.
The main findings of this study are listed below:

- The accuracy of the turn rate estimate is not very important as far as the quality of the position, speed, and heading estimates are concerned. What is important is the correct and timely detection of the maneuver and the fast response of the filter to this detection.

- The linear model design (Configuration 1) does obtain acceptable results for maneuvers with turn rates up to $3°/s$. It does not, however, yield good estimates for faster turns.

- The LS estimate of the turn rate fluctuates severely around the true rate: it first underestimates the turn rate at the beginning of the maneuver and then, to correct this mistake, it yields an overestimated turn rate.

- The design with the LS fit (Configuration 2) does not provide better estimates, especially at the onset of the maneuver, than Configuration 1 with linear models.

Next, consider the following IMM configurations:

3. One nearly constant velocity model with noise level accounting for slow turns and small linear accelerations and one nearly coordinated turn model with the turn rate considered as a random variable of a certain mean and variance. This is designated as IMM2.

4. One nearly constant velocity model with noise level accounting for slow turns and small linear accelerations and *two* nearly coordinated turn models (CT1 and CT2, defined later in Table 11.6.8-1) with the turn rate assumed a random variable. The two maneuver models are different only in their assumed expected turn rates and variances. This is designated as IMM3.

To accelerate the detection and response of the IMM algorithm to the maneuver, the standard IMM configuration is modified as follows. The expected turn rate with an appropriate variance is *instantaneously* fed into the coordinated turn model(s) at each algorithm cycle in the above two designs (IMM2 and IMM3). This is done right after the interaction step, that is, after the estimates given by the filter that is based on the linear nearly constant velocity model are obtained.

With this instantaneous feeding, when the aircraft is in the uniform motion, the estimates given by the filter(s) based on the nearly coordinated turn model(s) have little effect on the overall (combined) estimate of the IMM algorithm (and on the interaction estimates for the filter based on the linear model at the next cycle) because their estimates have small weights. Consequently, while the aircraft is indeed in uniform motion, the filter matched to the uniform motion model is dominant and yields accurate estimates. Once the aircraft starts to maneuver, the instantaneous feeding of the expected turn rate facilitates the filters matched to the coordinated turn models to take over rapidly, leading to a significant reduction in the peak estimation errors.

Results and Discussion

Several sets of values of the expected turn rate $\bar{\omega}$ and the associated variance σ_ω have been examined for both IMM2 and IMM3. Results obtained from 100 Monte Carlo simulations are presented here with $\bar{\omega}$ and σ_ω listed in Table 11.6.8-1 below. These are estimator design parameters, which should be selected based on the expected scenarios and the designer's experience.

	IMM3		IMM2	IMM2
	CT1	CT2	Version 1	Version 2
Expected turn rate $\bar{\omega}$ (°/s)	3.0	9.0	6.0	3.0
Standard deviation σ_ω (°/s)	2.0	3.0	4.0	3.0

Table 11.6.8-1: Expected turn rates and their accuracies used in different models.

In Tables 11.6.8-2, 11.6.8-3, and 11.6.8-4, all the position estimation errors are for the ξ and η coordinates *combined*. The *maneuver detection delay* is defined to be the latency, measured in sampling periods, from the maneuver onset time to the time that the probability of the uniform motion mode falls below 0.5. The "UM probability error" stands for the (steady-state) probability of the maneuver modes during the uniform motion. In other words, this is the estimator-calculated probability that the UM mode is not in effect while, in truth, it is.

	Turn rate		
	3°/s	5°/s	10°/s
Peak position error (m)	132	133	144
UM position error (m)	80	80	80
Peak speed error (m/s)	6.4	10.8	27.0
UM speed error (m/s)	1.8	1.8	1.8
UM heading error (deg)	2.6	2.6	2.6
Maneuver detection delay (scans)	1	1	1
UM probability error (%)	5.0	5.0	4.5

Table 11.6.8-2: Estimation errors in IMM3 (raw position measurement error 141m).

477

	Turn rate		
	3°/s	5°/s	10°/s
Peak position error (m)	146	148	145
UM position error (m)	82	82	82
Peak speed error (m/s)	9.6	11.1	23.0
UM speed error (m/s)	2.0	2.0	2.0
UM heading error (deg)	2.8	2.8	2.8
Maneuver detection delay (scans)	2	1	1
UM probability error (%)	2.6	2.6	2.6

Table 11.6.8-3: Estimation errors in IMM2 with $\bar{\omega} = 6°/s$ and $\sigma_\omega = 4°/s$.

	Turn rate		
	1°/s	3°/s	5°/s
Peak position error (m)	122	131	131
UM position error (m)	84	84	84
Peak speed error (m/s)	3.5	6.0	13.7
UM speed error (m/s)	1.8	1.9	1.9
UM heading error (deg)	2.0	2.0	2.0
Maneuver detection delay (scans)	6	2	1
UM probability error (%)	5.0	5.0	5.0

Table 11.6.8-4: Estimation errors in IMM2 with $\bar{\omega} = 3°/s$ and $\sigma_\omega = 3°/s$.

As presented in Table 11.6.8-2, Configuration 4 (i.e., IMM3) yields a peak RMS position error during the mode of flight change that is almost 10% below the raw measurement error (except for the 10°/s case, for which it is about the same) and 45% RMS error [2] reduction during the uniform motion. The speed estimation error during the uniform motion is 1.5% of the aircraft speed. The detection of the maneuver is quick: in two scans (i.e., one scan delay) for all three turns.

---

[2]Square root of the sample average (mean) of the squared error from the Monte Carlo simulations.

11.6.8 Example — Design of an IMM Estimator for ATC Tracking

Since the coordinate-combined raw measurement error (1σ) is $100\sqrt{2} = 141m$, the angle deviation from the straight line in one scan (sampling period) during the uniform motion due to 1σ measurement error can be as large as

$$\tan^{-1} \frac{141m}{120m/s \cdot 5s} = 13.2° \qquad (11.6.8\text{-}13)$$

This is even slightly higher than the expected heading change caused by a $5°/s$ turn in one scan, which is $\frac{1}{2}(5°/s) \cdot (5s) = 12.5°$. In view of this, the UM probability error of only 5% seems quite low. The rapid detection of the maneuver, together with this small probability error, verifies the good reliability of this design in terms of providing the correct and timely information of the flight mode.

The results from the simpler version, IMM2, are listed in Tables 11.6.8-3 and 11.6.8-4 for two different choices of $\bar{\omega}$ and σ_ω. These results are similar to those in Table 11.6.8-2. For the easier case of a $1°/s$ turn, the peak errors are below the $3°/s$ case for all these three designs.

To cover turns with turn rates up to $10°/s$, the use of two coordinated turn models in IMM3 with different expected turn rates is clearly superior to having a single expected turn rate as in IMM2.

Allowing more latitude for the peak error during the $10°/s$ turn, would lead to even more reduction during the uniform motion as well as lower peak errors for both the $3°/s$ and $5°/s$ turns. This is illustrated in Table 11.6.8-4 for the second version of the IMM2 with the expected rate and the associated standard deviation listed in Table 11.6.8-1.

The accuracy of the IMM3 and IMM2 can be improved further if Doppler (range rate) and/or turn rate data is available. These error reductions cannot be achieved by any single-model filter, switching filter, or other adaptive scheme currently available.

If the sampling period is shorter, one can achieve more reduction of the errors in the state estimate compared to the unfiltered radar measurements during the maneuver and substantially more noise reduction can be achieved during the uniform motion.

Figure 11.6.8-1 presents the RMS position errors (ξ and η coordinates *combined*) for the IMM3 and the first version of the IMM2 for a 90° turn with a $3°/s$ rate starting at $k = 8$ (which can be noticed in the absence of noise for the first time at $k = 9$) as well as a baseline Kalman filter (tuned specially for this case). The initial large estimation errors are due to the fact that the initial probability of each model was set to be equal to account for the worst case of ignorance.

The following observations can be made:

IMM3 has the peak RMS position error of about $130m$, superior to the single measurement position RMS error, which is $100\sqrt{2} = 141m$. Moreover, when the aircraft is not maneuvering, the RMS error is around $80m$, that is, a reduction of about 45% in the RMS error (70% in MSE). The Kalman filter has a peak RMS position error of $140m$. During the uniform motion, however, its RMS error stays at about $120m$. This estimation error of the specially tuned Kalman filter for the uniform motion can be reduced only slightly at a cost of a much higher peak error. This is the typical behavior of a single-model based Kalman filter.

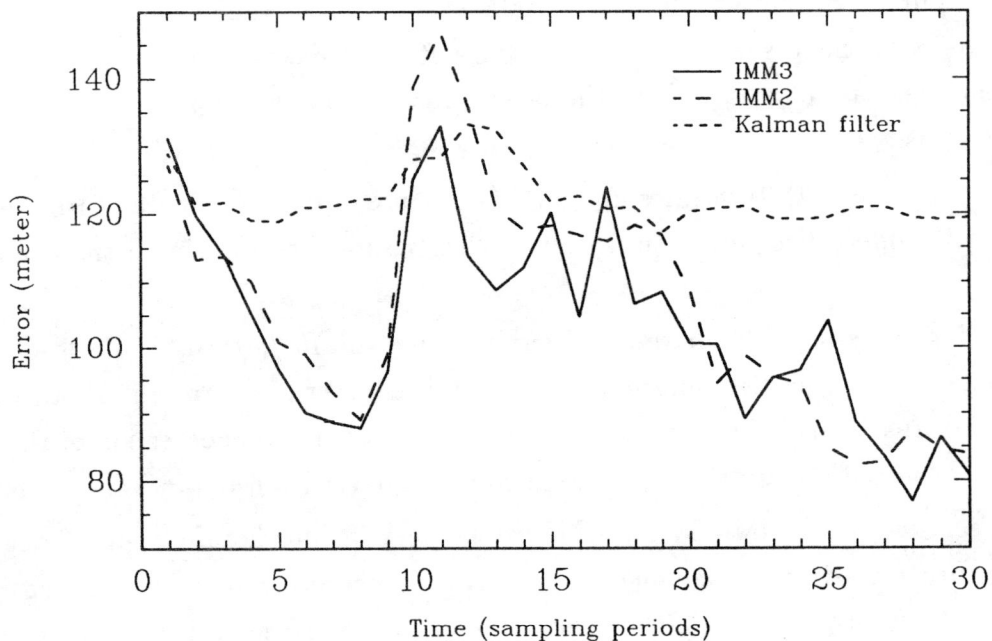

Figure 11.6.8-1: RMS position estimation errors for $3°/s$ turn rate.

Figure 11.6.8-2 shows the evolution of the average mode 2 (turning mode) probabilities in the first version of the IMM2. This indicates rapid "detection" — in two samples ($10s$) of the faster turns and in three samples ($15s$) of the standard ($3°/s$) turn. The more sophisticated IMM3 detects all the turns in two samples. Also, the less stringently designed IMM2 (version 2), with more error allowance for the $10°/s$ turn, has slightly better capability of maneuver detection.

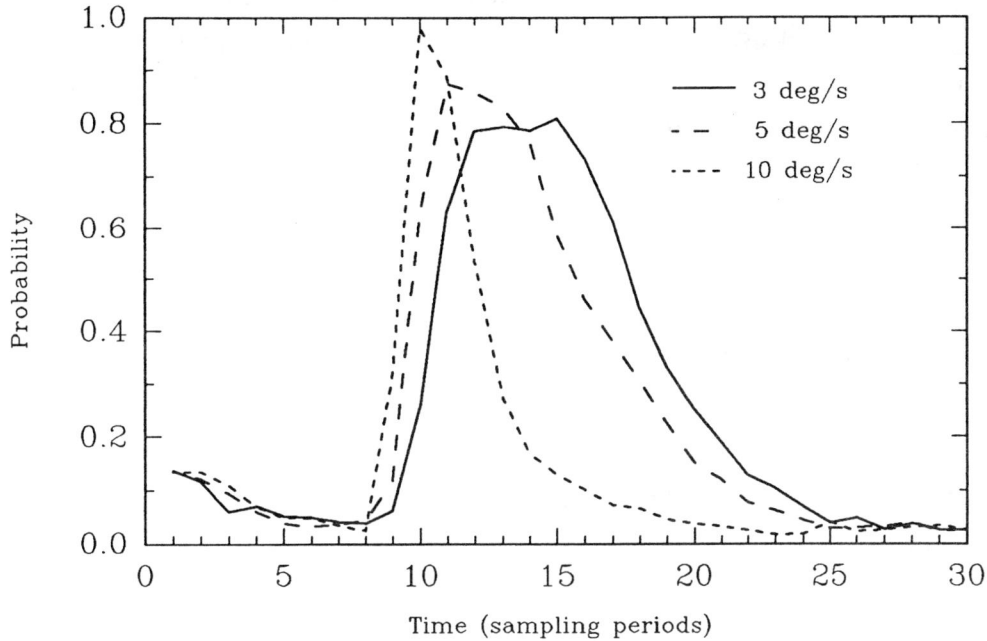

Figure 11.6.8-2: Mode 2 probabilities in IMM2 version 2 for a 90° turn starting at $k = 8$.

In view of the fact that the aircraft position measurements are usually available in polar coordinates (range and azimuth or bearing), the recently developed debiased consistent transformation technique [LB93a] can be used here. In this technique, the polar measurements are converted into the Cartesian coordinates in such a way that the consistence and unbiasedness of the converted measurements is guaranteed for all practical situations. Consequently, the mode-matched filters corresponding to the uniform motion in the IMM configuration can work in a linear setting, that is, on the basis of the (optimal) Kalman filter instead of the (suboptimal) extended Kalman filter.

No left turns are involved in the scenario considered here. In practical situations, both right and left turns are to be considered. This requires the use of additional models, which may lead to some deterioration in performance. It is possible that the variable structure technique of [LB92], where the mode set is changed in real time, can be used to improve the performance in this case.

11.6.9   The Multiple Model Approach — Summary

The *multiple model* or *hybrid system* approach assumes the system to be in one of a finite number of modes (i.e., that it is described by one out of a finite number of models).

Each model is characterized by its parameters — the models can differ in, for instance, the level of the process noise (its variance), a deterministic input, and/or in any other parameter (different dimension state vectors are also possible).

For the *fixed model* case the estimation algorithm consists of the following:

- For each model a filter "matched" to its parameters is yielding *model-conditioned estimates and covariances*.

- A *mode probability calculator* — a *Bayesian model comparator* — updates the probability of each mode using

  - the likelihood function (innovations) of each filter;
  - the prior probability of each model.

- An *Estimate combiner* computes the overall estimate and the associated covariance as the weighted sum of the model-conditioned estimates and the corresponding covariance — via the (Gaussian) mixture equations.

Ad hoc modifications of the fixed model (static) approach to handle switching models:

- Finite length window or fading memory for the likelihood functions;
- Imposing a lower bound on the probability of each model.

For systems that undergo changes in their mode during their operation — *mode jumping (model switching)* — one can obtain the *optimal multiple model estimator* which, however, consists of an exponentially increasing number of filters.

This is because the optimal approach requires conditioning on each *mode history* and their number is increasing exponentially. Thus, suboptimal algorithms are necessary for the (realistic) mode transition situation.

The *first order generalized pseudo-Bayesian* (GPB1) MM approach computes the state estimate accounting for each possible current model.

The *second order generalized pseudo-Bayesian* (GPB2) MM approach computes the state estimate accounting for

- each possible current model;
- each possible model at the previous time.

The *interacting multiple model* (IMM) approach computes the state estimate that accounts for *each possible current model* using a suitable mixing of the previous model-conditioned estimates depending on the current model.

These algorithms are **decision free** — no maneuver detection decision is needed — the algorithms undergo a **soft switching** according to the latest updated mode probabilities.

Table 11.6.9-1 presents a comparison of the complexities (in terms of the functions to be performed) of the static algorithm and the three algorithms presented for switching models.

	Static	*GPB1*	*GPB2*	*IMM*
Number of filters	r	r	r^2	r
Number of combinations of r estimates and covariances	1	1	$r+1$	$r+1$
Number of probability calculations	r	r	r^2+r	r^2+r

Table 11.6.9-1: Comparison of complexities of the MM algorithms.

As can be seen from the above, the static algorithm has the same requirements as the GPB1. The IMM has only slightly higher requirements than the GPB1, but clearly significantly lower than GPB2.

In view of this, the modifications of the static algorithm for the switching situation are considered obsolete.

In view of the fact that the IMM performs significantly better than GPB1 and almost as well as GPB2 [BB88], the IMM is considered to be the best compromise between complexity and performance.

Furthermore, the IMM has been shown to be able to keep the position estimation error not worse than the raw measurement error during the critical maneuver periods (onset and termination) and provide significant improvement (noise reduction) at other times.

11.7 USE OF EKF FOR SIMULTANEOUS STATE AND PARAMETER ESTIMATION

11.7.1 Augmentation of the State

The extended Kalman filter can be used for suboptimal state estimation in nonlinear dynamic systems.

The situation of linear systems with unknown parameters that are continuous-valued can be put in the framework of nonlinear state estimation by augmenting the base state. The **base state** is the state of the system with the parameters assumed known.

Denoting the unknown parameters as a vector θ, the **augmented state** will be the **stacked vector** consisting of the base state x and θ

$$y(k) \triangleq \begin{bmatrix} x(k) \\ \theta \end{bmatrix} \tag{11.7.1-1}$$

The linear dynamic equation of x, with the known input u and process noise v,

$$x(k+1) = F(\theta)x(k) + G(\theta)u(k) + v(k) \tag{11.7.1-2}$$

and the "dynamic equation" of the parameter vector (assumed time invariant)

$$\theta(k+1) = \theta(k) \tag{11.7.1-3}$$

can be rewritten as a nonlinear dynamic equation for the augmented state

$$y(k+1) = f[y(k), u(k)] + v(k) \tag{11.7.1-4}$$

The resulting nonlinear equation (11.7.1-4) is then used for an EKF to estimate the entire augmented state.

The same technique can be used if the dynamic equation of the base state is nonlinear.

The model represented by (11.7.1-3) for the parameter dynamics assumes it to be constant. Therefore, the covariance yielded by the EKF will, asymptotically, tend to zero, and, consequently, the filter gain for these components will tend to zero. This is because there is no process noise entering into these components of the augmented state — the **controllability condition** (2) from Subsection 5.2.5, pertaining to the Riccati equation for the state estimation covariance, is not satisfied.

11.7.1 Augmentation of the State

Since the EKF is not an optimal estimation algorithm, it will in general not yield consistent estimates for the parameters — the estimates will not converge to the true values. Thus the situation where the parameter variances tend to zero is undesirable because it will lead in practice to estimation errors much larger than the filter-calculated variances.

This can be remedied (to some extent) by assuming an **artificial process noise** (or **pseudo-noise**) entering into the parameter equation. This amounts to replacing (11.7.1-3) by

$$\theta(k+1) = \theta(k) + v_\theta(k) \tag{11.7.1-5}$$

where the parameter process noise is assumed zero mean and white. Thus the parameter is modeled as a (discrete time) Wiener process.

Any nonzero variance of this process noise will prevent the filter-calculated variances of the parameter estimates from converging to zero. Furthermore, this also gives the filter the ability to estimate **slowly varying parameters**.

The choice of the variance of the artificial process noise for the parameters — the **tuning of the filter** — can be done as follows:

1. Choose the standard deviation of the process noise as a few percent of the (estimated/guessed) value of the parameter.
2. Simulate the system and the estimator with random initial estimates (for the base state as well as the parameters) and monitor the normalized estimation errors.
3. Adjust the noise variances until, for the problem of interest, the filter is consistent — it yields estimation errors commensurate with the calculated augmented state covariance matrix. The criteria to be used are those from Section 5.4 — the estimation bias and the normalized estimation error squared (NEES).

The example of the coordinated turn motion with an unknown rate from (11.6.8-5) to (11.6.8-7) falls into this catefory.

11.7.2   An Example of Use of the EKF for Parameter Estimation

Consider the scalar system, that is, its base state x is a scalar

$$x(k+1) = a(k)x(k) + b(k)u(k) + v_1(k) \qquad (11.7.2\text{-}1)$$

where $v_1(k)$ is the base state process noise and the two unknown parameters are $a(k)$ and $b(k)$, possibly time varying.

The observations are

$$z(k) = x(k) + w(k) \qquad (11.7.2\text{-}2)$$

Following the procedure of the previous subsection, the augmented state is

$$y(k) \triangleq \begin{bmatrix} y_1(k) \\ y_2(k) \\ y_3(k) \end{bmatrix} \triangleq \begin{bmatrix} x(k) \\ a(k) \\ b(k) \end{bmatrix} \qquad (11.7.2\text{-}3)$$

With this the nonlinear dynamic equation corresponding to (11.7.2-1) can be written as

$$y_1(k+1) = f^1[y(k), u(k)] + v(k) \triangleq y_1(k)y_2(k) + y_3(k)u(k) + v_1(k) \qquad (11.7.2\text{-}4)$$

and the "dynamic equation" of the parameters is

$$y_i(k+1) = y_i(k) + v_i(k) \qquad i = 2, 3 \qquad (11.7.2\text{-}5)$$

The augmented state equation is then

$$y(k+1) = f[y(k), u(k)] + v(k) \qquad (11.7.2\text{-}6)$$

with the augmented process noise

$$v(k) \triangleq \begin{bmatrix} v_1(k) \\ v_2(k) \\ v_3(k) \end{bmatrix} \qquad (11.7.2\text{-}7)$$

assumed zero mean and with covariance

$$Q = \text{diag}(q_1, q_2, q_3) \qquad (11.7.2\text{-}8)$$

With the linear measurements (11.7.2-2), the only place in the EKF where linearizations are to be carried out is the base state prediction.

The second order EKF will use the following augmented state prediction equations. From (11.7.2-4) the base state prediction can be obtained directly as

$$\hat{y}_1(k+1|k) = \hat{y}_2(k|k)\hat{y}_1(k|k) + \hat{y}_3(k|k)u(k) + P_{21}(k|k) \qquad (11.7.2\text{-}9)$$

since

$$E[a(k)x(k)|Z^k] = E[y_2(k)y_1(k)|Z^k] = \hat{y}_2(k|k)\hat{y}_1(k|k) + \text{cov}[y_2(k), y_1(k)|Z^k] \quad (11.7.2\text{-}10)$$

Note that the last term in (11.7.2-9) is the only second order term.

The predicted values of the remaining two components of the augmented state, which are the system's unknown parameters, follow from (11.7.2-5) as

$$\hat{y}_i(k+1|k) = \hat{y}_i(k|k) \qquad i = 2,3 \qquad (11.7.2\text{-}11)$$

Equations (11.7.2-9) and (11.7.2-11), in general, have to be obtained using the series expansion of the EKF, which requires evaluation of the Jacobian of the vector f and the Hessians of its components. In the present problem, the Jacobian is

$$F(k) = \begin{bmatrix} \hat{y}_2(k|k) & \hat{y}_1(k|k) & u(k) \\ 0 & 1 & 0 \\ 0 & 0 & 1 \end{bmatrix} \qquad (11.7.2\text{-}12)$$

and the Hessians are

$$f_{yy}^1 = \begin{bmatrix} 0 & 1 & 0 \\ 1 & 0 & 0 \\ 0 & 0 & 0 \end{bmatrix} \qquad f_{yy}^2 = 0 \qquad f_{yy}^3 = 0 \qquad (11.7.2\text{-}13)$$

Then it can be easily shown (see problem 11-3) that using (10.3.2-4) for the augmented state, that is,

$$\hat{y}(k+1|k) = F(k)\hat{y}(k|k) + \frac{1}{2}\sum_{i=1}^{n_x} e_i \text{tr}[f_{yy}^i P(k|k)] \qquad (11.7.2\text{-}14)$$

yields (11.7.2-9) and (11.7.2-11).

The prediction covariance of the base state can be obtained, using (10.3.2-6) as

$$\begin{aligned} P_{11}(k+1|k) =\ & \hat{y}_2(k|k)^2 P_{11}(k|k) + 2\hat{y}_2(k|k)\hat{y}_1(k|k)P_{21}(k|k) + 2\hat{y}_2(k|k)u(k)P_{13}(k|k) \\ & + \hat{y}_1(k|k)^2 P_{22}(k|k) + 2\hat{y}_1(k|k)u(k)P_{23}(k|k) + u(k)^2 P_{33}(k|k) \\ & + P_{21}(k|k)^2 + P_{22}(k|k)P_{11}(k|k) \end{aligned} \qquad (11.7.2\text{-}15)$$

11.7.3 EKF for Parameter Estimation — Summary

The EKF can be used to estimate simultaneously:

- the base state, and
- the unknown parameters

of a system.

This is accomplished by stacking them into an augmented state and carrying out the series expansions of the EKF for this augmented state.

Since the EKF is a suboptimal technique, significant care has to be exercised to avoid filter inconsistencies. The filter has to be tuned with artificial process noise so that its estimation errors are commensurate with the calculated variances.

11.8 NOTES, PROBLEMS, AND TERM PROJECT

11.8.1 Bibliographical Notes

The "adjustment" of the process noise covariance, presented in Section 11.2, has become part of the Kalman filtering folklore. It can be found in [Jaz69, Jaz70]. Among its applications, the one presented in [CWA77] deals with state and parameter estimation for maneuvering reentry vehicles. In [THS77, TBM77] similar ideas are used for passive tracking of maneuvering targets including reinitialization of the filter upon detection of the maneuver. In [Cas80] a continuous update of the process noise covariance is presented based on a fading memory average of the residuals. An adaptation of the filter gain based on the residuals' deviation from orthogonality is presented in [HC73]. In [SW72] a least squares approach with exponential discount of older measurements is used for maneuvering targets. Estimation of the noise covariances has been discussed in, for example, [LB94] and [MT76].

The switching from one model to another with different parameters is discussed in [MD73] in a manner somewhat similar to Subsection 11.2.2.

The input estimation method of Section 11.3 is based on [CHP79]. Estimation of the input and the measurement noise covariance has been discussed in [MK86]. The generalized likelihood ratio technique [WJ76, KGW82] deals simultaneously with the estimation of the input and its onset time. Such an approach was taken by [Bog87] who augmented the IE technique of [CHP79] to include maneuver onset time estimation; however, the resulting algorithm was very costly and performed less well than the IMM [BCB89]. In [Dem87] an approach consisting of hypothesis testing via dynamic programming is proposed.

The variable dimension approach to tracking maneuvering targets from Section 11.4 is from [BB82], which also presented the comparison of algorithms discussed in Section 11.5. An extensive comparison of various maneuver detection approaches is described in [Woo85].

A weighted state estimate, along the lines of Section 11.6, is presented in [Tho73]. The multiple model (MM) approach was originally presented in [Mag65]. The Markov switching of models is discussed in [MW73, Moo75, MVM79, MVM80, GM77]. The generalized pseudo-Bayesian (GPB) algorithms are proposed by [AF70, JG71b, JG71a]. The GPB2 MM approach is based on [CA78]. A survey of the MM techniques and their connection with failure detection is given in [Tug82]. Related work in the area of failure detection is [Wil76, Cag80]; the connection between failure detection, multiobject tracking and general modeling uncertainties is discussed in [PS83]. Detection-estimation algorithms that approximate the optimal algorithm of Subsection 11.6.3 are discussed in [HS78, TH79]. The Interacting MM algorithm is from [Blo84, BB88]. The application of the IMM to air traffic control can be found in [Bar92](Chapters 1 and 2). Subsection 11.6.8 is from [LB93b]. Reference [LB93c] discusses how one can predict the performance of an IMM algorithm without Monte Carlo simulations and illustrates this technique on examples, including a simplified ATC problem.

An adaptive sampling technique for a maneuvering target is presented in [BBC81].

The use of target orientation measurements for tracking maneuvering targets is explored in [KMR81, Lef84, AKG86, SH89].

11.8.2 Problems

11-1 CRLB for a two-model parameter estimation problem. Given an estimation problem where the noise behaves according to one of two models. Using the binary random variable α, the observation can be written as

$$z = x + \alpha w_1 + (1 - \alpha) w_2$$

where x is the (unknown constant) parameter to be estimated,

$$P\{\alpha = 1\} = p_1 \qquad\qquad P\{\alpha = 0\} = p_2 = 1 - p_1$$

with $w_i \sim \mathcal{N}(0, \sigma_i^2)$ independent of each other and of α. In other words, the measurement has with probability p_1 accuracy σ_1 and with probability p_2 accuracy σ_2.

1. Write the likelihood function of x, $\Lambda(x) = p(z|x)$.

2. Find $\hat{x}^{\mathrm{ML}}$.

3. Evaluate the corresponding MSE.

4. Evaluate the CRLB for this estimation problem assuming $p_1 = 0.5$, $\sigma_1 = 1$, $\sigma_2 = 100$. (Hint: use the fact that $\sigma_1 \ll \sigma_2$ to approximate the integral that requires otherwise numerical evaluation).

5. Compare the results of (3) and (4) for the above numbers. Comment on the usefulness of the bound in this problem.

11-2 Two-model parameter estimation problem with a prior. Given the random parameter $x \sim \mathcal{N}(x_0, \sigma_0^2)$ to be estimated based on the same measurement as in problem 11-1.

1. Find the MMSE estimate of x given z.

2. Find the conditional variance of x given z.

Assume x, α and w_i are all independent.

11-3 EKF prediction equations for a system with unknown parameters.

1. Derive the predicted augmented state from (11.7.2-14).

2. Prove (11.7.2-15) and derive the remaining terms of this covariance.

11-4 EKF simulation for state and parameter estimation. Consider the system (11.7.2-1) with $x(0) = 0$, $a(k) = 0.5$, $b(k) = 0$, process noise zero mean with variance $q_1 = 0.09$. The observations are given by (11.7.2-2) with measurement noise zero mean with variance $r = 0.09$. The covariance matrix of the initial augmented state estimate is $P(0|0) = \mathrm{diag}(0.09, 0.04)$.

1. Implement an EKF for this problem (to estimate x, a, and b) without an artificial process noise. Initialize with random initial estimates according to $P(0|0)$. List the normalized estimation error squared (NEES) for each component of the augmented state for 100 steps.

2. Perform 100 Monte Carlo runs for the above. List the averages of the normalized estimation error for bias check and the NEES for each component for 100 steps. Indicate the corresponding 95% confidence regions.

3. If the filter needs tuning, do it and present the results. Compare also the absolute RMS estimation errors with those from (2).

11-5 Fading memory average. Prove (11.2.2-7).

11-6 Sojourn time in a state for a Markov chain. Prove

$$E[\tau_i] = \frac{1}{1 - p_{ii}}$$

where τ_i is the expected sojourn time of a Markov chain in state i (mode i) and p_{ii} is the transition probability from state i to state i.

11.8.3 Term Project

Implement a two-model IMM for the following air traffic control problem:

The Ground Truth is a target moving with a constant speed of $150m/s$ with initial state in Cartesian coordinates

$$x = [\xi \ \dot{\xi} \ \eta \ \dot{\eta}]' = [0 \ 0 \ 0 \ 250]'$$

The sampling period is $T = 5s$. At $k = 20$ ($t = 100s$) it starts a (left) turn of $3°/s$ for $30s$, then continues straight until $k = 40$. (The turn is not known to the filter!)

Measurements are made starting from $k = 0$ on the position of this target according to the following two options:

(A) the Cartesian position (ξ, η) is observed with additive white Gaussian noise with covariance $R = \text{diag}[10^4, 10^4]$;

(B) in polar coordinates (range r and azimuth θ) by a radar located at $[\xi_0, \eta_0] = [-10^4, 0]$

$$r = \sqrt{(\xi - \xi_0)^2 + (\eta - \eta_0)^2} \qquad \theta = \tan^{-1}\left(\frac{\eta - \eta_0}{\xi - \xi_0}\right)$$

with additive white Gaussian noise with covariance $R = \text{diag}[2500m^2, (1°)^2]$.

The IMM estimator is initialized from the measurements at $k = 0$ and $k = 1$ and starts running from $k = 2$. Each model has initial probability 0.5 and the same initial estimate.

The two models to be used are second order kinematic (Section 6.3.2), with process noise (assumed zero-mean white Gaussian) with variances $\sigma_{v_m}^2$, $m = 1, 2$. These, together with the Markov chain transition matrix between the models, are the IMM estimator design parameters.

1. Calculate the true state according to the above specifications for $k = 0, \ldots, 40$ (the truth evolves without noise).

2. Generate the noisy measurements along the trajectory.

3. Implement the IMMKF for this problem. Indicate the rationale for the choice of the estimator design parameters.

4. Calculate the following average results for $N = 100$ runs for 2 possible designs (try to have the estimated position RMS errors not to exceed the single measurement position RMS

error during the maneuver — a few percent excess is OK — while having as much as possible "noise reduction" in the straight portions of the trajectory):

$$\text{NORXE} \triangleq \text{AVE}\{\tilde{x}_1(k|k)/\sqrt{P_{11}(k|k)}\}$$
$$\text{FPOS} \triangleq \sqrt{\text{AVE}\{P_{11}(k|k) + P_{33}(k|k)\}}$$
$$\text{FSPD} \triangleq \sqrt{\text{AVE}\{P_{22}(k|k) + P_{44}(k|k)\}}$$
$$\text{RMSPOS} \triangleq \text{RMS position error (Note: both coordinates combined)}$$
$$\text{RMSSPD} \triangleq \text{RMS speed error (Note: speed} \triangleq \text{magnitude of velocity vector)}$$
$$\text{NEES} \triangleq \text{AVE}\{\tilde{x}(k|k)'P(k|k)^{-1}\tilde{x}(k|k)\}$$
$$\text{MODPR} \triangleq \text{AVE}\{\text{model 2 probability}\}$$

where

$$\text{RMS}(y) \triangleq \sqrt{\frac{1}{N}\sum_{j=1}^{N}(y^j)^2} \qquad\qquad \text{AVE}(y) \triangleq \frac{1}{N}\sum_{j=1}^{N}y^j$$

and y^j is the outcome of y in run j. Provide the expressions you used for calculating RMSPOS and RMSSPD. Indicate which is your final ("best") design.

5. Indicate the distributions of NORXE and NEES and their 95% probability regions. Justify the final choice of the design parameters. Summarize the effect of the Markov chain transition matrix on the final results based on the designs illustrated in item 4.

6. Plot for both designs RMSPOS and FPOS; RMSSPD and FSPD; NORXE with its probability region; NEES with its probability region; and MODPR. Comment on the estimator bias and consistency.

7. Repeat item 4 for the best single-model filter (a standard KF) of your choice (a single design). Indicate how the design parameter was chosen and comment on its performance as compared to the best IMM.

DELIVERABLES:

A concise report is due.

Each student will have to make a 10–15 min. presentation in class. It is suggested to prepare 7–8 viewgraphs for this presentation. The most important plots are:

- A single run with the true and estimated trajectories (best IMM). Plot the true position (×), the measurement (□), and the updated position (o) as evolve in time. Provide a magnified plot of the turn portion.

and the following comparison plots:

- RMSPOS for the two IMMs and the KF (together, for comparison); also, indicate the (baseline) raw unfiltered position error.
- RMSSPD for the two IMMs and the KF (together, for comparison).
- NEES for the two IMMs and the KF (together, for comparison).
- MODPR for the two IMMs (together, for comparison).

Bibliography

[AF70]   G. A. Ackerson and K. S. Fu. On State Estimation in Switching Environments. *IEEE Trans. Automatic Control*, AC-15(1):10–17, Jan. 1970.

[AH83]   V. J. Aidala and S. E. Hammel. Utilization of Modified Polar Coordinates for Bearings-Only Tracking. *IEEE Trans. Automatic Control*, AC-28:283–293, March 1983.

[AKG86]   D. Andrisani, F. P. Kuhl, and D. Gleason. A Nonlinear Tracker Using Attitude Measurements. *IEEE Trans. Aerospace and Electronic Systems*, AES-22:533–539, Sept. 1986.

[AM79]   B. O. D. Anderson and J.B. Moore. *Optimal Filtering*. Prentice-Hall, Englewood Cliffs, NJ, 1979.

[AWB68]   M. Athans, R. P. Wishner, and A. Bertolini. Suboptimal State Estimation for Continuous-Time Nonlinear Systems from Discrete Noisy Measurements. *IEEE Trans. Automatic Control*, AC-13:504–514, Oct. 1968.

[Bal84]   A. V. Balakrishnan. *Kalman Filtering Theory*. University Series in Modern Engineering, Springer, Berlin, Germany, 1984.

[Bar85]   Y. Barniv. Dynamic Programing Solution to Detecting Dim Moving Targets. *IEEE Trans. Aerospace and Electronic Systems*, AES-21:144–156, Jan. 1985.

[Bar90]   Y. Bar-Shalom, editor. *Multitarget-Multisensor Tracking: Advanced Applications*. Artech House, Norwood, MA, 1990.

[Bar91]   Y. Bar-Shalom. BEARDAT — Bearing and Frequency Data Association Tracker 2.0. 1991. Interactive software.

[Bar92]   Y. Bar-Shalom, editor. *Multitarget-Multisensor Tracking: Applications and Advances*. Volume II, Artech House, Norwood, MA, 1992.

[BB62]   T. R. Benedict and G. W. Bordner. Synthesis of an Optimal Set of Radar Track-While-Scan Smoothing Equations. *IRE Trans. Automatic Control*, AC-7:27–32, July 1962.

[BB82]   Y. Bar-Shalom and K. Birmiwal. Variable Dimension Filter for Maneuvering Target Tracking. *IEEE Trans. Aerospace and Electronic Systems*, AES-18(5):621–629, Sept. 1982.

[BB83]   Y. Bar-Shalom and K. Birmiwal. Consistency and Robustness of PDAF for Target Tracking in Cluttered Environments. *Automatica*, 19:431–437, July 1983.

[BB86]   M. Basseville and A. Benveniste, editors. *Detection of Abrupt Changes in Signals and Dynamical Systems*. Volume 77 of *Lecture Notes in Control and Information Sciences*, Springer-Verlag, Berlin, 1986.

BIBLIOGRAPHY

[BB88] H. A. P. Blom and Y. Bar-Shalom. The Interacting Multiple Model Algorithm for Systems with Markovian Switching Coefficients. *IEEE Trans. Automatic Control*, AC-33(8):780–783, Aug. 1988.

[BBC81] S. S. Blackman, T. J. Broida, and M. F. Cartier. Applications of a Phased Array Antenna in a Multiple Maneuvering Target Environment. In *Proc. 20th IEEE Conf. Decision and Control*, San Diego, CA, Dec. 1981.

[BC75] R. E. Bogner and A. C. Constantinides. *Introduction to Digital Filtering*. Wiley, New York, 1975.

[BCB89] Y. Bar-Shalom, K. C. Chang, and H. A. P. Blom. Tracking a Maneuvering Target Using Input Estimation vs. the Interacting Multiple Model Algorithm. *IEEE Trans. Aerospace and Electronic Systs*, AES-25:296–300, March 1989.

[Bel61] R. Bellman. *Adaptive Control Processes: A Guided Tour*. Princeton University Press, Princeton, NJ, 1961.

[BF88] Y. Bar-Shalom and T. E. Fortmann. *Tracking and Data Association*. Academic Press, New York, 1988.

[BH75] A. Bryson and Y. C. Ho. *Applied Optimal Control*. Hemisphere, Cambridge, MA, 1975.

[BH92] R. G. Brown and P. Y. C. Hwang. *Introduction to Random Siganls and Apllied Kalman Filtering*. Wiley, New York, 2nd edition, 1992.

[Bie77] G. J. Bierman. *Factorization Methods for Discrete Sequential Estimation*. Academic Press, New York, 1977.

[BL90] R. S. Baheti, A. J. Laub, and D. M. Wiberg (Eds.). Bierman Memorial Issue on Factorized Estimation Applications. *IEEE Trans. Automatic Control*, AC-35:1282–1319, Dec. 1990.

[Bla86] S. S. Blackman. *Multiple Target Tracking with Radar Applications*. Artech House, Norwood, MA, 1986.

[Blo84] H. A. P. Blom. An Efficient Filter for Abruptly Changing Systems. In *Proc. 23rd IEEE Conf. Decision and Control*, Las Vegas, NV, Dec. 1984.

[BN93] M. Basseville and I. V. Nikiforov, editors. *Detection of Abrupt Changes: Theory and Applications*. Prentice-Hall, Englewood Cliffs, NJ, 1993.

[Bog87] P. L. Bogler. Tracking a Maneuvering Target Using Input Estimation. *IEEE Trans. Aerospace and Electronic Systems*, AES-23(3):298–310, May 1987.

[Boz79] S. M. Bozic. *Digital and Kalman Filtering*. E. Arnold, 1979.

[BR87] R. E. Bethel and R. G. Rahikka. An Optimum First-Order Time Delay Tracker. *IEEE Trans. Aerospace and Electronic Systems*, AES-23:718–725, Nov. 1987.

[BS78] Y. Baram and N. R. Sandell, Jr. Consistent Estimation on Finite Parameter Sets with Application to Linear Systems Identification. *IEEE Trans. Automatic Control*, AC-23(3):451–454, June 1978.

[CA78] C. B. Chang and M. Athans. State Estimation for Discrete Systems with Switching Parameters. *IEEE Trans. Aerospace and Electronic Systems*, AES-14(5):418–425, May 1978.

[Cag80] A. K. Caglayan. Simultaneous Failure Detection and Estimation in Linear Systems. In *Proc. 19th IEEE Conf. Decision and Control*, Albuquerque, NM, Dec. 1980.

[Cas80] F. R. Castella. An Adaptive Two-Dimensional Kalman Tracking Filter. *IEEE Trans. Aerospace and Electronic Systems*, AES-16:822–829, Nov. 1980.

494

BIBLIOGRAPHY

[Cas81] F. R. Castella. Tracking Accuracies with Position and Rate Measurements. *IEEE Trans. Aerospace and Electronic Systems*, AES-17:433–437, May 1981.

[CDY84] C. B. Chang, K. P. Dunn, and L. C. Youens. A Tracking Algorithm for Dense Target Environments. In *Proc. 1984 American Control Conf.*, San Diego, CA, June 1984.

[Che84] C. T. Chen. *Linear System Theory and Design*. Holt, New York, 1984.

[CHP79] Y. T. Chan, A. G. C. Hu, and J. B. Plant. A Kalman Filter Based Tracking Scheme with Input Estimation. *IEEE Trans. Aerospace and Electronic Systems*, AES-15(2):237–244, March 1979.

[COD91] E. Cortina, D. Otero, and C. E. D'Attelis. Maneuvering Target Tracking Using EKF. *IEEE Trans. Aerospace and Electronic Systs.*, AES-27:155–158, Jan. 1991.

[CWA77] C. B. Chang, R. H. Whiting, and M. Athans. On the State and Parameter Estimation for Maneuvering Reentry Vehicles. *IEEE Trans. Automatic Control*, AC-22(2):99–105, Feb. 1977.

[Dem87] K. Demirbas. Maneuvering Target Tracking with Hypothesis Testing. *IEEE Trans. Aerospace and Electronic Systems*, AES-23:757–766, Nov. 1987.

[DF93] O. E. Drummond and G. Frenkel. Glossary of Tracking Terms of SDI Panels on Tracking. In *Proc. SPIE Conf. on Signal and Data Processing of Small Targets, Vol. 1954*, April 1993.

[Eks83] B. Ekstrand. Analytical Steady State Solution for a Kalman Tracking Filter. *IEEE Trans. Aerospace and Electronic Systems*, AES-19:815–819, Nov. 1983.

[Elb84] T. F. Elbert. *Estimation and Control of Systems*. Van Nostrand, New York, 1984.

[Eyk74] P. Eykhoff. *System Identification: Parameter and State Estimation*. Wiley, Chichester, England, 1974.

[FH77] T. E. Fortmann and K. L. Hitz. *An Introduction to Linear Control Systems*. Dekker, New York, 1977.

[Fit80] R. J. Fitzgerald. Simple Tracking Filters: Filtering and Smoothing Performance. *IEEE Trans. Aerospace and Electronic Systems*, AES-16:860–864, Nov. 1980.

[Fit81] R. J. Fitzgerald. Simple Tracking Filters: Closed Form Solutions. *IEEE Trans. Aerospace and Electronic Systems*, AES-17:781–785, Nov. 1981.

[Fri69] B. Friedland. Treatment of Bias in Recursive Filtering. *IEEE Trans. Automatic Control*, AC-14:359–367, Aug. 1969.

[Fri73] B. Friedland. Optimum Steady State Position and Velocity Estimation Using Noisy Sampled Position Data. *IEEE Trans. Aerospace and Electronic Systems*, AES-9:906–911, Nov. 1973.

[FS85] A. Farina and F. A. Studer. *Radar Data Processing, Vol. I: Introduction and Tracking, Vol. II: Advanced Topics and Applications*. Research Studies Press, Letchworth, Hertfordshire, England, 1985.

[Gel74] A. Gelb. *Applied Optimal Estimation*. MIT Press, Cambridge, MA, 1974.

[GM77] N. H. Gholson and R. L. Moose. Maneuvering Target Tracking Using Adaptive State Estimation. *IEEE Trans. Aerospace and Electronic Systems*, AES-13:310–317, May 1977.

[GM93] J. Gray and W. Murray. A Derivation of an Analytic Expression for the Alpha-Beta-Gamma Filter. *IEEE Trans. Aerospace and Electronic Systems*, AES-29:1064–1065, July 1993.

[GS84] G. C. Goodwin and K. S. Sin. *Adaptive Filtering, Prediction, and Control*. Prentice-Hall, Englewood Cliffs, NJ, 1984.

BIBLIOGRAPHY

[Hay86] S. Haykin. *Adaptive Filter Theory*. Prentice-Hall, Englewood Cliffs, NJ, 1986.

[HC73] R. L. T. Hampton and J. R. Cooke. Unsupervised Tracking of Maneuvering Vehicles. *IEEE Trans. Aerospace and Electronic Systems*, AES-9:197–207, March 1973.

[Hel91] C. W. Helstrom. *Probability and Stochastic Processes for Engineers*. Macmillan, New York, 2nd edition, 1991.

[How71] R. A. Howard. *Dynamic Probabilistic Systems, I: Markov Models, II: Semi-Markov and Decision Processes*. Wiley, New York, 1971.

[HS78] M. T. Hadidi and S. C. Schwartz. Sequential Detection with Markov Interrupted Observations. In *Proc. 16th Allerton Conf. on Communication, Control and Computing*, Univ. of Illinois, Oct. 1978.

[IB89] H. R. Itzkowitz and R. S. Baheti. Demonstration of Square Root Kalman Filter on WARP Parallel Computer. In *Proc. 1989 American Control Conf.*, Pittsburgh, PA, June 1989.

[Jaz69] A. H. Jazwinski. Adaptive Filtering. *Automatica*, 5(4):475–485, July 1969.

[Jaz70] A. H. Jazwinski. *Stochastic Processes and Filtering Theory*. Academic Press, New York, 1970.

[JB90] C. Jauffret and Y. Bar-Shalom. Track Formation with Bearing and Frequency Measurements in the Presence of False Detections. *IEEE Trans. Aerospace and Electronic Systems*, AES-25:999–1010, Nov. 1990.

[JD91] D. Jaarsma and R. A. Davis. Joint Estimation of Time Delay and Doppler Ratio from Broadband Using Discrete Estimation Techniques. In *Proc. 1991 International Conf. Acoustics, Speech, and Signal Processing*, Toronto, Canada, May 1991.

[JG71a] A. G. Jaffer and S. C. Gupta. Optimal Sequential Estimation of Discrete Processes with Markov Interrupted Observations. *IEEE Trans. Automatic Control*, AC-16:417–475, Oct. 1971.

[JG71b] A. G. Jaffer and S. C. Gupta. Recursive Bayesian Estimation with Uncertain Observation. *IEEE Trans. Information Theory*, IT-17:614–616, Sept. 1971.

[Joh72] J. Johnston. *Econometric Methods*. McGraw-Hill, New York, 1972.

[Kai74] T. Kailath. A View of Three Decades of Linear Filtering Theory. *IEEE Trans. Information Theory*, IT-20(2):146–181, March 1974.

[Kai81] T. Kailath. *Lectures on Wiener and Kalman Filtering*. Springer-Verlag, New York, 1981.

[Kal60] R. E. Kalman. A New Approach to Linear Filtering and Prediction Problems. *Trans. ASME, J. Basic Engineering*, 82:34–45, March 1960.

[Kal84] P. R. Kalata. The Tracking Index: A Generalized Parameter for Alpha-Beta-Gamma Target Trackers. *IEEE Trans. Aerospace and Electronic Systems*, AES-20:174–182, March 1984.

[Kay93] Steven M. Kay. *Fundamentals of Statistical Signal Processing: Estimation Theory*. PTR Prentice-Hall, Englewood Cliffs, NJ, 1993.

[KB61] R. E. Kalman and R. Bucy. New Results in Linear Filtering and Prediction Theory. *Trans. ASME, J. Basic Engineering*, 83:95–108, March 1961.

[KGW82] J. Korn, S. W. Gully, and A. S. Willsky. Application of the Generalized Likelihood Ratio Algorithm to Maneuver Detection and Estimation. In *Proc. 1982 American Control Conf.*, Arlington, VA, June 1982.

[Kle89] D. L. Kleinman. Private Communication, 1989.

BIBLIOGRAPHY

[KMR81] J. D. Kendrick, P. S. Maybeck, and J. G. Reid. Estimation of Aircraft Target Motion Using Orientation Measurements. *IEEE Trans. Aerospace and Electronic Systems*, 17:254–260, March 1981.

[KS87] S. C. Kramer and H. W. Sorenson. Bayesian Parameter Estimation. In *Proc. 1987 American Control Conf.*, Minneapolis, MN, June 1987.

[LB92] X. R. Li and Y. Bar-Shalom. Mode-Set Adaptation in Multiple Model Approach to Hybrid State Estimation. In *Proc. 1992 American Control Conf.*, pages 1794–1799, Chicago, IL, June 1992.

[LB93a] D. Lerro and Y. Bar-Shalom. Tracking with Debiased Consistent Converted Measurements vs. EKF. *IEEE Trans. Aerospace and Electronic Systems*, AES-29(3):1015–1022, July 1993.

[LB93b] X. R. Li and Y. Bar-Shalom. Design of an Interacting Multiple Model Algorithm for Air Traffic Control Tracking. To appear in *IEEE Trans. Control Systems Technology*, 1(3), Sept. 1993. Special issue on Air Traffic Control.

[LB93c] X. R. Li and Y. Bar-Shalom. Performance Prediction of the Interacting Multiple Model Algorithm. *IEEE Trans. Aerospace and Electronic Systems*, AES-29(3):755–771, July 1993. A brief version was published in *Proc. 1992 American Control Conf.*, Chicago, IL, June 1992, 2109-2113.

[LB94] X. R. Li and Y. Bar-Shalom. A Recursive Multiple Model Approach to Noise Identification. To appear in *IEEE Trans. Aerospace and Electronic Systems*, AES-30(3), July 1994. A brief version was published in *Proc. 1st IEEE Conf. Control Applications*, Dayton, OH, Sept. 1992, 847-852.

[Lef84] C. C. Lefas. Using Roll-Angle Measurement to Track Aircraft Maneuvers. *IEEE Trans. Aerospace and Electronic Systems*, AES-20:672–681, Nov. 1984.

[Leh83] E. L. Lehmann. *Theory of Point Estimation*. Wiley, New York, 1983.

[Leh86] E. L. Lehmann. *Testing Statistical Hypotheses*. Wiley, New York, 2nd edition, 1986.

[Lew86] F. L. Lewis. *Optimal Estimation*. Wiley, New York, 1986.

[Lju87] L. Ljung. *System Identification*. Prentice-Hall, Englewood Cliffs, NJ, 1987.

[LP66] R. E. Larson and J. Peschon. A Dynamic Programming Approach to Trajectory Estimation. *IEEE Trans. Automatic Control*, AC-11:537–540, July 1966.

[Mag65] D. T. Magill. Optimal Adaptive Estimation of Sampled Stochastic Processes. *IEEE Trans. Automatic Control*, AC-10:434–439, 1965.

[May79] P. S. Maybeck. *Stochastic Models, Estimation and Control*, Vol. I. Academic Press, New York, 1979.

[May82] P. S. Maybeck. *Stochastic Models, Estimation and Control*, Vols. II, III. Academic Press, New York, 1982.

[MC78] J. L. Melsa and D. L. Cohn. *Decision and Estimation Theory*. McGraw-Hill, New York, 1978.

[MD73] R. J. McAulay and E. J. Denlinger. A Decision-Directed Adaptive Tracker. *IEEE Trans. Aerospace and Electronic Systems*, AES-9(2):229–236, March 1973.

[Med69] J. Meditch. *Stochastic Linear Estimation and Control*. McGraw-Hill, New York, 1969.

[Men73] J. M. Mendel. *Discrete Techniques of Parameter Estimation*. Marcel Dekker, New York, 1973.

[Men87] J. M. Mendel. *Lessons in Digital Estimation Theory*. Prentice-Hall, Englewood Cliffs, 1987.

BIBLIOGRAPHY

[MK86]   A. Moghaddamjoo and R. L. Kirlin. Robust Adaptive Kalman Filtering with Unknown Inputs. In *Proc. 1986 American Control Conf.*, Seattle, WA, June 1986.

[Moo75]   R. L. Moose. An Adaptive State Estimator Solution to the Maneuvering Target Problem. *IEEE Trans. Automatic Control*, AC-20(3):359–362, June 1975.

[Mor69]   N. Morrison. *Introduction to Sequential Smoothing and Prediction*. McGraw-Hill, New York, 1969.

[MT76]   K. A. Myers and B. D. Tapley. Adaptive Sequential Estimation with Unknown Noise Statistics. *IEEE Trans. Automatic Control*, AC-21:520–523, Aug. 1976.

[MVM79]   R. L. Moose, H. F. VanLandingham, and D. H. McCabe. Modeling and Estimation of Tracking Maneuvering Targets. *IEEE Trans. Aerospace and Electronic Systems*, AES-15(3):448–456, May 1979.

[MVM80]   R. L. Moose, H. F. VanLandingham, and D. H. McCabe. Application of Adaptive State Estimation Theory. In *Proc. 19th IEEE Conf. Decision and Control*, Albuquerque, NM, Dec. 1980.

[MW73]   R. L. Moose and P. L. Wang. An Adaptive Estimator with Learning for a Plant Containing Semi-Markov Switching Parameters. *IEEE Trans. Systems, Man and Cybernetics*, SMC-3:277–281, May 1973.

[NA81]   S. C. Nardone and V. J. Aidala. Observability Criteria for Bearings-Only Target Motion Analysis. *IEEE Trans. Aerospace and Electronic Systems*, AES-17:162–166, March 1981.

[Nav77]   A. M. Navarro. General Properties of Alpha-Beta and Gamma Tracking Filters. 1977. Report PHL 1977-02, National Defense Research Organization, Physics Lab., The Hague, The Netherlands, Jan. 1977 (US Dept. of Commerce, National Technical Information Servies, Report 77-24347).

[NL83]   L. C. Ng and R. A. LaTourette. Equivalent Bandwidth of a General Class of Polynominal Smoothers. *J. Acoustical Society of America*, 74:814–816, Sept. 1983.

[OB88]   D. R. O'Halloran and R. K. Baheti. Fast Mapping of a Kalman Filter on WARP. GE-CRD, Schenectady, NY, 1988.

[Pap84]   A. Papoulis. *Probability, Random Variables, and Stochastic Processes*. McGraw-Hill, New York, 1984.

[Par71]   H. Parkus. *Optimal Filtering*. Springer-Verlag, New York, 1971.

[Poo88]   H. V. Poor. *An Introduction to Signal Detection and Estimation*. Springer-Verlag, New York, 1988.

[PS83]   K. R. Pattipati and N. R. Sandell, Jr. A Unified View of State Estimation in Switching Environments. In *Proc. 1983 American Control Conf.*, pages 458–465, 1983.

[Rab89]   L. R. Rabiner. A Tutorial on Hidden Markov Models and Selected Applications in Speech Recognition. *IEEE Proc.*, 77(2):257–286, Feb. 1989.

[Rho71]   I. B. Rhodes. A Tutorial Introduction to Estimation and Filtering. *IEEE Trans. Automatic Control*, AC-16(6):608–704, Dec. 1971.

[Rog88]   S. R. Rogers. Steady State Performance of Decoupled Kalman Filter. In *Proc. NAECON*, Dayton, OH, May 1988.

[RS72]   H. Raiffa and R. Schlaifer. *Applied Statistical Decision Theory*. M.I.T. Press, Cambridge, MA, 1972.

498

[RTS65]   H. E. Rauch, F. Tung, and C. T. Striebel. Maximum Likelihood Estimation of Linear Dynamic Systems. *AIAA Journal*, 3:1445–1450, Aug. 1965.

[Sch73]   F. Schweppe. *Uncertain Dynamic Systems.* Prentice-Hall, Englewood Cliffs, NJ, 1973.

[SH89]   D. D. Sworder and R. G. Hutchins. Image-Enchanced Tracking. *IEEE Trans. Aerospace and Electronic Systems*, AES-25:701–710, Sept. 1989.

[Sin70]   R. A. Singer. Estimating Optimal Tracking Filter Performance for Manned Maneuvering Targets. *IEEE Trans. Aerospace and Electronic Systems*, AES-6:473–483, July 1970.

[Skl57]   J. Sklansky. Optimizing the Dynamic Parameters of a Track-While-Scan System. *RCA Review*, 18:163–185, June 1957.

[SM71]   A. Sage and J. Melsa. *Estimation Theory with Applications to Communications and Control.* McGraw-Hill, New York, 1971.

[SM91]   D. L. Snyder and M. I. Miller. *Random Point Processes in Time and Space.* Springer-Verlag, New York, 2nd edition, 1991.

[Sor80]   H. W. Sorenson. *Parameter Estimation.* Marcel Dekker, New York, 1980.

[Sor85]   H. W. Sorenson, editor. *Kalman Filtering: Theory and Application.* IEEE Press, New York, 1985.

[SS85]   T. L. Song and J. L. Speyer. A Stochastic Analysis of a Modified Gain Extended Kalman Filter with Applications to Estimation with Bearings Only Measurements. *IEEE Trans. Automatic Control*, AC-30:940–979, July 1985.

[Sta87]   D. V. Stallard. An Angle-Only Tracking Filter in Modified Spherical Coordinates. In *Proc. 1987 AIAA Guid., Navig. and Control Conf.*, pages 542–550, 1987.

[Str65]   C. Striebel. Sufficient Statistics in the Optimum Control of Stochastic Systems. *J. Math. Anal. and Appl.*, 12:576–592, Dec. 1965.

[SW72]   K. Spingarn and H. L. Weidemann. Linear Regression Filtering and Prediction for Tracking Maneuvering Aircraft Targets. *IEEE Trans. Aerospace and Electronic Systems*, AES-8:800–810, Nov. 1972.

[Swe59]   P. Swerling. First Order Error Propagation in a Stagewise Smoothing Procedure for Satellite Observations. *J. Astronautical Sciences*, 12:46–52, Autumn 1959.

[TBM77]   R. R. Tenney, T. B. Ballard, and L. E. Miller. A Dual Passive Tracking Filter for Maneuvering Targets. In *Proc. ONR Passive Target Tracking Conf.*, Naval Postgrad. School, Monterey, CA, May 1977.

[TH79]   J. K. Tugnait and A. H. Haddad. A Detection-Estimation Scheme for State Estimation in Switching Environments. *Automatica*, 15(4):477–481, July 1979.

[Tho73]   J. S. Thorp. Optimal Tracking of Maneuvering Targets. *IEEE Trans. Aerospace and Electronic Systems*, AES-9:512–519, July 1973.

[THS77]   R. R. Tenney, R. S. Hebbert, and N. R. Sandell. A Tracking Filter for Maneuvering Sources. *IEEE Trans. Automatic Control*, AC-22:246–251, April 1977.

[Tug82]   J. K. Tugnait. Detection and Estimation for Abruptly Changing Systems. *Automatica*, 18(5):607–615, Sept. 1982.

[Van68]   H. L. Van Trees. *Detection, Estimation, and Modulation Theory*, Part I. Wiley, New York, 1968.

[WH85]   E. Wong and B. Hajek. *Stochastic Processes in Engineering Systems.* Springer-Verlag, New York, 1985.

[Wil76]   A. S. Willsky. A Survey of Design Methods for Failure Detection in Dynamic Systems. *Automatica*, 12(6):601–611, Nov. 1976.

[WJ76]   A. S. Willsky and H. L. Jones. A Generalized Likelihood Ratio Approach to the Detection and Estimation of Jumps in Linear Systems. *IEEE Trans. Automatic Control*, AC-21:108–112, Feb. 1976.

[WLA70]   R. P. Wishner, R. E. Larson, and M. Athans. Status of Radar Tracking Algorithms. In *Proc. Symp. Nonlinear Estimation*, San Diego, CA, Sept. 1970.

[WM80]   H. Weiss and J. B. Moore. Improved Extended Kalman Filter Design for Passive Tracking. *IEEE Trans. Automatic Control*, AC-25:807–811, Aug. 1980.

[Woo85]   M. S. Woolfson. An Evaluation of Manoeuvre Detector Algorithms. *GEC J. of Research (Chelmsford, England)*, 3(3):181–190, 1985.

[ZD63]   L. A. Zadeh and C. A. Desoer. *Linear System Theory.* McGraw-Hill, New York, 1963.

Index

ERRATA

for

ESTIMATION AND TRACKING
Yaakov Bar-Shalom and Xiao-Rong Li
Artech House, 1993

p. xvii, l. 4: "lead" should be "led"

p. 16, l. 13: remove sentence "In ..."

p. 16, l. 14: at end add sentence "In the case of a full rank $n \times n$ matrix, its columns (and rows) span the n-dimensional space."

p. 17, l. -2: "a_i" should be "a_1"

p. 18, l. 9: "*symmetric*" should be "*symmetric* (with non-zero eigenvalues)"

p. 33, l. 7: "$p_x(\xi) \triangleq$" should be "$p_x(\xi) = \lim_{\xi_i \to 0}$"

p. 33, l. -2: "parantheses" should be "parentheses"

p. 62, l. -2: "F" should be "$F(k)$"

p. 62, l. -1: "process" should be "sequence"

p. 73, l. 5: "(a) ... H_1" should be "(a) ... H_0"

p. 79, l. 3: ".139" should be "1.39"; ".277" should be "2.77"

p. 82, l -13: "constant" should be "constant unknown"

p. 82, l. -9: "σ" should be "σ^2"

p. 99, l. 13: "idependent" should be "independent"

p. 103, l. 3: "(2.4.2-2)" should be "(2.4.2-3)"

p. 104, ll. 8, -3: outside summations should be over j

p. 106, l. 7: "variance" should be "MSE"

p. 120, l. -13: outside summation should be over j

p. 120, l. 3: "1" should be "-1"

p. 120, l. 6: "$1 \triangleq$" should be "$\mathbf{1} \triangleq$"

p. 121, l. 9: "in terms of w" should be "in terms of z"

p. 135, l. -1: "diag$[R(k)]$" should be "diag$[R(i)]$"

p. 153, l. 10: second "$-$" should be "$+$"

p. 153, l. -5: "$-$" should be a blank space

p. 153, l. -4: "initiation" should be "continuation"

p. 168, l. -5: eliminate "$'$"

p. 174, l. 5: "1000" should be "10000"; "3000" should be "20000"

p. 175, figure parts (b) of the two scenarios should be switched

p. 179, l. -11: "accelerations" should be "velocities"

p. 199, l. -3: add transpose "$'$" at end of Eq. (4.3.5-7)

p. 200, l. 6: add transpose "$'$" at end of Eq. (4.3.5-8)

p. 201, l. 7: both "H" should be "$\mathcal{H}$"

p. 214, l. -9: "(5.2.1-2)" should be "(5.2.1-3)"

p. 214, l. -5: "(5.2.1-2)" should be "(5.2.1-3)"

p. 217, l. 2: "(5.2.3-17)" should be "(5.2.3-11)"

p. 218, l. -2: "state" should be "state and measurements"

p. 219, right column of figure, block 3 from top: "$P(k|k)$" should be "$P(k+1|k)$"; "$R(k)$" should be "$R(k+1)$"

p. 225, l. -7: "$i = 1$" should be "$j = 1$"

p. 239, Eq. (5.4.2-9), p. 240, Eq. (5.4.2-12): should be for a single innovation component at a time

p. 239, l. -9, p. 240, l. -7: "problem 4-4" should be "problem 5-4"

p. 256, l. 11: "confidence" should be "probability"

p. 256, l. -11: "$\tilde{x}(k|k)$" should be "$\tilde{x}(k+1|k)$"

p. 270, ll. 3, -2: all references "(1.*)" should be "(3.*)"

p. 271, ll. 8, 14: all references "(1.*)" should be "(3.*)"

p. 276, l. 6: "T/β" should be "β/T"

p. 276, l. -10: "(6.5.2-4)" should be "(6.3.2-4)"

p. 276, l. -9: "(6.5.3-14)" should be "(6.5.3-13)"

p. 284, l. -11: the equation should be broken into two separate equations with the text line "and" between them

p. 308, l. 5: "$\bar{L}\bar{D}\bar{L}$" should be "$\bar{L}\bar{D}\bar{L}'$"

p. 329, l. 5: "n" should be "N"

p. 343, l. -5: "(9.2.1-2)" should be "(9.2.1-3)"

p. 345, l. -3: "$C(t)$" should be "$C(t)'$"

p. 347, l. -9: "$P(t) =$" should be "$\dot{P}(t) =$"

p. 350, ll. -5, -10: "R" should be "$\tilde{R}$"

p. 358, l. -1: "W" should be "L"

p. 377, l. 11: "*smoothing ... (1.4.12-4)*" should be "*total probability theorem (1.4.10-10)*"

p. 377, l. 13: eliminate the middle term

p. 384: all equation references "(6.*)" should be "(10.*)"

p. 387, right column of figure: add "down arrow" from block 2 to block 3

p. 387, right column of figure, block 4 from top: "$P(k|k)$" should be "$P(k+1|k)$"; "$R(k)$" should be "$R(k+1)$"

p. 415, l. 15: "$[x_1 x_2]$" should be "$[x_1 \quad x_2]$"

p. 421, l. 1: "LEEL" should be "LEVEL"

p. 427, l. 1: "$\sum$" should be "$\prod$"

p. 432, l. 1: " ARIABLE" should be "VARIABLE"

p. 458, between the filter blocks' line and the merging blocks' line: all $k-1$ should be k

p. 464, l. 11: "x" should be "z"

p. 466, l. 8: "0.316^2" should be "0.0316^2"

p. 480, l. 8: "140" should be "133"

p. 486, l. 12: in Eq. (11.7.2-4) "$v(k)$" should be "$v_1(k)$"

p. 486, l. 14: in Eq. (11.7.2-5) "=" should be "$= f^i[y(k), u(k)] + v_i(k) \triangleq$"

p. 487, l.- 6: in Eq. (11.7.2-14) "$F(k)\hat{y}(k|k)$" should be "$f[\hat{y}(k|k), u(k)]$"

p. 487, l.- 1: in Eq. (11.7.2-15) add at end "$+q_1$"

p. 491, l. 15: "250" should be "150"

In the software DynaEst 2.0 add in `start40.200` between lines 2 and 3:

```
cd(PATH_TO_MATLAB)
delete startup.m
```

(This correction has been incorporated into later versions)